Compositional Grading in Oil and Gas Reservoirs

Compositional Grading in Oil and Gas Reservoirs

Rogério Oliveira Espósito
Senior Process Engineer, Petrobras, Brazil

Pedro Henrique Rodrigues Alijó
Rio de Janeiro State University, Rio de Janeiro, Brazil

José Antonio Scilipoti
Federal University of Rio de Janeiro, Rio de Janeiro, Brazil

Frederico Wanderley Tavares
Federal University of Rio de Janeiro, Rio de Janeiro, Brazil

ELSEVIER

Gulf Professional Publishing
An imprint of Elsevier

Gulf Professional Publishing is an imprint of Elsevier
50 Hampshire Street, 5th Floor, Cambridge, MA 02139, United States
The Boulevard, Langford Lane, Kidlington, Oxford, OX5 1GB, United Kingdom

Notices

Knowledge and best practice in this field are constantly changing. As new research and experience broaden our understanding, changes in research methods, professional practices, or medical treatment may become necessary.

Practitioners and researchers must always rely on their own experience and knowledge in evaluating and using any information, methods, compounds, or experiments described herein. In using such information or methods they should be mindful of their own safety and the safety of others, including parties for whom they have a professional responsibility.

To the fullest extent of the law, neither the Publisher nor the authors, contributors, or editors, assume any liability for any injury and/or damage to persons or property as a matter of products liability, negligence or otherwise, or from any use or operation of any methods, products, instructions, or ideas contained in the material herein.

British Library Cataloguing-in-Publication Data
A catalogue record for this book is available from the British Library

Library of Congress Cataloging-in-Publication Data
A catalog record for this book is available from the Library of Congress

ISBN: 978-0-12-812452-9

Publishing Director: Joe Hayton
Senior Acquisition Editor: Katie Hammon
Editorial Project Manager: Mariana Kuhl
Production Project Manager: Mohana Natarajan
Designer: Vicky Pearson

Typeset by MPS Limited, Chennai, India

Contents

Biography

Frederico W. Tavares—B.S. in Chemical Engineering (1981), M.Sc. in Physical Chemistry (1984), and D.Sc. in Chemical Engineering (1992) at the Federal University of Rio de Janeiro (UFRJ), Brazil. He started his career in 1983 at the Federal University of Uberlândia, Minas Gerais, Brazil. Now, he is a full professor at UFRJ, at School of Chemistry (EQ-UFRJ) and at the Chemical Engineering Program of COPPE (PEQ-COPPE-UFRJ). His researches are related with Statistical Thermodynamics, Non-Equilibrium Thermodynamics, and Molecular Simulations applied for product and process design in Chemical and Petroleum Engineering. During sabbatical leaves, he spent 2 years with Prof. Stanley I. Sandler at University of Delaware and one year with Prof. John M. Prausnitz at University of California, Berkeley. He published more than 120 scientific papers and advisor more than 44 dissertation (M.Sc.) and 17 Doctor theses (D.Sc.).

José A. Scilipoti is B.S. in Chemical Engineering (2005) at National University of Córdoba (UNC—Córdoba, Argentine) and D.Sc. (2014) in Chemical Engineering at the National University of South (UNS—Bahía Blanca, Argentine). He worked as a process engineer at TEKSID Argentina (2005–06) and at MATERFER (2006–08). He was a professor (2009–14) at UNC teaching programming for undergraduate courses of Chemical Engineering. In 2014, he carried out research tasks focused on enzymatic reactions in two-phase systems at l'Institut National des Sciences Appliquées (INSA—Toulouse, France). His research interests are related with phase equilibria, separation process, engineering of chemical reactions, molecular design, and Applied Thermodynamics for product and process design in Chemical and Petroleum Engineering.

Pedro H. R. Alijó is B.S. in Chemical Engineering at Federal University of Bahia (UFBA, 2009), M.Sc. (2011) and D.Sc. (2014) in Chemical Engineering at the Federal University of Rio de Janeiro, Brazil. Between 2012 and 2016, he has worked as a research assistant at COPPETEC Foundation, in a project funded by Petrobras S.A., involving compositional grading calculations in oil and gas reservoirs. He started his academic career in 2015 as an adjunct Professor at the Rio de Janeiro State University (UERJ), Brazil, teaching Thermodynamics and Physical Chemistry for undergraduate courses of Chemical Engineering and Chemistry, at the

Institute of Chemistry of UERJ (IQ-UERJ). His research interests are related with colloidal systems modeling, electrokinetic phenomena, and Applied Thermodynamics for product and process design in Chemical and Petroleum Engineering.

Rogério O. Espósito—B.S. in Chemical Engineering at the Federal University of Rio de Janeiro (UFRJ, 1997). His M.Sc. (1999) and D.Sc. theses (2004) were about equilibrium calculations under the influence of both gravitational field and rock-fluid interactions. He joined Petrobras in 2000, acting on both propane deasphalting and product-blending optimization processes in Duque de Caxias Refinery (REDUC). In 2004, after concluding D.Sc., he moved on to the PVT laboratory of Petrobras Research Center (CENPES), also developing a postdoctoral work about thermodynamic modeling of CO_2 coreflooding processes using carbonate plugs in the Reservoir Engineering Research Institute (RERI) under Prof. Abbas Firoozabadi advising. This work yielded a publication commended with the Cedric Ferguson medal in SPE's Annual Technical Conference and Exhibition, Amsterdam, 2014. Currently, he is a technical consultant for reservoir-fluid thermodynamic modeling in the Ultradeep-Water Department of Petrobras.

Prefaces

Wanna shout an unhuman scream out loud which is a way to be listened to

Invoking the symbol poet from times even harder than nowadays, we here propose another way...

Feeling sorry for the context of both eras, this work, at least in this moment, certainly marches offhand the recent history of absurd merit-and-value switching and complete failure of education, culture, science, and technology. This book does not only have the objective of teaching its readers. We expect it to be as a real hope for unstimulated grad students. A viable alternative for engineers who are trying to qualify for their jobs; and finally, a source of options in research lines for both students and advisors in postgrad, increasing the possibilities of international publication in our overwhelmed institutions. We are not the owners of the truth in compositional grading area. We are investigators searching for answers to our doubts, hoping they can generate high-level discussions, new publications and also new master, doctoral, and postdoctoral scholarships. We hope that this work helps frame important decisions along the development of petroleum fields, independently of the international scenario, we know to be cyclic. Crises come and go, with or without the bias of market opportunists. But the knowledge acquired and settled is eternal and ensures employability, even though moral. We hope the shout proclaimed by this work represents that damped minority who still believes that a country is built up with people and books. Among so much lying and brute force, we humbly look for understanding the natural laws acting on the deep reservoir rocks, helping our contribution to take off from there the richness we wanted to see divided with justice. Let us move ahead, for "if things reduce us simply to nothing, simply from nothing we have to start." We greet those who will look for a future in these pages, admitting that it is not at all pleasant to start over (and over and over) from zero mark due to mere opportunity destruction. Here is an opportunity, at least we hope so. Write to us. Put your courses together. Enrich your programs. Make your critics and suggestions for future editions. Why should you do it? Well... a regarded director of a Brazilian institution told me once that "the world has only 500 people. The rest are supporting actors." We don't want to switch sides. We want the side to change.

Rio de Janeiro, November 2016
Rogério Espósito

Among the technological challenges arising from petroleum exploration at severe conditions (high pressures and water depths, strong thermal gradients, porous media of complex geomorphology, etc.), the thermodynamic modeling of reservoir fluids is one of the most important, providing valuable information during exploration and formation evaluation periods. Under conditions of deep wells containing considerable amounts of asphaltenes and carbon dioxide, the existent models are limited, and the data are scarce (when available) and costly. The lack of consistent information hinders the development of new models and the improvement of reliable and robust methodologies for the calculation of thermodynamic properties that are useful for the design and evaluation stages of reservoirs.

In this scenario, the interaction between the academy and the oil and gas industry is necessary and beneficial for both sides. Although the academy has the responsibility of leading the discussions with the scientific rigor required for the development of thermodynamic models, moving away from the real industrial problems makes models and methodologies developed with low technological impact. In Brazil, one finds very often many obstacles to the interaction of academy and industry, success examples being rare. Offhand this tendency, one can say that this book is an encouraging example.

Identifying the necessity of improving the modeling tools to face the new challenges arising from new reservoirs containing high contents of carbon dioxide, Dr. Rogério O. Espósito (Petrobras S.A.) came to the ATOMS ("Applied Thermodynamics and Molecular Simulation") laboratory, at the Federal University of Rio de Janeiro (UFRJ), to propose an ambitious academy—industry joint venture project, within the framework of the Brazilian National Petroleum Agency (ANP)—Petrobras program, entitled "Theoretical Modeling of the Compositional Grading in Reservoirs." The project involved members of the technical staff of Petrobras, teachers and postgraduate students of UFRJ, and a strategic partnership with RERI (Reservoir Engineering Research Institute), directed by Prof. Abbas Firoozabadi.

Initially, this book was not in the horizon of this academy—industry cooperation project. It arose stimulated by the discussions among the authors before the scientific tournaments that showed along the path. The information available in the literature on specific studies about compositional grading in oil and gas reservoirs are scattered, lacking for a consensual and unified theoretical treatment. In this sense, we figured that publishing a textbook would give an important contribution to both academic and industrial audience as it presents the main concepts in a more explaining way, systematically discussing the new approaches proposed. Therefore, we hope that we can stimulate new research lines (theoretical and applied) in a subject that still needs deeper investigations.

Finally, we hope this book confirms that partnerships between academy and industry can be successful. The technological innovation needed to face the current challenges demands more dialog among different sectors of the society. To sum up competences and skills, divide responsibilities. This is a path we believe to be indispensable. We expect to encourage the reader to join us in this journey.

Rio de Janeiro, November 2016

Pedro H.R. Alijó
José A. Scilipoti
Frederico W. Tavares

Production from hydrocarbon reservoirs is based on integration of various disciplines and topics. Most advanced concepts in physics, chemistry, and mathematics are integrated in geological description of subsurface formations to optimally produce hydrocarbon fluids. One of the basic elements in efficient hydrocarbon production is fluid itself. The reservoir fluid is composed of various species, ranging from methane to asphaltenes, resins, water, and salts. These species are not homogeneously distributed in the subsurface due to various mechanisms. In many reservoirs, the liquid content of the produced fluids may vary orders of magnitude with depth. Depending on the position of the well both in the $x-y$ plane and in the $z-y$ and $z-x$ planes, the composition of the produced fluids may be very different. Knowledge of species distribution is a key to well placement.

The proposed book is the only book that is devoted to compositional variation in hydrocarbon reservoirs. Petrobras is perhaps the only major energy company in the world which has a large number of huge reservoirs with large variation in species distribution. The offshore fields in Brazil show significant variation of composition both in the $x-y$ plane and in the z-direction. The first author is an experienced engineer and scientist in Petrobras. The fourth author has contributed significantly to thermodynamics which is one of the foundations in interpretation of data and modeling. Many years ago, the first author did his PhD thesis on the subject of the book.

The material in the book is much broader than the variation of species in hydrocarbon reservoirs. The authors present novel methodologies for prediction of phase behavior of hydrocarbon fluids comprehensively. The first few chapters of the book are devoted to phase behavior calculation and measurements. The techniques devoted to phase behavior modeling are used later in the book. The second important aspect of the book is vast amount of data on giant offshore fields which show compositional variation. The data plus analysis of published data from a unified approach are unique aspects of the book.

The book also covers association theory for water and other species in a unique engineering approach and incorporates the theory in the computation of species distribution.

As a whole, the book is very valuable addition to the array of books related to thermodynamics and the first book devoted to species distribution in hydrocarbon reservoirs. A large number of students, professors, and scientists active in oil and gas production will be the audience for the book. The examples at the end of many chapters will increase the possibility that part of the book may be used as text.

Abbas Firoozabadi

Among the technological challenges faced since the discovery and development startup of our petroleum fields, one of the most intrigant and of difficult solutions was that of understanding and modeling the fluid behavior, not only at the reservoir original conditions, but also at the dynamic ones during the production periods.

Simply speaking, the reservoir fluid is consisted by a mixture of both hydrocarbons and contaminants, distributed at different proportions among the various accumulations, subject to gravity, diffusive and natural convection effects. The information acquired as one drilled the wells and carried out extended well tests were debris of a puzzle that looked incomplete, obliging one to line up phenomena normally not considered in the common approaches of fluid modeling.

Multidisciplinary groups of geosciences provided interpretations and elaborated models that go through origin, generation, and migration routes of hydrocarbons that interacted with different amounts of contaminants previously occurring in the accumulations. The integration of such information made possible to identify and understand the phenomena which, gradually, were incorporated in the equations of state used in the flow simulators.

I had the privilege of following up the technical evolutions, supported by envying technical and scientific rigor in this relentless and creative journey endeavored by Dr. Rogério Espósito.

The knowledge and experience to attend the challenge of compositional modeling of reservoir fluids are condensed in this book, which is doubtless a mark in fluid engineering. The conceptual and theoretical basis is presented in an elegant and precise way, attending practical consults and fulfilling the curiosity of the most exigent students, engineers, and scientists.

In the former chapters, one presents several modeling approaches depending on the complexity of the accumulations and the composition of their fluids, adding up a brief description of laboratory assays to obtain the parameters necessary to the fluid representation. In Chapter 3, Phase Equilibrium Under the Influence of the Gravitational Field, Chapter 4, Irreversible Thermodynamics Applied to Reservoir Engineering, and appendices, one goes through the phenomena of gravity segregation, thermal diffusion, and steady-state natural convection based on heat-of-transport parameters. Classic examples from the literature are rediscussed in Chapter 5, Classic Examples From Literature. New approaching proposals are shown, leaving the way open for future studies, which will certainly be thought of for the so far unpublished case studies of Chapter 6, Case Studies and Chapter 7, The Influence of Molecular Association.

Petrobras outstands for its excellence and pioneering in all disciplines, keeping and stimulating in its staff professionals of the very highest degree, like our colleague Dr. Rogério Espósito.

Saulo de Tarso Cerqueira Lima,
Petroleum Engineer—Petrobras S.A.

Acknowledgments

We are grateful to Petrobras for thoroughly carrying out its social responsibility, contributing for the enhancement of the applied research in our universities, generating academic production at international level, stimulation and hope for young investigators.

Rogério O. Espósito thanks both collaboration and review of E&P-PRESAL, E&P-CORP, E&P-EXP, E&P-Libra, CENPES, and RH-UP. We will not cite names, there would be many. The authorization for publishing the examples of Chapters 6 and 7 and the focusing of worktime for the conclusion of this book is much appreciated.

Rogério O. Espósito acknowledges Calsep Inc. team, leaded by Dr. Karen Schou Pedersen, for their hospitality and valuable discussions in various examples of this book.

We also thank Japan Petroleum Exploration Co., LTD (JAPEX), especially Dr. Toshiyuki Anraku, for providing the PVT analyses of some of the main correlation wells in Yufutsu. This was crucial for the quality assurance of our PR-EoS fitting, which parameters, once estimated based on these reports, will remain the same everywhere in the reservoir.

Our special thanks to Prof. Abbas Firoozabadi, for valuable discussions and powerful contributions, not only in the confection of the case studies, but also reviewing the main chapters. The partnership of both Petrobras and UFRJ with RERI is the great driving force of this work.

We thank our wives, sons, and relatives, including those who are gone, hoping that one day we will celebrate all together what we planted here.

Our special thanks to

anp
Agência Nacional
do Petróleo,
Gás Natural e Biocombustiveis

BR
PETROBRAS

Introduction

Full representation of petroleum reservoir-fluid phase behavior remains a challenge. Thermodynamic behavior of confined fluids within the pores of a rock is influenced by various phenomena originated in solid–fluid and fluid–fluid microscopic interactions, and not all of them are completely understood. Although thermodynamic modeling is a powerful tool for calculation of fluid properties along field production life, some of these phenomena still cannot be predicted with an unified model that represent them all simultaneously.

For example, after oil migration from source to reservoir rock, we would expect petroleum to have enough geological time to reach thermodynamic equilibrium state as if it was a simple, closed system not acted by external fields or chemical reactions other than those already occurred during genesis process. Some issues complicate this scenarium: (1) the action of Earth's gravitational field, which tends to segregate heavier components to the bottom of the structure (Schulte, 1980; Creek and Schrader, 1985; Firoozabadi, 1999; Espósito et al., 2000); (2) the effect of capillarity, which tends to change the interface regions between the phases present in the reservoir, depending on rock–fluid interactions of inert kind, like wettability or adsorption (Wheaton, 1991); and (3) noninert kind rock–fluid interactions, e.g., chemical reactions between reservoir fluids and carbonate rocks (Nghiem et al., 2011). None of these effects prevent the reservoir to reach thermodynamic equilibrium, only changes species distribution within the pore space, if the system can be considered isothermal with reasonable approximation. Gravitational and capillary effects can change fugacity equations by adding new terms and chemical reactions that could be assumed to advance on local equilibrium a priori.

Rigorously, no reservoir can be considered isothermal, since the heat flux from Earth's core to surface generates entropy uninterruptedly inside its geological limits, which, due to rock thermal conductivity, will never grow adiabatic. So, if geothermal gradient is really to be considered, the equilibrium equations that lead to minimization of thermodynamic potentials like Gibbs and Helmholtz free energies are locally valid but are also insufficient to describe the global system.

Several approaches were made to solve this problem, with different levels of complexity (Montel and Gouel, 1985; Shukla and Firoozabadi, 1998;

Ghorayeb and Firoozabadi, 2000; Pedersen and Lindeloff, 2003). Irreversible Thermodynamics (Fitts, 1962; De Groot and Mazur, 1962, Haase, 1969) remains as an essential mathematical tool. In this context, the system of equations no longer takes fugacity equality, no matter if they are modified or not by the gravitational field. As it will be seen along this text, one suggests expressions for diffusive fluxes in microscopic terms based on local equilibrium hypothesis at each point of the flow. In geological time scale, the reservoir fluid rearranges itself according to both geothermal gradient and the gravitational field acting on it and will reach steady state, when, despite inequity among fugacities, no property will vary with time. Accurate values of diffusion coefficients are essential for the determination of steady-state condition, in which all diffusive fluxes are null.

Thermal diffusion (or *Soret effect*), i.e., the appearance of a component flux as a consequence of a thermal gradient, is not completely dominated yet, at least for complex mixtures containing a high number of components with different sizes and polarities, like petroleum systems. Although gravity tends to segregate the heavies to the bottom of the reservoir and the lights to its top, thermal diffusion can both attenuate or enhance this effect, depending on the affinity of each component for the hot (bottom) or cold (top) side. In the Yufutsu field (Ghorayeb et al., 2000), located in Hokkaido, Japan, and discussed in detail in Chapter 5, the small difference between phase densities present in the reservoir has reduced the influence of the gravitational field, so the tendency of lighter hydrocarbons (mainly methane) of diffusing to the hot side made gas phase to lie below the liquid phase on a critical transition. As a consequence, density increases upwards and the gas—oil ratio decreased with production time due to the flow of heavier top components towards the producer wells.

The main goal of considering all these phenomena is to obtain a trustable component distribution along all the reservoir extension, being it or not at thermodynamic equilibrium. In doing so, the initialization of the flux simulator will be as much representative of the reservoir initial and boundary conditions as one wants it to be, enhancing the reliability of the production forecast.

It is necessary to determine the contribution of each phenomenon on the fluid, confirming or dropping a priori the isothermal-system hypothesis, using well-fitted parameters in the diffusion models as well. First of all, a fine-tuned fluid model for property calculation at any pressure, temperature, and composition is crucial.

The parameters of the equations of state (EOS) must be estimated on the basis of experimental PVT data from representative reservoir sampling. The cubic EOS introduced in the 1970s (Soave, 1972; Peng and Robinson, 1976) are still the most used thermodynamic models for reservoir fluid representation. In some cases, critical transitions, organic precipitation, and contamination by filtrate of perforation fluids may have already been occurred at reservoir conditions, and their complete representation through tuning of these equations has not been accomplished so far.

Microscopically, short-range interactions called "associations" change significantly the thermodynamic behavior of some fluids. Molecular associations like hydrogen bonds play an important role on changing properties, and mainly on how these properties vary with temperature. In reservoir fluids, asphaltenes are the components that present more polar sites due to heteroatoms and can associate with other asphaltene molecules, resins or even low molar-mass contaminants, like CO_2 and H_2S.

There are suitable thermodynamic models to treat this type of molecular interaction, e.g., the EOS derived from the so-called Statistic Association Fluid Theory (SAFT). See, for example, Jackson et al. (1988) and Chapman et al. (1990).

In these equations, a reference fluid of hard-sphere type, which presents only a repulsion term up on contact, undergoes a series of perturbations in its molecular interactions, generating additional terms that represent the possibility of interactions with other molecules (dispersion), chain formation and association between them. The basic idea of this equation is to consider that these contributions are additive in free-energy terms. The terms of a SAFT-based EOS are mathematically complex, and its parameter estimation is much more cumbersome than that needed for the traditional cubics. So, the latter still predominate in the literature due to their simplicity and cost–benefit ratio on predicting thermodynamic properties.

Of special interest in modeling associating fluids is the work of Kontogeorgis et al. (1996), which linked repulsion and dispersion terms of a cubic EOS to the association term from SAFT, proposing a hybrid model so-called CPA (*Cubic Plus Association*). A number of works were published since then, seeking a CPA-parameter set estimation for several polar systems, from binary aqueous mixtures to reservoir fluids exhibiting asphaltene precipitation (Li and Firoozabadi, 2009, 2010; Santos et al., 2015).

The association phenomenon must be considered in an independent way from thermal diffusion. Its understanding might be of value for the petroleum industry, mainly in reservoir simulation and its interfaces with flow assurance and surface facilities.

The main objective of this book is to establish separately the contributions of the isothermal gravitational field, thermal diffusion, and molecular association phenomena to the compositional grading observed in petroleum fields. Starting from the experience gathered from other fields in the literature where these phenomena are noticeable, new premises, methods of calculation and parameter estimation were then applied to new case studies not published so far. Nothing prevents the thoughts employed in the examples presented here to be extended to other sedimentary basins, since one can always perform a parameter reestimation for local contexts.

We start, in Chapter 1, by classifying reservoir fluids (liquids or gases) as a function of the reservoir conditions of temperature and pressure relative to the critical point conditions in its phase envelope. Then we proceed to a description of the experimental procedures for determination of PVT properties of a reservoir fluid. These properties represent the volumetric behavior of this complex hydrocarbon-plus-contaminant mixture as a function of temperature and pressure. Therefore, they can be simulated through thermodynamic equilibrium algorithms using EOS, once these are duly calibrated to reproduce such experimental data.

In Chapter 2, some basic concepts of Classic Thermodynamics will be revisited to derive equilibrium conditions in closed simple systems. The algorithms necessary to determine compositions and properties of the phases in equilibrium within a PVT cell (multiphase flash and saturation pressure) will be shown in detail. So will be the parameter estimation techniques for Peng–Robinson EOS, which numerical solution allows for updating important intermediate variables, like phase densities and fugacity coefficients.

As petroleum contains a very high number of components, some lumping techniques for generating pseudocomponents are necessary to reduce computation time in commercial simulators without loss of thermodynamic information. Some of these techniques will be revisited according to the cumulated experience of the authors on their utilization.

In Chapter 3, we will proceed to deriving the thermodynamic potentials modified by the gravitational field. Gravity-modified fugacity equations will then be applied to reservoirs considered isothermal a priori, without molecular association effects. The solution of this system of equations allows us to obtain composition at any point of the reservoir from measured data in a reference level. Depths related to the interfaces between the phases present in the reservoir (gas–oil contacts) can also be pointed with good agreement and, therefore, be used for model calibration. Some classic examples of the literature will be revisited (Schulte, 1980; Creek and Schrader, 1985).

Chapter 4 describes the microscopic equations of fluid transport, which support Irreversible Thermodynamics. Different contributions for internal entropy generation, like heat conduction, molecular diffusion, and viscous dissipation will be derived. Soret and Duffour effects are discussed, and the methods for calculation of thermal diffusion parameters for compositional grading proposed by two main investigation groups in the literature will also be presented (Firoozabadi et al., 2000; Pedersen and Lindeloff, 2003). The systems of equations proposed independently by these two groups to eliminate the diffusive fluxes at steady state will be compared to those obtained by the constant-temperature hypothesis along reservoir depth. The latter characterizes the gravity-modified thermodynamic equilibrium state. Finally, the equations governing fluid moving by natural convection in two-dimensional (2D) and three-dimensional (3D) systems with lateral temperature gradients will also be derived, including a numerical solution sketched specifically for 2D domains.

Chapter 5 presents the application of this theory to examples of the literature, from very simple systems (binaries without gravitational field) to more complex ones, like actual reservoirs influenced by thermal diffusion and gravity simultaneously. Examples showing simulations of laboratory experiments splitting the components of simple mixtures by thermal diffusion will also be discussed, testing some model parameters. The current practical limitations of such experiments, along with their importance in modeling more complex thermal separation phenomena, open a wide and promising research line. Finally, the models proposed by both research groups cited in Chapter 4 are also tested in case studies for which temperature gradient has contributed for attenuation or enhancement of the compositional grading predicted by isothermal hypothesis.

The concepts used in Chapter 5 will support our interpretation of the conjugated influence of both gravity and thermal diffusion in the reservoirs studied in Chapter 6. Simulation of compositional grading from punctual fluid samples will allow us to obtain composition (and consequently any PVT property) at any other point in the reservoir, since hydraulic connection is confirmed. Only 1D and 2D domains are considered. Simulations in 3D nonstructured grids along geological time are still under research in high-precision numerical method area.

In Chapter 7, the incorporation of SAFT's association term on cubic EOS, proposed by Kontogeorgis et al. (1996), will be discussed to explain the physical meaning of the new CPA parameters to be estimated. An example will be shown, in which asphaltene precipitation takes place in the reservoir, by simple depletion or evidenced by an asphaltic-oil region covered by a lighter oil zone and/or even a gas cap at steady state.

Finally, in the epilogue, we discuss some extensions of the research lines explored here and their applications in the mean term, which, we hope, may keep on attracting young professionals seeking specialization in one of the most promising areas of petroleum industry, i.e., the compositional grading in oil and gas reservoirs.

REFERENCES

Chapman, W.G., Gubbins, K.E., Jackson, G., Radosz, M., 1990. New reference equation of state for associating liquids. Ind. Eng. Chem. Res. 29, 1709–1721.

Creek, J.L. and Schrader, M.L., East Painter reservoir: an example of a compositional gradient from a gravitational field, SPE 14411, 60th Annual Technical Conference and Exhibition, Las Vegas-NV, September/1985.

De Groot, S.R., Mazur, P., 1962. Nonequilibrium Thermodynamics. North-Holland Publishing Co., Amsterdam.

Espósito, R.O., Castier, M., Tavares, F.W., 2000. Calculations of thermodynamic equilibrium in systems subject to gravitational fields. Chem. Eng. Sci. 55, 3495–3504.

Firoozabadi, A., 1999. Thermodynamics of Hydrocarbon Reservoirs. McGraw Hill, New York, NY.

Firoozabadi, A., Ghorayeb, K., Shukla, K., 2000. Theoretical model of thermal diffusion factors in multicomponent mixtures. AIChE J. 46 (5), 892–900.

Fitts, D.D., 1962. Nonequilibrium Thermodynamics. McGraw-Hill Series in Advanced Chemistry, New York, NY.

Ghorayeb, K., Firoozabadi, A., 2000. Modeling multicomponent diffusion and convection in porous media. SPE J. 5 (2), 158–171, June.

Ghorayeb, K., Anraku, T and Firoozabadi, A., *Interpretation of the fluid distribution and GOR behavior in the Yufutsu fractured gas-condensate field*, SPE 59437, SPE Asia Pacific Conference, Yokohama, Japão, April/2000.

Haase, R., 1969. Thermodynamics of Irreversible Processes. Addison-Wesley, Reading, MA.

Jackson, G., Chapman, W.G., Gubbins, K.E., 1988. Phase equilibria of associating fluids. Spherical molecules with multiple bonding sites. Mol. Phys. Vol. 65 (No 1), 1–31.

Kontogeorgis, G.M., Voutsas, E.C., Yacoumis, I.V., Tassios, D.P., 1996. An equation of state for associating fluids. Ind. Eng. Chem. Res. 35, 4310–4318.

Li, Z., Firoozabadi, A., 2009. Cubic-plus-association equation of state for water-containing mixtures: is 'cross association' necessary? AIChE J. Vol. 55 (7).

Li, Z., Firoozabadi, A., 2010. Cubic plus association equation of state for asphaltene precipitation in live oils. Energy Fuels 24, 2956–2963.

Montel, F. and Gouel, P.L., *Prediction of compositional grading in a reservoir fluid column*, paper SPE 14410, presented at the SPE Annual Technical Conference and Exhibition, Las Vegas, Nevada, September, 22–25, 1985.

Nghiem, L., Shrivastava, V. and Kohse, B., *Modeling aqueous phase behavior and chemical reactions in compositional simulation*, SPE 141417, Reservoir Simulation Symposium, The Woodlands, Texas, EUA, February, 2011.

Pedersen, K.S., and Lindeloff, N., *Simulations of compositional gradients in hydrocarbon reservoirs under the influence of a temperature gradient*, SPE 84364, SPE Annual Technical Conference and Exhibition, Denver, Colorado, Outubro/2003.

Peng, D.Y., Robinson, D.B., 1976. A new two-constant equation of state. Ind. Eng. Chem. Fund. 15, 59.

dos Santos, L.C., Abunahman, S.S., Tavares, F.W., Ahón, V.R.R., Kontogeorgis, G.M., 2015. Cubic plus association equation of state for flow assurance projects. Ind. Eng. Chem. Res. 54 (26), 6812–6824. Available from: http://dx.doi.org/10.1021/acs.iecr.5b01410.

Schulte, A.M., 1980. Compositional variations within a hydrocarbon column due to gravity, SPE 9235. Dallas-Texas, 21–24.

Shukla, K., Firoozabadi, A., 1998. A new model of thermal diffusion coefficients in binary hydrocarbon mixtures. Ind. Eng. Chem. Res. 37, 3331–3342.

Soave, G., 1972. Equilibrium constants from a modified Redlich–Kwong equation of state. Chem. Eng. Sci. 27, 1197–1203.

Wheaton, R.J., 1991. Treatment of variations of composition with depth in gas-condensate reservoirs (includes associated papers 23549 and 24109). Soc. Pet. Eng. Available from: http://dx.doi.org/10.2118/18267-PA.

Chapter 1

Reservoir Fluids and PVT Analysis

Reservoir fluids are classified in five types, depending on the light-, intermediate,- and heavy-hydrocarbon contents in the mixture. Thermodynamic behavior and the physical state in which they show within the rock pores are suggested by their phase envelope (Pressure and Temperature (PT) diagram), together with reservoir temperature and pressure. Depending on its composition, one can infer oil and gas relative proportions obtained at the surface after separation, as well as the yields of distilled fractions to be used upon refining. The fluid submitted to the temperature and pressure conditions within the pore space of the rock, still containing "dissolved gas" (light components like N_2, CO_2, CH_4, C_2H_6, etc.) is called "live oil," whereas "dead oil" means the produced fluid after surface separation processing (no longer containing the lights), which is ready to be sent to downstream refineries. The dead oil is stored at the so-called stock-tank (or standard) conditions, i.e., 60°F and 1.01325 bar. For example, the "American Petroleum Institute (API) degree," a density measurement scale, is defined for the dead oil as follows:

$$\mathrm{API} = \frac{141.5}{d_{60/60}} - 131.5 \tag{1.1}$$

where $d_{60/60}$ is the specific gravity of the oil, i.e., the ratio of its own density at 60°F and the one of pure water (999 kg/m^3) at the same temperature.

There are other quantities relating certain live- and dead-oil characteristics, which are called Pressure, Volume and Temperature (PVT) properties. In this chapter, we will present the reservoir-fluid types and their main characteristics, detailing also the concepts and procedures to obtain the PVT properties.

1.1 RESERVOIR FLUIDS

1.1.1 "Black Oil"

A hydrocarbon mixture classified as *black oil* is liquid at reservoir conditions, which are, by definition, far from the critical region of the phase envelope. When taken to standard conditions, it shows relatively low initial

Compositional Grading in Oil and Gas Reservoirs. DOI: http://dx.doi.org/10.1016/B978-0-12-812452-9.00001-0

gas–oil ratio (GOR), i.e., the volume of produced gas over volume of residual dead (gas-free) oil is normally below 400 m^3 std/m^3 std. An isothermal depressurization carried on a sample of this kind of fluid starting from reservoir static pressure would lead necessarily to a bubble point when saturation was reached.

This kind of fluid has relatively low content of "dissolved gas," i.e., light components like N_2, CO_2, CH_4, and C_2H_6. Below saturation pressure, gas liberation causes low shrinkage (i.e., volume reduction) in the oil due to its low compressibility. Fig. 1.1 shows a phase envelope typical of a *black oil* fluid. Reservoir temperature is much lower than its critical one, that is, the temperature and pressure conditions are far to the left of the critical point on the phase envelope. Quality lines (relative liquid and vapor contents inside the envelope) are quite sparse, which causes a relatively small liberation of gas below the bubble point. Trajectory 1-2-3 shows an isothermal depressurization at the reservoir temperature (around 200°F), starting by static pressure, reaching bubble point (around 2800 psia), and ending below 1500 psia, with approximately 65% of liquid. Surface separator conditions are about 300 psia and 100°F. The trajectory of a fluid particle from static pressure

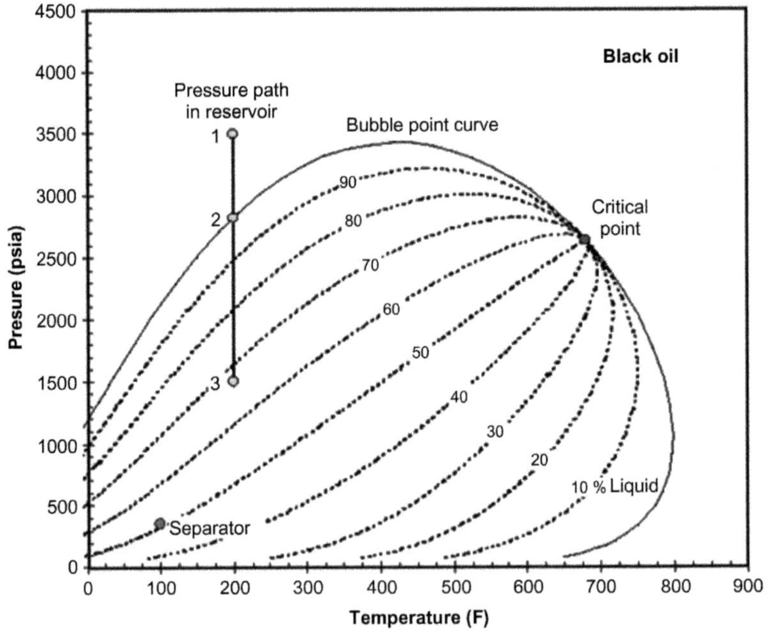

FIGURE 1.1 Phase diagram of a black oil fluid. Quality lines vary from bubble point (100% liquid) to dew point (0% liquid), all of them converging at the critical point. Trajectory 1-2-3 is an isothermal depressurization from static pressure. Surface separator conditions at low pressure are also shown. *Adapted from McCain Jr., W.D., The Properties of Petroleum Fluids, Pennwell Publishing Company, 1990.*

toward the separator is not isothermal (depends on the temperature profile inside the column of production (CoP), from reservoir to surface) and therefore is not shown in Fig. 1.1.

1.1.2 Volatile Oil

As reservoir conditions in Fig. 1.2 get close to the critical point, reservoir temperature being lower than the critical one, one notes an increase of the light and intermediate fractions relative to the heavy ones, which increases fluid compressibility and, consequently, its shrinkage. At the vicinity of the critical point, the space among the quality lines reduces considerably. So, the difference between volatile and black oil is that, although both are liquids at reservoir conditions, the volume of gas evolved below saturation pressure for a small depletion (or pressure reduction) on a volatile oil is much greater than that observed for black oils. Normally, the initial GOR for a volatile oil overlaps 400 std m^3/std m^3 and its yields in more valuable intermediate fractions are greater than the ones for black oils. When managing a volatile oil reservoir, one has to consider the advantages that the closeness to the critical point provides: low viscosity and easier miscibility with injected gases, which tends

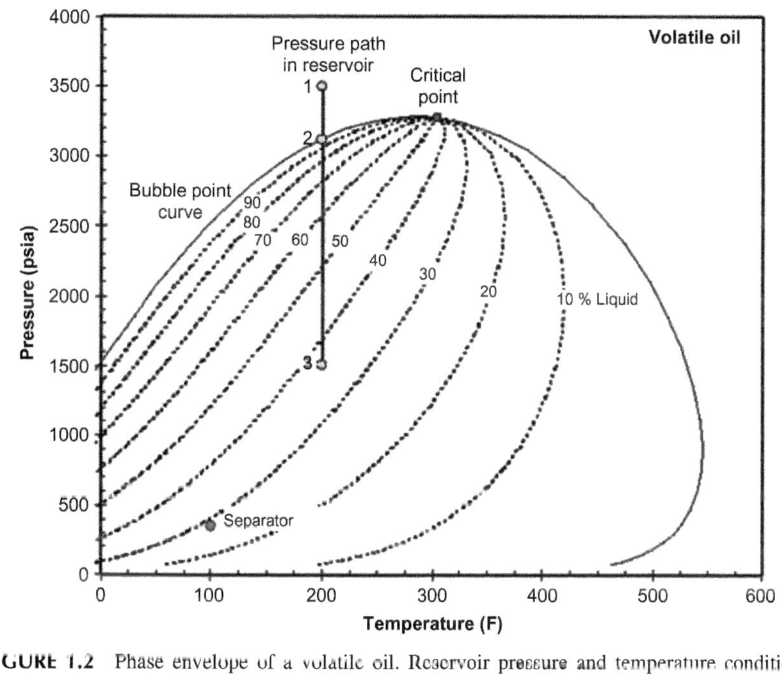

FIGURE 1.2 Phase envelope of a volatile oil. Reservoir pressure and temperature conditions are close to the critical point ones. Less sparse quality lines. Note the trajectory of isothermal depressurization, with more gas liberation, and the surface separator conditions. *Adapted from McCain Jr., W.D., The Properties of Petroleum Fluids, Pennwell Publishing Company, 1990.*

to increase recovery. Nevertheless, pressure maintenance is essential for gas (lights) to be kept in solution, avoiding the so-called *breakthrough* (premature gas irruption at the producer wells), and running down of the driving force for oil production. Fig. 1.2 shows a typical phase envelope of a volatile oil. One can note the proximity of the critical point, which leads to both higher gas liberation and shrinkage of the oil below saturation pressure (which is still a bubble point), due to the shorter space among the quality lines.

1.1.3 Gas Condensate

Gases subject to retrograde condensation show pressure and temperature conditions greater than the critical temperature and lower than the cricondentherm (maximum temperature of the phase envelope). As its temperature is greater than the critical temperature, its original state is gaseous, the lighter hydrocarbons being able to solubilize heavier ones, which would be liquid if were alone at these conditions. As one lowers the pressure, the solvency power of these lighter hydrocarbons decreases. When one reaches saturation pressure, the heavies come out of solution forming a liquid phase of density as greater than the remaining gas phase as we are far from the critical point. This phenomenon, apparently unexpected, of condensation by pressure reductions is known as "retrograde condensation." That is why this kind of fluid is also called "retrograde gas." Fig. 1.3 shows a typical envelope. Below dew point, the amount of liquid dropped out by isothermal pressure reduction increases till some extent; in the case of Fig. 1.3, a little bit below 30%. Proceeding with depletion, the pressure gets low enough for some intermediate hydrocarbons to revaporize (return point), which characterizes the way back to the "normal vaporization zone." When one reaches the dew point curve again, this time at its bottom-part limit, this is the so-called "lower dew-point." In practice, the lower dew point will likely be located below atmospheric pressure; therefore, some condensate will always remain.

1.1.4 Wet Gas

The wet gas shows a critical temperature greater than the cricondentherm, which was already defined as the highest temperature two phases can coexist at. So, the isothermal pressure reduction does not lead to liquid drop out at reservoir temperature, although some liquid can be observed at the surface separator conditions (located within the envelope). Fig. 1.4 shows a typical phase envelope of a wet gas.

1.1.5 Dry Gas

This is the simplest among all reservoir fluids, being made up by methane (more than 90%), ethane and traces of contaminants like CO_2, H_2S, and N_2.

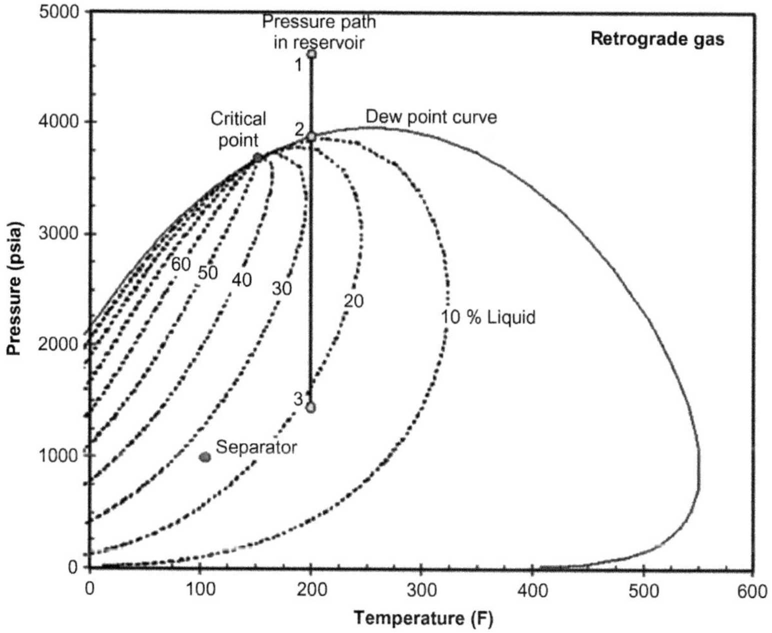

FIGURE 1.3 Phase envelope of a gas condensate. Reservoir temperature is already greater than the critical one, which causes the phenomenon of retrograde condensation, that is, the liquid dropout when one reaches saturation pressure (dew-point 2 in trajectory 1-2-3). *Adapted from McCain Jr., W.D., The Properties of Petroleum Fluids, Pennwell Publishing Company, 1990.*

There is no retrograde condensation at any point of the production system, from reservoir to surface. Therefore, global composition is always constant along time everywhere. It is the easiest type of fluid to be modeled either by empirical correlations or equation of state. Even so, some tuning is still necessary to determine its thermodynamic properties accurately. This technique will be detailed in Chapter 2. Fig. 1.5 shows a typical phase envelope of a dry gas. Note that, reservoir and production-system conditions are totally to the one-phase gas right side of the envelope. Both reservoir and separator temperatures are far away from cricondentherm. The GOR of a dry gas is theoretically infinite, but, in practice, one can consider a limit of say 20,000 std m^3/std m^3.

In general, fixing T and P reservoir coordinates, the phase envelope moves to the left as one increases the proportion of light hydrocarbons relative to heavier ones. Fig. 1.6 shows this tendency qualitatively. Table 1.1 shows some criteria for classifying a reservoir fluid sample on the laboratory bench (McCain, 1990). These numbers are not rigorous, and it is up to the field engineer to define the classification from an integrated analysis of the fluid properties. It is important to note that, among the five reservoir-fluid types, the volatile oil and the gas condensate are the closest ones to the critical point.

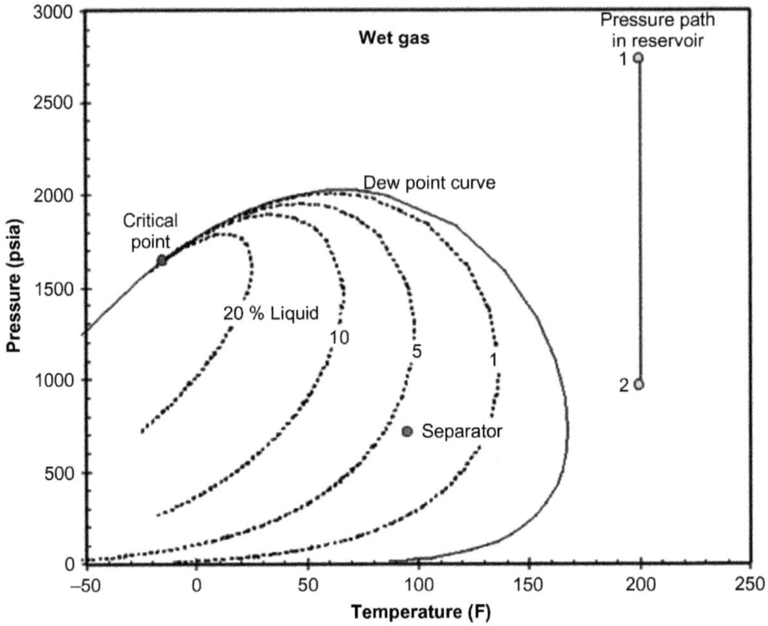

FIGURE 1.4 Phase envelope of a wet gas. Reservoir temperature is already beyond cricondentherm, so there is no liquid dropout in the reservoir due to isothermal depletion from static pressure. Separator conditions may lie inside the envelope with little liquid dropout though. *Adapted from McCain Jr., W.D., The Properties of Petroleum Fluids, Pennwell Publishing Company, 1990.*

In a hypothetical experiment, it is possible to turn a volatile oil into a gas condensate by isobaric heating above cricondenbar without seeing two phases, since the pressure will be always above both bubble and dew points.

1.2 PVT ANALYSIS AND BLACK OIL MODELING

PVT analysis is related to the way properties that depend intrinsically or directly on the volume vary with temperature and pressure. Note that GOR and Bo, already mentioned in this text, are nothing but ratios of oil and gas volumes at different T and P conditions relative to oil produced at tank conditions. From now on, we are going to detail the laboratory assays from which one can obtain the properties that make up the PVT analysis of the reservoir fluid. Such assays are chosen to be carried out depending on its previous classification. All of them are volume measurements as a function of temperature and pressure, keeping certain conditions invariant.

As reservoir fluids are confined within the pore space of the rock often under high-pressure conditions (frequently above 500 bar), it is expected that there is a suitable technology for collecting these fluids in situ with representativeness, keeping such conditions inside the sampler in order to avoid leakage with loss of volatile fractions. Service companies that developed such

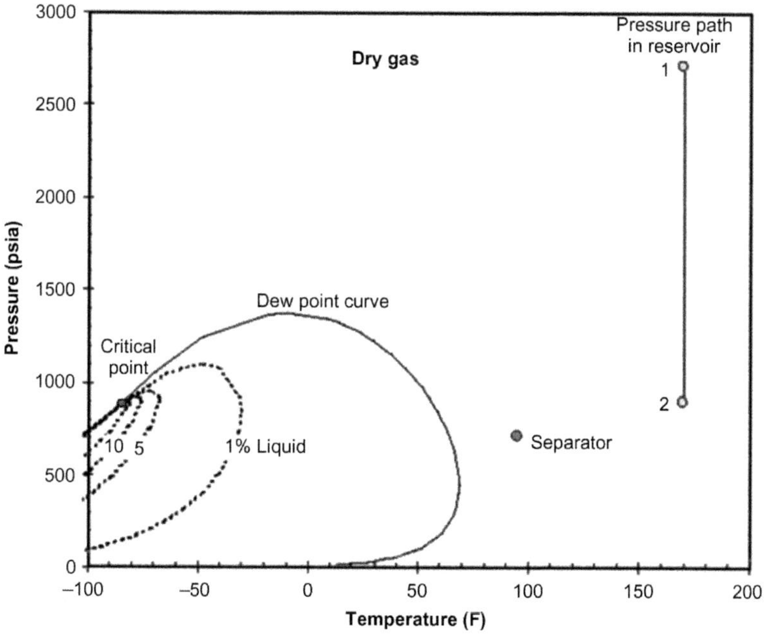

FIGURE 1.5 Phase envelope of a dry gas. Reservoir temperature is beyond cricondentherm. There is no condensation at all. *Adapted from McCain Jr., W.D., The Properties of Petroleum Fluids, Pennwell Publishing Company, 1990.*

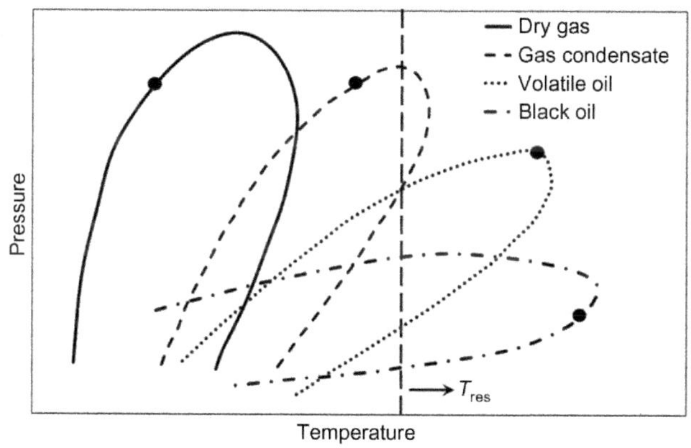

FIGURE 1.6 Qualitative comparison between phase envelopes of different kinds of reservoir fluids. Temperature T_{res} shows reservoir conditions. *Adapted from McCain Jr., W.D., The Properties of Petroleum Fluids, Pennwell Publishing Company, 1990.*

technologies have rigid internal safety procedures and quality control, disposing of tools that allow for bottom-hole fluid sampling during well testing (before or after casing), taking advantage of induced flow to storage the live oil practically exempted of water, sand, grit, and contaminant fluids.

TABLE 1.1 Reservoir-Fluid Classification

Property	Black Oil	Volatile Oil	Gas Condensate	Wet Gas	Dry Gas	Recommended Criterium?
GOR (m^3 std/m^3 std)	>400	300–600	>600	>3000	>20,000	Yes
°API	<45	>40	>40	<70	<70	No
Color	Dark	Variable	Slightly colored	Light	Light	No
Phase change	Bubble	Bubble	Dew	Dew	Dew	Yes
C_{7+} (mol%)	>20	12 to 20	<12	<4	<0.7	Yes
Bo (m^3/m^3 std)[a]	<2.0	>2.0	–	–	–	Yes

[a] *Bo = Ratio between the Live-oil volume at reservoir conditions and the dead-oil volume at standard (std) conditions. See the definition in Eq. (1.2).*
Adapted from McCain Jr., W.D., The Properties of Petroleum Fluids, Pennwell Publishing Company, 1990.

Fig. 1.7 shows the sketch of a PVT bottle (which was already filled up by transferring the reservoir fluid from the field sampler) in use at the laboratory, i.e., duly connected to some basic equipment.

The first assay the reservoir-fluid sample is submitted to the so-called *flash liberation*. Part of this sample is roughly depressurized from a high-pressure condition (generally even higher than the reservoir static pressure) to ambient pressure by passing through a micrometric valve. In this depressurization, the evolved gas is computed in a gasometer and the dead oil (or condensate) is weighted in an Erlenmeyer held in a thermal bath kept at 60°F. After weighting, part of the oil has its density measured in a bench densitometer at 60°F, from which one can easily calculate its API. The gas volume (converted from ambient to standard conditions) divided by the dead-oil volume (mass over density at standard conditions) provides the GOR of the fluid. In the vase of a black oil sample, the live-oil volume displaced at high-pressure bottle conditions, compared to the dead-oil volume at the end of the test, provides the shrinkage of the sample.

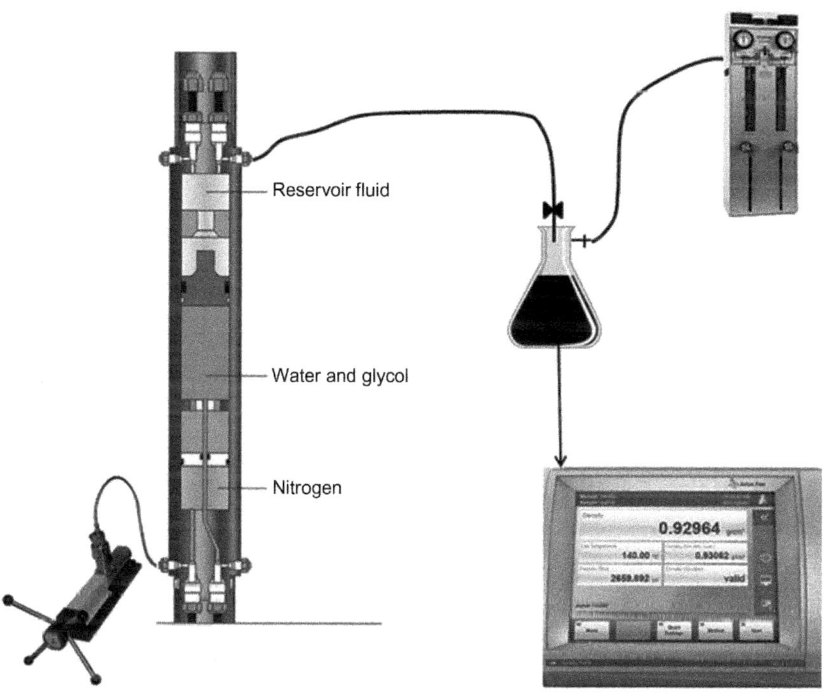

FIGURE 1.7 Sketch of a PVT bottle in use at the laboratory for flash liberation assay: part of the live-oil sample is pushed out by a positive-displacement pump and passes to the micrometric valve, where pressure is abruptly dropped down to ambient. Evolved gas is computed in a gasometer and dead-oil density is measured in a densitometer. *Parts are copyright of Anton-Paar Inc., Schlumberger Ltd., Vinci Technologies S.A., and Chandler Engineering. Used with permission.*

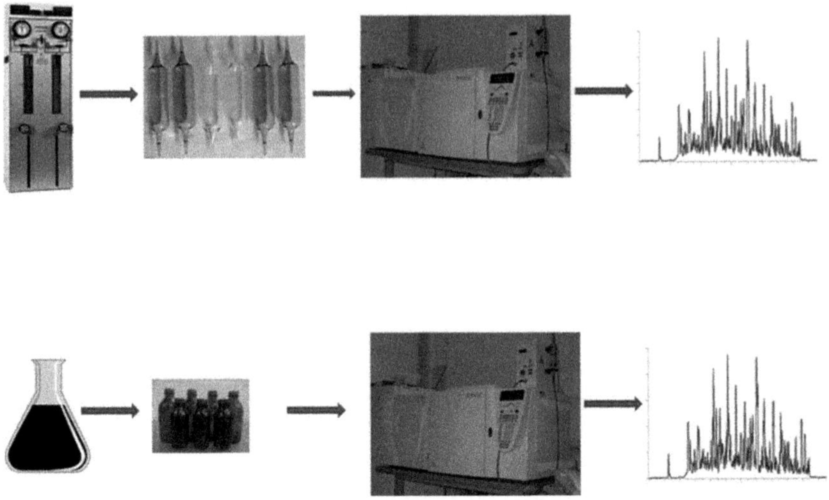

FIGURE 1.8 Sample-taking sketch after flash liberation for chromatography of both gas (glass ampoules) and dead oil (flasks). *Gasometer is copyright of* Chandler Engineering. *Used with permission. Other clipboards reproduced from* Wikipedia.org.

In order to conclude flash liberation, Fig. 1.8 shows the sample-taking sketch for chromatography of both gas and dead oil. The pressure necessary for live oil to get into the chromatographer makes its direct compositional analysis impossible. Chromatograms of dead oil and gas from the flash provide its global composition through a simple mass balance.

Once GOR and API of the sample are determined, the procedure is repeated for duplicate PVT bottles of the same fluid to check data consistency. Once repeatability is verified, one chooses the bottle that will proceed to the complete PVT analysis, which, in the case of the reservoir fluid be a black oil, is composed by three main assays: Constant Composition Expansion (CCE), Differential Liberation (DL), and Viscosity. The former two are carried out in cylinder of steel-made walls, thick enough to bear the required high pressures, named PVT cell (Fig. 1.9). The steel structure of the PVT cell is normally involved in a thermal jacket by which the internal temperature is kept constant (equal to the reservoir temperature) using a Proportional-Integral-Derivative control scheme. Alternatively, the cell might be immersed in an air bath.

CCE is a set of precise measurements of the volume occupied by the live oil within the PVT cell as a function of pressure for a fixed temperature, generally, the reservoir one. In conventional cells, where the internal pressure of the live oil is exerted by an auxiliary fluid (e.g., liquid mercury), these measurements must take into account the expansion and contraction of the steel lines, the cell walls, and the mercury itself, the latter both at the reservoir (it is in direct contact with live oil within the cell) and room

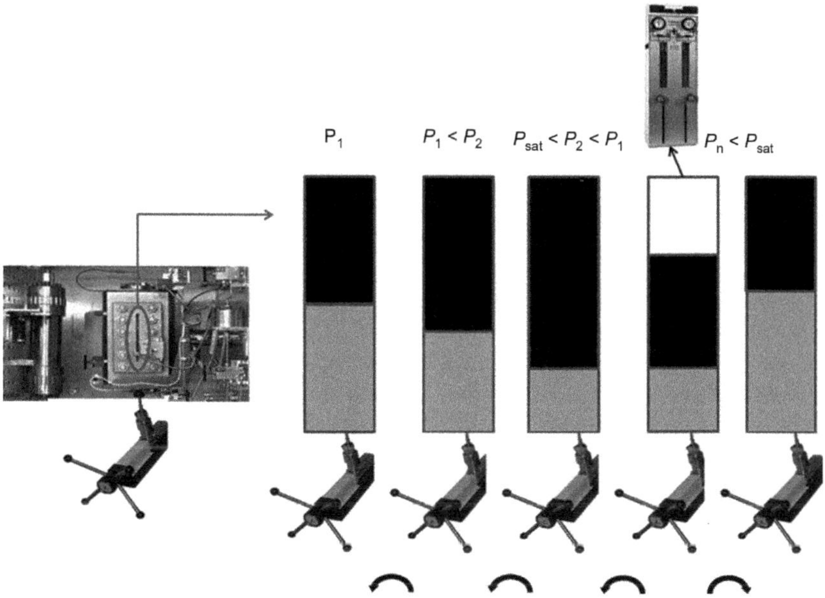

FIGURE 1.9 Another part of the chosen bottle is fed to the PVT cell (left side). The right side shows the live-oil sample kept at high pressures within the cell by a mercury seal. One retreats the pump, so mercury expands, releasing the pressure on the oil, which increases its volume. When pressure is stabilized, volume is written down, ending up the differential liberation stage. Note the first gas bubble at the saturation pressure (P_{sat}). Below it, a gas cap is formed and pushed off the cell by advancing the pump at constant pressure. Gas is computed in the gasometer. *Parts are copyright of Schlumberger Ltd., Vinci Technologies S.A., and Chandler Engineering. Used with permission.*

temperature (inside the external flexible lines). The description of these calculations is out of the scope of this book, but they are part of the daily routine of the lab engineer, who obtains PVT properties from the nominal measurements of pump-displaced volumes at each stage, corrected by all influences cited above. For more details, see for example, Whitson (1998).

To feed the PVT cell, initially full of mercury at room temperature, about 100 mL of live oil are pumped from the bottle, mercury being displaced and collected in an external recipient for safe wasting. The cell temperature control is then activated and the fluid reaches the assay condition, that is, reservoir temperature and pressure.

Then, the pump piston is retreated in predefined pressure steps and the corresponding displaced volume, read at the pump body scale. Once the already cited corrections are done, the definite volume occupied by the oil within the cell maybe calculated. Below saturation pressure, each equilibrium stage is characterized by an evolved gas cap which lies above the oil. Here is the main difference between CCE and DL: if, from static to saturation pressure, both assays are the same, below the bubble point, the evolved gas is pushed out *only in the* DL, keeping the temperature and pressure of that stage constant, till only the saturated liquid of the stage remains within the cell. The n_{th} stage below P_{sat}, in the very right side of Fig. 1.9, shows schematically the gas being pushed off the cell. Note that pump wheel is now rotated clockwise to advance the mercury seal. The gas is computed in a gasometer, and part of this volume is collected in an ampoule to have its composition determined by chromatography. The procedure is repeated till one reaches atmospheric pressure, when, by definition, one has no dissolved gas anymore. The residual oil volume, i.e., the volume of DL's "dead oil," is then measured not only at the assay's temperature but also at 60°F. In CCE, there is no gas expulsion below P_{sat}; therefore, only total volume can be measured within the cell. The assay has this name because the global composition of the mixture is always the same, once there is no mass withdrawing in the cell till the end of the test. In the case of DL, with the "nominal" measurements of displaced volume done at the pump scale and the gas volumes measured in the gasometer, it is then possible to compute the volumes of oil and gas in equilibrium within the cell at each stage, so obtain the PVT properties. Let us describe them further in more detail.

1.2.1 Oil Formation Volume Factor (FVF or Bo)

By definition, the oil formation volume factor (FVF or Bo) is the volume of live oil at any T and P (not necessarily the original reservoir conditions) divided by the volume of the respective dead oil at standard conditions, 1 atm and 60°F:

$$\text{Bo} = \frac{\text{Oil volume at } P \text{ and } T}{\text{Oil volume at } P_{\text{std}} \text{ and } T_{\text{std}}} \tag{1.2}$$

where $[\text{Bo}] = m^3/m^3_{\text{std}}$.

Note that the numerator of this fraction (Eq. (1.2)), the live oil, does not show the same composition of the denominator (dead oil), because, from T and P of interest to standard conditions, the gas comes out of solution. The inverse of Bo is known as *shrinkage*, because generally the dead-oil volume is lower than the live-oil one.

As composition varies, it is expected that Bo is a function of the fluid trajectory from live- to dead-oil conditions. Fig. 1.10 shows the qualitative behavior of Bo along DL. Naturally, none of these values is necessarily equal to the Bo of the flash liberation.

In practice, Bo can only be calculated at the end of DL because we need the residual oil volume at 60°F to obtain its denominator. As to the numerator, the live-oil volume inside the cell above P_{sat} is obtained from nominal-volume readings on the pump-body scale, applying the due corrections already cited. Below P_{sat}, the free-gas volume in equilibrium with the oil

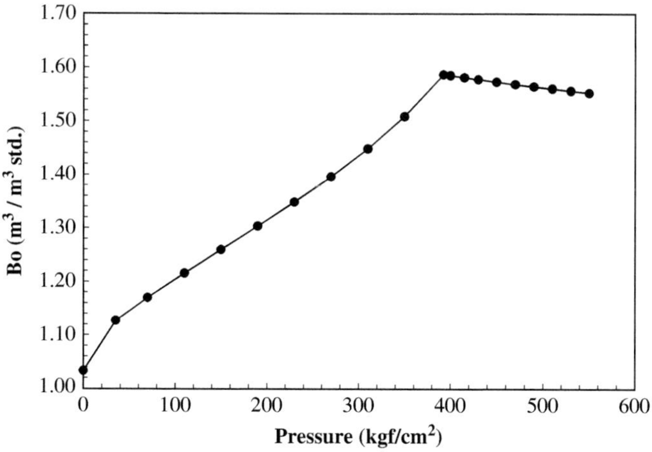

FIGURE 1.10 Example of a Bo graphic in a differential liberation assay.

(given by the difference on the pump reading before and after the gas expulsion) must be subtracted from total volume of the sample inside the cell in an order that only saturated oil is computed.

Starting from static pressure and reducing it to P_{sat} (bubble point), the increase in Bo shown in Fig. 1.10 might be explained by expansion due to oil compressibility. Such expansion might be more pronounced, depending mainly on the intermediate fractions in the mixture. Generally, the leaner the dissolved gas (and the heavier the dead oil), the lower the live-oil compressibility; then the increase in Bo tends to be moderated and linear. Below P_{sat}, however, the oil shrinkage due to gas evolving tends to overcome its expansion by pressure reduction, so one observes a steeper decline in Bo. Finally, the last reported value (that is atmospheric pressure) is greater than one because of thermal expansion, since the assay (reservoir) temperature is always greater than 60°F.

1.2.2 Solubility Ratio (Rs)

It is the volume of gas still dissolved in the live oil at T and P, but reported at standard conditions and divided by the residual-oil volume, the latter also reported at 1 atm and 60°F:

$$\text{Rs} = \frac{\text{Volume of gas still dissolved in the oil at } T \text{ and } P, \text{ reported at std. conditions}}{\text{Residual oil volume at std. conditions}} \tag{1.3}$$

where $[\text{Rs}] = m^3_{std}/m^3_{std}$.

The denominator of Rs is exactly the same as Bo's, and therefore, it can only be computed at the end of the experiment. Starting from static pressure,

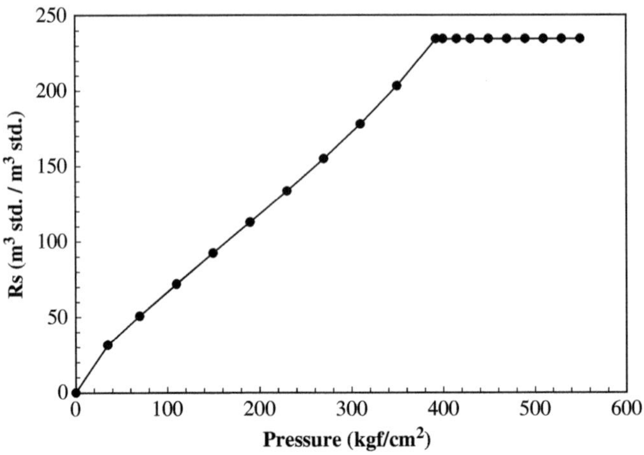

FIGURE 1.11 Example of an Rs graphic in a differential liberation.

till P_{sat}, the Rs is constant and maximum because all the gas is dissolved in the oil, as one can see in Fig. 1.11. This maximum value is conventionally called the GOR of DL. It differs from flash's GOR due to two opposite factors which are as follows:

1. Gradual pressure reduction: The bigger number of steps tends to retain the dissolved gas in the oil. The more stages DL has, the less the amount of gas evolved until the end of the assay will be.
2. The temperature of DL: Normally much higher than the flash's, it tends to produce more gas from the live-oil sample than the single-stage low-temperature pressure drop.

The volume of gas evolved at each pressure step is computed in the gasometer, converted to standard conditions, and the sum up of them all, divided by the residual-oil volume, gives the DL GOR. Of this total gas volume, one discounts the amount liberated at each stage to obtain the volume still dissolved below P_{sat}. By hypothesis, the last reported Rs value (at atmospheric pressure) is zero.

1.2.3 Oil Density (ρ_o)

Analogously to Bo and Rs, the live-oil density can only be calculated at the end of the experiment too because it is necessary to know the total mass of the sample fed to the PVT cell. Total mass is equal to the residual-oil volume multiplied by its density (both at 60°F) plus the sum of evolved gases mass. Each one of the latter is given by the volume computed in the gasometer multiplied by its density (both at standard conditions), which, in

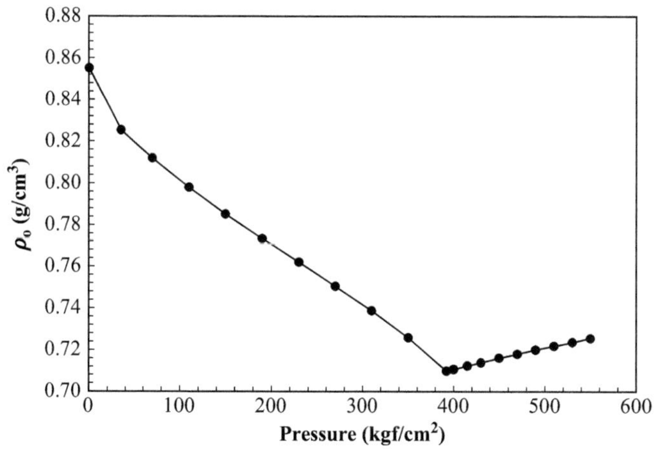

FIGURE 1.12 Example of an oil-density graph in a differential liberation.

turn, comes from chromatography composition. The total mass, divided by the sample volume inside the cell at each pressure step above P_{sat}, gives the live-oil density until the bubble point. Below it, the evolved gas mass at each step must be subtracted from the total mass, this result being divided by the volume of the saturated oil remaining in the cell at that step. Fig. 1.12 presents a typical oil-density graph in a DL. It decreases from static pressure to P_{sat} due to expansion; however, from then on, it starts increasing due to the higher concentration of heavier hydrocarbons, a consequence of light-component liberation with the evolved gas.

1.2.4 Oil Viscosity (μ_o)

Paralelly to the DL assay, another part of the live-oil sample is charged into an equipment named *rolling ball* or *falling ball* viscometer (Fig. 1.13). In such device, the time taken for a ball to roll (or fall) within a tube filled with live oil, preferably at the same conditions as DL steps, is measured. Using a calibration curve that relates the times with actual viscosities of reference fluids at various conditions, it is possible to infer the live-oil viscosity at the DL steps. The gas evolved below P_{sat} is stored and eliminated in a chamber at the upper part of the viscometer, so the ball can roll (or fall) through the oil without bubbles. The plot of measured viscosity points is similar to density: starting from static pressure, viscosity tends to decrease with pressure reduction as one releases the contact and friction among oil molecules. However, below P_{sat}, the liberation of light components to the gas phase increases the concentration of heavies in the oil, which in turn increases its viscosity (Fig. 1.14).

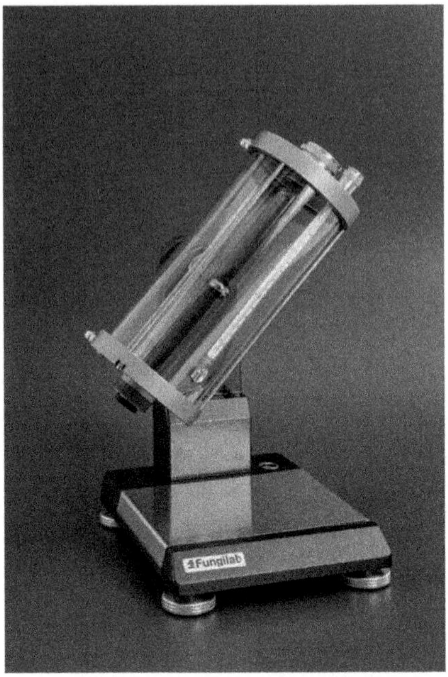

FIGURE 1.13 Falling Ball viscometer with internal pressurized chamber in which the ball falls through the oil sample in a standardized-length and angle glass tube. The falling time of the ball at each pressure step is measured through edge-located light-barrier detectors and then used in a calibration curve to calculate viscosity. Image is copyright of Fungilab Inc. Used with permission.

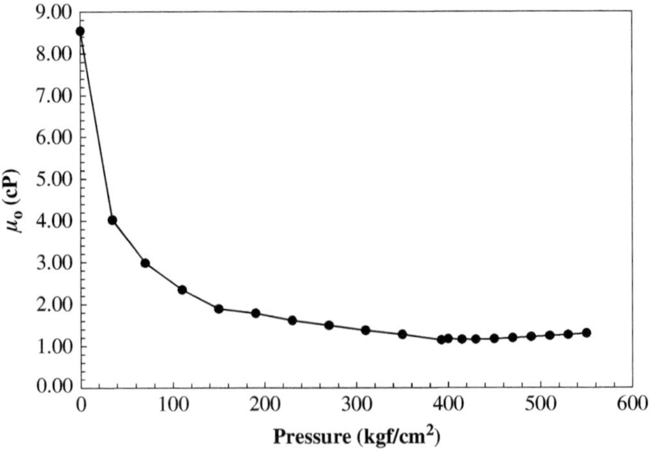

FIGURE 1.14 Example of an oil-viscosity graph in a differential liberation.

1.2.5 Gas Formation Volume Factor (Bg)

It is the volume occupied by the gas at the temperature and pressure of any stage below P_{sat}, divided by the volume of the same gas at standard conditions. This is the main difference between Bo and Bg: in the former, the denominator represents a fluid which is different from that one computed in the numerator. In the latter, by hypothesis, the gas is necessarily the same. Fig. 1.15 shows a typical graph of Bg in a DL. The volume occupied by the gas at interest conditions inside the cell might be calculated from the difference between the pump readings before and after its expulsion. The respective standard-condition volume is obtained from that one measured in the gasometer, assuming that the ideal-gas hypothesis is valid.

1.2.6 Gas Gravity (γ_g)

The so-called gravity of the evolved gas is a dimensionless property given by the ratio between its density and the one of the air, both measured at standard conditions. Assuming both are ideal gases in such conditions, the gas gravity collapses to the ratio of these two molar masses. Gas molar mass can be obtained from chromatography analysis. Fig. 1.16 shows a typical graph of gas gravity in the DL.

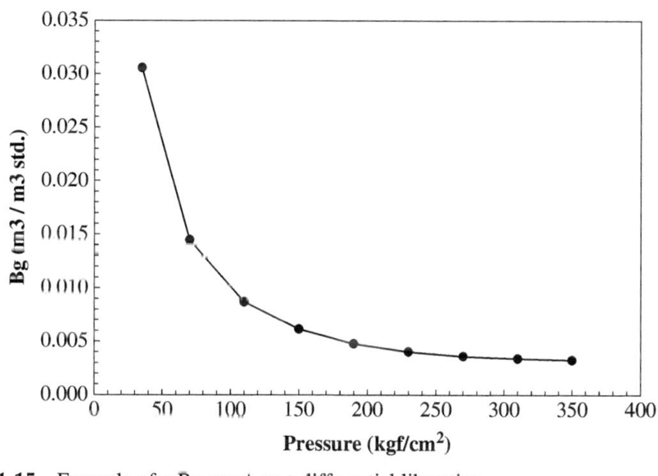

FIGURE 1.15 Example of a Bg-graph in a differential liberation.

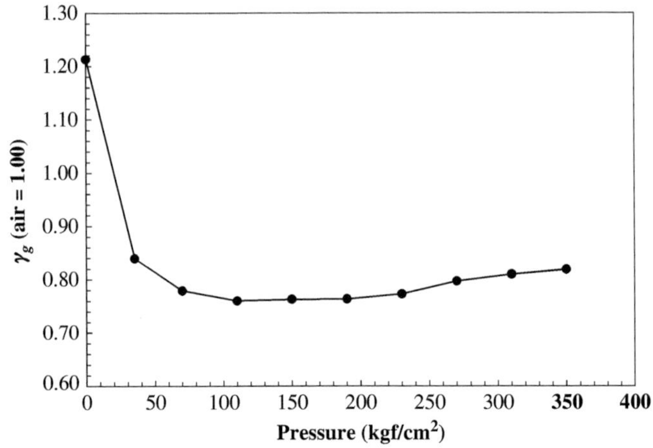

FIGURE 1.16 Example of a gas gravity graph in the differential liberation.

1.2.7 Compressibility Factor of the Gas (Z)

This property is directly proportional to Bg, once it indicates the ideality deviation as we move away from standard conditions, for which the relation $P\overline{V} = RT$ (where $\overline{V} = V/N$ is the molar volume) is valid:

$$Z = \frac{PV}{\mathrm{NRT}} = \frac{P\overline{V}}{\mathrm{RT}} \tag{1.4}$$

As $Z = 1$ at standard conditions, we have

$$Z = 1 = \frac{P_{\mathrm{std}} V_{\mathrm{std}}}{NRT_{std}} \Rightarrow Z = \frac{(PV/T)}{(P_{\mathrm{std}} V_{\mathrm{std}}/T_{\mathrm{std}})} = \frac{P}{T}\frac{T_{\mathrm{std}}}{P_{\mathrm{std}}} B_g \tag{1.5}$$

Fig. 1.17 shows a typical graph of compressibility factor Z in the DL.

It is important to mention a detail that can influence the measurement of gas properties: depending on the liberation condition at each step and on the live-oil composition, the evolved gas can contain heavy components that can drop out as a condensate at lab conditions. The amount of such condensate must have both its mass and composition measured so one can estimate its molar mass. Then, its number of moles might be calculated and multiplied by 23.69 $\mathrm{cm^3/mol}$ (molar volume of the ideal gas at standard conditions) to provide an equivalent gas volume to be added on Bg's denominator. For heavy oils liberating lean gases, this contribution might not be very important. Even so, for the sake of simulation of dew-point pressures and density, the global composition of the gas that leaves the cell must be corrected using the condensate information. For light oils and rich gases, the contribution might be strongly significant and may never be neglected.

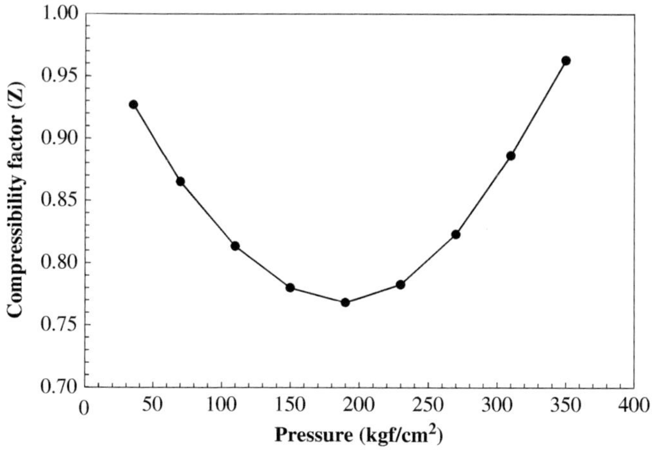

FIGURE 1.17 Example of a graph of gas compressibility factor in a differential liberation.

Once we have the PVT properties, the so-called *black oil modeling* (that is the ability of obtaining PVT properties at any T and P conditions using fitted empirical correlations based on experimental data) might be applied in order to obtain the relative in situ amounts of oil and gas during the production life of the reservoir, as well as inside the CoP and on surface facilities.

In this way of modeling the fluid, one is not interested in its composition but in its "consequences," i.e., shrinkage and swelling of the oil due to dissolved gas, oil and gas amounts obtained at surface conditions, determination of derived properties like relative permeabilities in the reservoir and thermal exchange coefficients, flow regimes within CoP, and many others that will be essential for the project of equipment upon field development. For reservoir simulations where gas reinjection is essential to the increase of the recovery factor, the so-called *compositional modeling* is recommended because it generally provides better results than *black oil*'s for both miscible and immiscible gases. For example, CO_2, the most widely used miscible gas, can even take the mixture to the critical region still at reservoir conditions, which makes *black oil modeling* practically inapplicable. Even under less severe conditions, the reinjection of lights changes the properties of the reservoir fluids, influencing directly the oil- and gas-production strategies.

The PVT properties obtained in the DL constitute the experimental database for equation of state parameter fitting, which resolution as a series of successive flashes (its feed being always the liquid phase of the previous one) must accurately provide, for each temperature and pressure condition, the volume occupied for each one of the phases in equilibrium inside the cell.

For gases, the CCE is carried out in visual cells, in which we can observe sample hazing as one reaches the dew-point pressure. The first condensate drops will also drip down on the glass inner walls of the cell. Below this

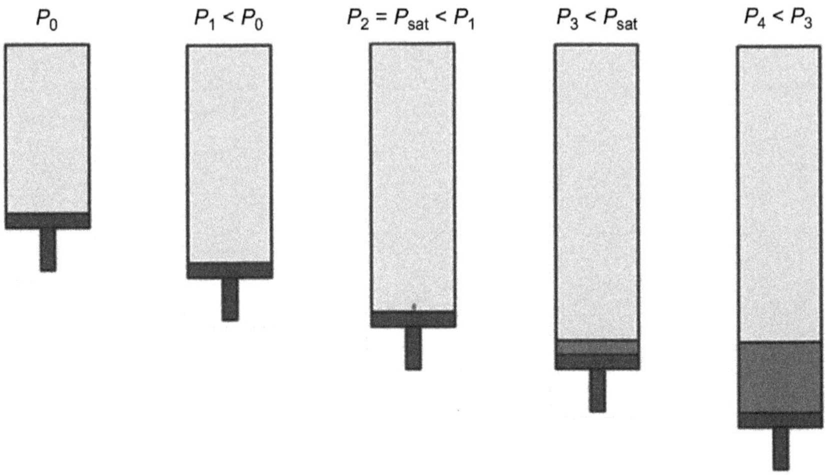

FIGURE 1.18 Constant Composition Expansion (CCE) of a gas condensate. Note the first condensate drop at the retrograde dew-point pressure and the gradual increase on the liquid volume settled at the bottom of the cell.

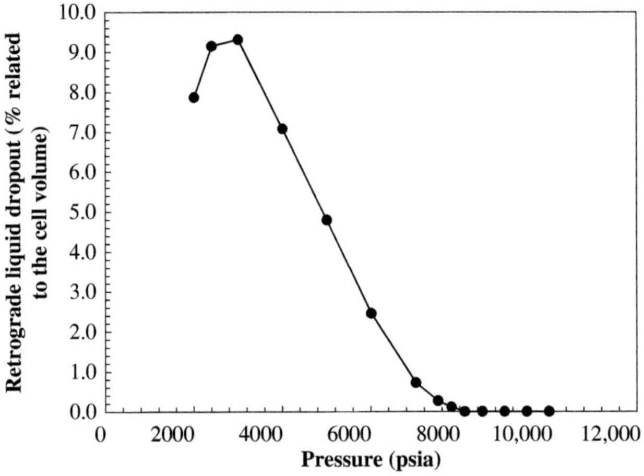

FIGURE 1.19 Retrograde liquid dropout as a function of pressure in a constant composition expansion of a gas condensate. Normally, cell's volume capacity is not sufficiently high to store the gas expansion at pressures low enough to reach the normal condensation zone in the phase envelope.

retrograde dew-point pressure, the liquid volume is computed by the height of the liquid—vapor interface relative to an arbitrary reference level. Fig. 1.18 shows schematically the steps of this assay.

Fig. 1.19 shows a typical graph of liquid dropout relative to the total volume of the cell as a function of pressure. It is important to emphasize that generally the volumetric capacity of PVT cells is not enough to store the strongly expanded gas at pressures low enough to reach the normal vaporization zone in the phase envelope. In other words, even that in volume basis the liquid dropout fraction starts decreasing (as is the case in Fig. 1.20, for example), most likely, *this occurs due to the great volumetric expansion of the gas and not by the revaporization of the condensate*. In molar or mass basis, as the depletion has not reached the normal condensation zone yet, the liquid fraction only increases.

For volatile oils, and mainly for gas condensates, besides the CCE, it is common to carry out another assay named Constant Volume Depletion (CVD), which simulates the production of gas as a result of the volume in excess caused by its expansion relative to a reference volume, normally the one at the saturation pressure. This thought is based on the hypothesis that the compressibility of the porous reservoir rock is negligible compared to the compressibility of the gas, which makes the available pore volume (PV) practically constant and therefore pressure-independent. So, for a discrete cell in a gas reservoir, pressure reduction implies in the expulsion of gas volume in excess. However, this move does not change the global composition of the reservoir fluid, except if one reaches the saturation pressure. If we assume the volume occupied by the gas within the PVT cell at the dew point as the reference volume, the excess of gas volume at lower pressures (excluding the retrograde condensate) is pushed off the cell at constant

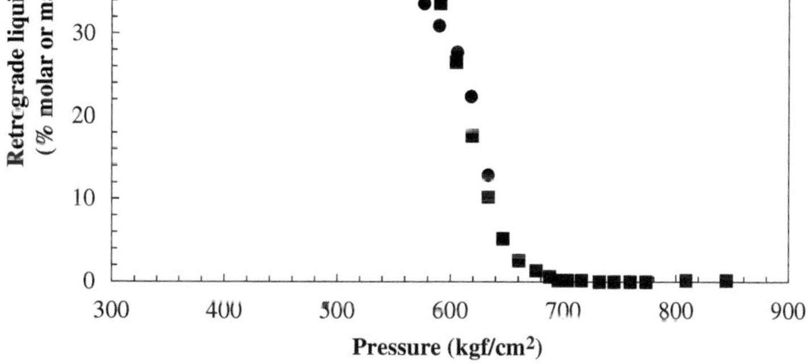

FIGURE 1.20 Percentages of retrograde liquid dropout in mass and volumetric basis in a constant composition expansion of a gas condensate. Volume-based curve's decline below 400 kgf/cm^2 does not necessarily mean that the system has reached the normal condensation zone. The more pronounced expansion of the gas makes its volume even greater than the liquid's, while the mass of the later is still increasing with no revaporization.

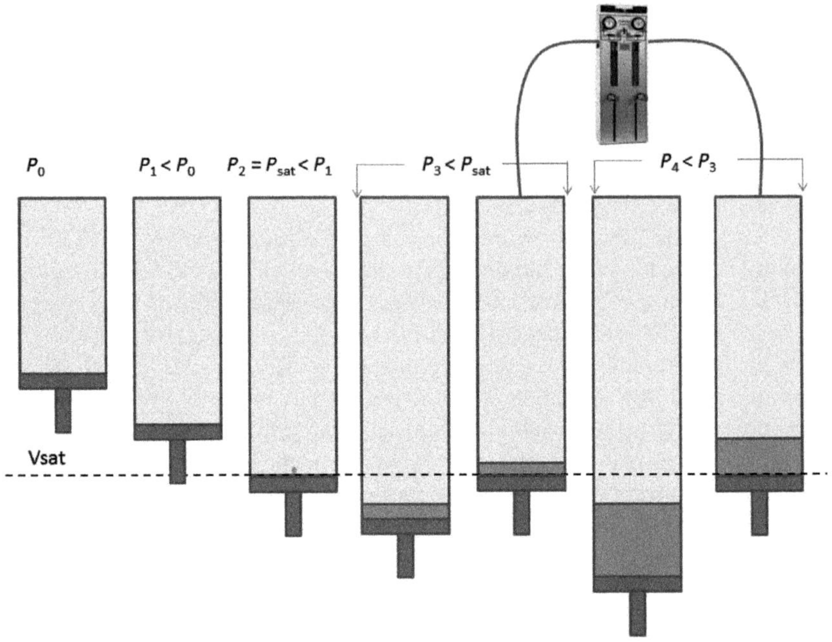

FIGURE 1.21 Constant Volume Depletion of a gas sample. Note the first retrograde condensate drop at dew-point pressure P_2. At pressures P_3 and P_4, the excess of volume relative to P_{sat} is pushed off the cell through the top (only gas), by piston displacement, back to saturation position, keeping the condensate dropout at the bottom of the cell.

pressure in order to recover the original volume at saturation. The evolved gas is computed in the gasometer to obtain its cumulative molar production relative to the number of moles charged into the cell in the beginning of the test. Its compressibility factor is proportional to the ratio between the pump reading before and after the expulsion and the gasometer volume (Bg). Fig. 1.21 illustrates the procedure. We emphasize that CVD starts, by hypothesis, from the saturation pressure, being identical to CCE for pressures higher than P_{sat}.

1.3 COMPLEMENTARY TESTS: THE CONCEPT OF MISCIBILITY

Besides the traditional assays of PVT analysis described before, there are two others, the *swelling test* and the measurement of the *minimum miscibility pressure* (MMP), used when one intends to inject gas in the reservoir to enhance the recovery factor. In these cases, the injected gas may dissolve in the oil, reducing its viscosity, and increasing its volume, forcing the excess to leave the porous media. In some cases, depending on the solubility of the gas in the oil at those temperature and pressure conditions, liquid–vapor equilibrium in the reservoir can occur mainly around the injector well, phases moving according to their relative permeabilities. The higher the gas mobility (and the lower its solubility in the oil), the worse for the project because the irruption of the gas (that is its arrival to the producer wells, or *breakthrough*) might be premature, leaving recoverable oil behind. On the other hand, in a totally opposite hypothetical situation, if the reservoir pressure was always higher than the saturation pressure of any gas–oil-proportion mixture, such mixture would always be monophasic, less and less viscous, and more and more swollen, maximizing recovery. In other words, depending on the injected gas and oil (originally a *black oil*, for example) proportions, final mixtures could turn into a volatile oil or even a gas condensate never splitting in two phases. What happens in practice is the occurrence of liquid–vapor equilibrium around the injector well, where there is gas in excess to oil. However, as the gas moves contacting more untouched oil, the mass transfer between them might be attenuated so that their compositions become increasingly different; or, what is desirable, their mutual solubility increases more and more, causing the phenomenon of *cosolvency*, for which the final mixture presents two phases increasingly alike, being finally soluble in all proportions in the further not-yet-visited oil. In this case, the critical point of the final mixture matches the temperature and pressure conditions at that region. *To the capacity of the gas-mixed reservoir fluid of reaching its critical point by cosolvency at any point of the reservoir by injection of a determined gas, is given the name of miscibility of the live oil to that injected gas*. The MMP is the lowest pressure to which the reservoir must be submitted in order that the miscibility to a determined gas is always achieved. It must be emphasized that the MMP of the reservoir fluid depends on the gas considered for injection.

The swelling test is a relatively simple assay that consists on injecting increments of a gas into a prefixed amount of live oil inside a PVT cell. For each one of the resulting mixtures one performs a CCE, writing down the increase in saturation pressure and the volume occupied by the oil at this new P_{sat} relative to the volume occupied at the original bubble point. We call this last quantity the *swelling factor*. From a certain amount of injected gas, global GOR and injection pressure are high enough for the reservoir fluid, originally a black oil, to become a gas condensate, which is visually noticed by the transition of P_{sat} from bubble to dew point. So to speak, the series of phase envelopes presented in Fig. 1.6 can also be considered (at least qualitatively) as a swelling test carried out at a fixed arbitrary temperature. Each gas injection displaces the critical point of the envelope to the left, finally taking the system to a gas condensate state. However, in the case of the swelling test, as the global mass of heavies is fixed and originally belonged to a black oil fluid, the dry gas state will never be achieved, at least within the limits of temperature and pressure of a conventional PVT cell (around 200°C and 700 atm). Fig. 1.22 shows qualitatively, in color scheme, the evolution of a swelling test in a PVT cell.

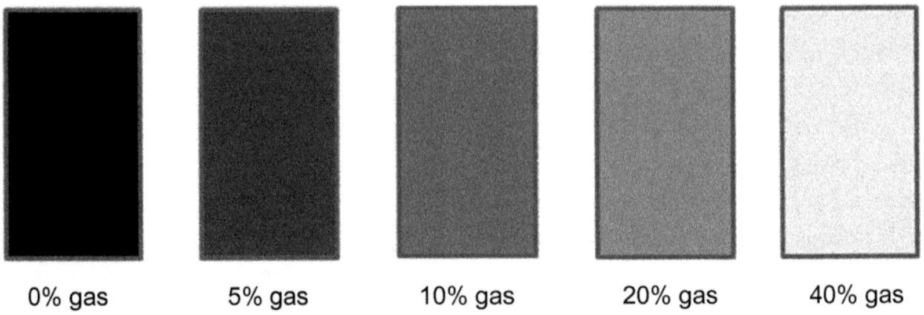

FIGURE 1.22 Sketch of a mixture evolution during a swelling test in a PVT cell. Values are arbitrary and merely illustrative.

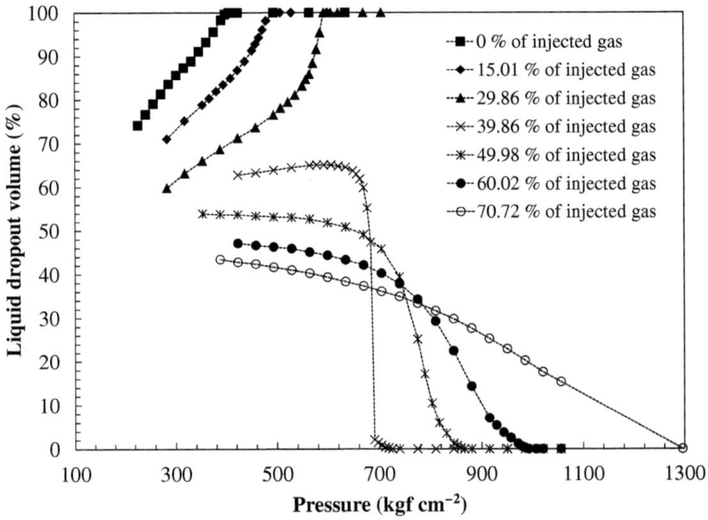

FIGURE 1.23 Liquid dropout curves in the CCEs of a swelling test. Transition from bubble to dew point occurs between 30 and 40 mole % of injected gas.

Fig. 1.23 shows the liquid dropout curves obtained in the CCEs of a real swelling test. Obviously, when the amount of injected gas is enough for the mixture to become a gas condensate, one notices that P_{sat} becomes a dew point, with liquid percentages starting from zero and not from 100 anymore. The most pronounced variations of these percentages in the last bubble point and first dew point are due to the proximity of the critical region in the phase envelopes of both last volatile oil and first rich gas condensate. As quality lines are too close, it is expected that a small pressure reduction causes great variation in the relative amounts of the phases in equilibrium.

The swelling test is a static assay that aims only to study the phase behavior of the reservoir fluid with the injected gas. The slim-tube test, by its turn, is a dynamic assay in which a certain amount of gas is injected in a coiled tube filled with a porous media of high permeability, saturated with a representative sample of live oil above bubble point. Normally, the injection is made in 0.2-PVs batches till one reaches a total amount of 1.2 PVs injected (Danesh, 1998). The oil volume recovered at the tube outlet after gas breakthrough is written down to compute the recovery factor. Then, the tube is cleaned and filled with a new oil sample; one increases the pressure and repeats the procedure. The MMP is conventionally defined as the one for which the oil recovery gets around 95%. The flow of fluids inside the coil is unidimensional and can be simulated in commercial softwares to validate fluid modeling, i.e., the Equation of State (EoS) fitting based on PVT analysis and swelling test. More details about this assay can be found in Ekundayo and Ghedan (2013). Pedersen et al. (2014) also presented a procedure with some conventional parameters used in commercial softwares.

Fig. 1.24 shows a real slim-tube equipment, the coil being on the right side of the chamber and the containers of oil, gas, and water standing on the left. Fig. 1.25 shows a typical recovery versus operation pressure curve in a real slim-tube test.

FIGURE 1.24 Slim tube apparatus. Coil within air bath together with fluid containers. *Image courtesy of Sanchez Technologies, Copyright 2016 Sanchez Technologies SAS and affiliates. Used with permission.*

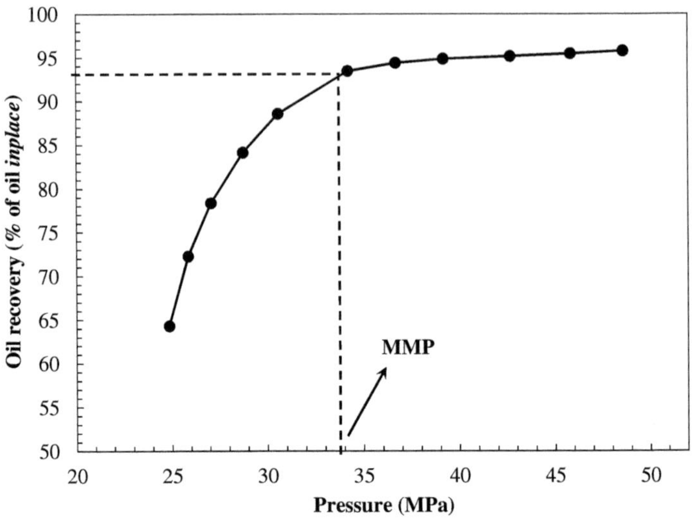

FIGURE 1.25 Oil recovery × pressure in a real slim-tube test. The slope breakpoint, from which there are no more considerable increments in recovery, characterizes the minimum miscibility pressure.

1.4 RESERVOIRS WITH PHASE TRANSITIONS

Such reservoirs are particular cases in which more than one fluid type coexist in the same porous formation. The most common case is the "*gas-cap reservoir*," where a Black Oil is already in equilibrium with a gas condensate, showing a well-defined interface between them known as *gas–oil contact (GOC)*. If one could neglect the compositional grading due to gravity, the bubble-point pressure of the oil would be equal to the dew-point pressure of the gas, and both would be the reservoir static pressure. If one considers gravity segregation, the static pressure itself increases with depth, with a more pronounced slope in the oil zone, which, due to its density considerably higher than the gas-phase one, shows a higher hydrostatic load. In this general case, the oil bubble-point pressure is equal to the dew point of the gas only at the *GOC*, that is, precisely at the interface. In the gas cap, dew point increases with depth down to the interface. Below it, the bubble point of the oil decreases due to the lower concentration of lighter components toward the bottom of the structure.

There is still a very interesting (and not very rare) particular case in which one observes a liquid–vapor transition without a definite interface. These are the *quasicritical* or *near-critical* fluids, for which the density difference between phases is low and, therefore, from bottom to top, liquid turns into gas without abrupt phase change or interface; only by the approach of the fluid's critical temperature toward the reservoir temperature along depth, as already shown in Fig. 1.6.

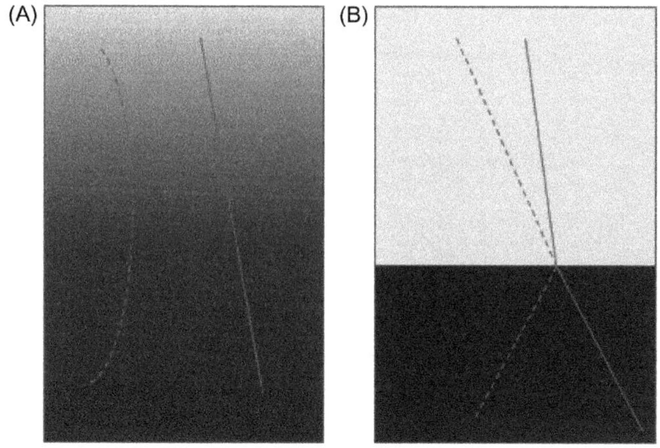

FIGURE 1.26 Phase transition, showing saturation-pressure profile (dashed line) and static-pressure one (full line), in a near-critical reservoir (A) and in a gas-cap reservoir (B).

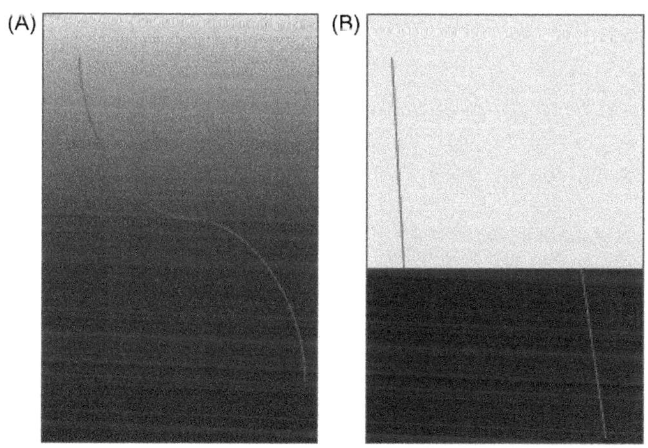

FIGURE 1.27 Phase transition, showing density profile along a near-critical reservoir (A) and a gas-cap one (B). Mole fractions of heavy hydrocarbons would follow this tendency. Lighter hydrocarbons would show inverted tendency, with symmetric curves.

In this case, the saturation pressure, in its gradual transformation from bubble do dew point, never touches the static-pressure curve. Fig. 1.26 shows both two-phase reservoirs we just introduced, emphasizing the saturation-pressure profile (dashed line) and static-pressure one (full line) along depth for the case of critical smooth transition (A) and the one of *GOC* (B). Fig. 1.27 shows how the tendency of density would be in both cases. Mole fractions of heavier hydrocarbons would follow the same tendency, the lights having a similar symmetric behavior. Fig. 1.28 shows qualitatively the phase envelopes of the oil and gas in equilibrium at the

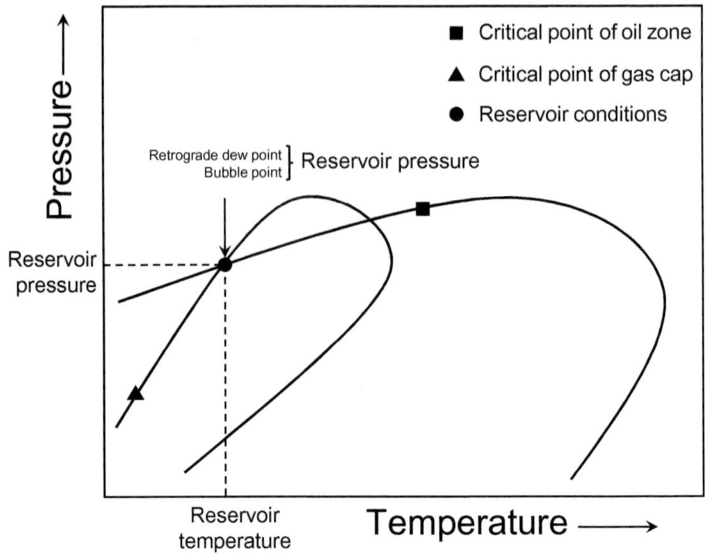

FIGURE 1.28 Phase envelopes of both oil and gas, exactly at the GOC depth.

GOC depth. Note the intersection between the bubble-point pressure of the oil and the dew-point pressure of the gas, both being the same as reservoir static pressure at that depth.

Finally, there are also the multiphase reservoirs, in which the segregation of heavier fractions, like paraffins and asphaltenes lead to the formation of a new phase below the conventional oil zone. Regarding thermodynamic modeling, if the new phase is rich in asphaltenes, it can be treated as an ultra-viscous fluid, still able to be described by the same equation of state used for the other fluid phases. In the case of paraffins, a model for solids should be applied, increasing even more the complexity of mathematical representation of the reservoir-condition equilibrium. Asphaltene segregation can occur by simple pressure reduction, which impairs their suspension in the oil, normally supported by other heavy components named *resins*. Or still by their low affinity with either lighter hydrocarbons or contaminant gases which act as solvent extractors, e.g., CO_2. In general, asphaltene content is not a determinant variable for its precipitation. Even at low concentrations, these low-compressibility macromolecules may turn unstable in the oil, mainly due to the volume increase of the lighter components when undergoing pressure reduction, which reduces the available space for its suspension. For a black or volatile oil, the pressure at which the asphaltenes become unstable is called the *upper onset pressure of asphaltene precipitation* (Pedersen et al., 2014). This pressure is determined with reasonable visual precision by this equipment known as high-pressure electronic microscope (Fig. 1.29). The zoomed image shows the formation of clusters of a darker

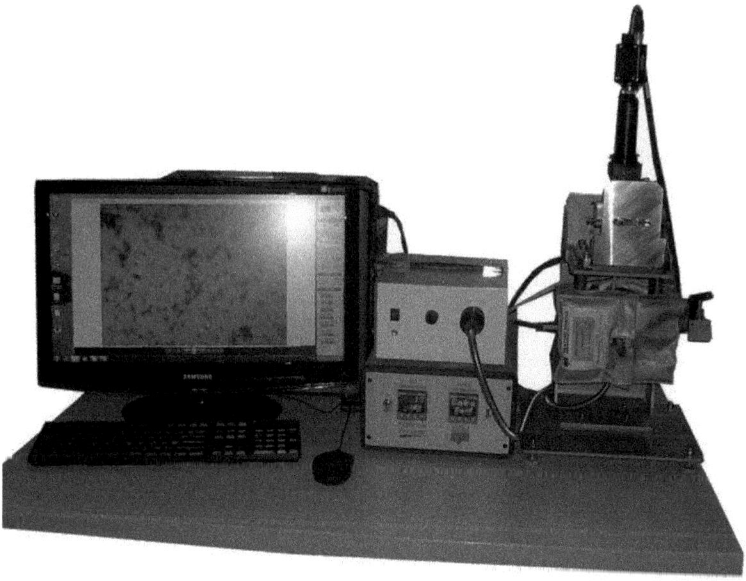

FIGURE 1.29 High-pressure electronic microscope capturing the segregation of an asphalthene-rich phase. *The image is a courtesy of Vinci Technologies S.A. Reproduced with permission.*

phase, which concentration clearly increases with pressure reduction still above the bubble point. With gas liberation, the remaining heavier fractions (specially the resins) become more concentrated, able back to support the asphaltene suspension, the latter being more soluble in the heavier oil. The pressure at which asphaltenes are totally resolubilized is called *lower onset pressure* (Pedersen et al., 2014). The kinetics of asphaltene resolubilization is not fully dominated in the literature and not always the lower onset curve is reported. Such discussion is beyond the scope of this book. Details can be found in Hammami et al. (1999). Fig. 1.30 shows an example of phase diagram of a live oil with some points of both upper-onset and bubble-point curves, experimentally measured (González et al., 2008). Single-phase reservoirs may exhibit asphaltene precipitation by pressure reduction far below the static pressure. However, for reservoirs showing two liquid phases in equilibrium, the production of the upper oil zone may cause deposition of the new phase just at reservoir conditions, due to the closeness to the onset curve determined in situ. As we will see in Chapter 7, there are gas-cap reservoirs (lights excess) over two distinct liquid phases. Reproduction of the compositional grading observed along the three phases is essential for the applied thermodynamic modeling to support the reliability of the field's production forecasts.

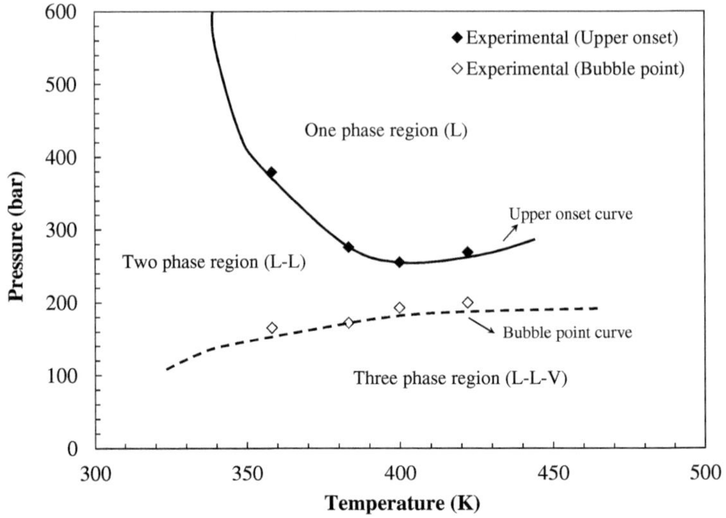

FIGURE 1.30 Live oil phase diagram showing upper onset and bubble point curves, with monophasic, biphasic and three-phase regions. *Adapted from Gonzalez, D.L., Vargas, F.M, Hirasaki, G.J. and Chapman, W.G., Modeling study of CO_2-induced asphaltene precipitation, Energy Fuels, 22, pp. 757–762, 2008.*

1.5 FINAL REMARKS

Predictability of compositional variation along depth, as a result of gravity segregation, depends, as we will see later, on the experimental determination of the PVT properties of a high-quality sample taken at a certain reference level, from which any other property can be simulated at any other point of the reservoir by the hypothesis of thermodynamic equilibrium or steady state. The calculation of compositional grading along depth is essential for the initialization of the composition in different fluid regions in commercial simulators, directly influencing the reliability of the production curves, once we are given a drainage plan, that is, a spatial configuration of producer and injector wells. It is important to emphasize that the real conditions of the reservoir during its production will not necessarily be the same for which the PVT analysis was done, since the gas reinjection or its liberation below P_{sat} change the original fluid composition in its trajectory toward the producer well. Therefore, a well-fitted equation of state is essential for simulating volumetric properties at any point of the reservoir at any time.

1.6 EXERCISES

1. (Adapted from Rosa et al., 2005): Given the sketch below, calculate, for the pressures of 2500, 2000, 1000, and 500 psia:
   a. The oil FVF (Bo)
   b. The solubility ratio (Rs)
   c. The gas FVF (Bg)

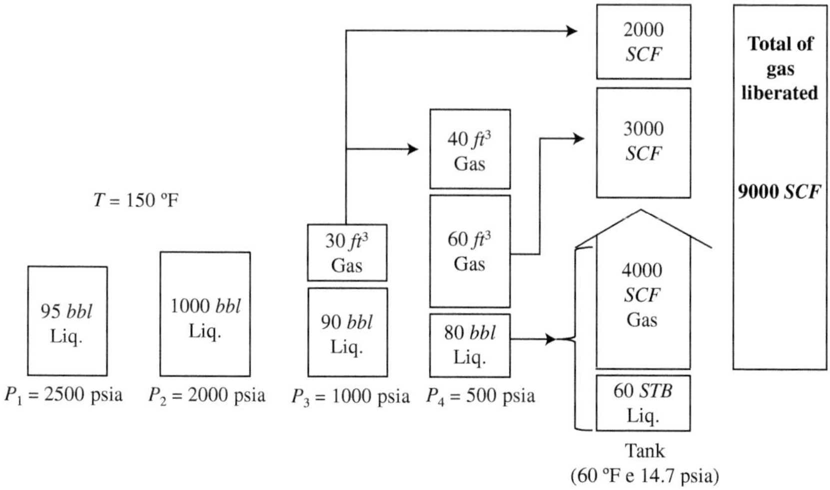

Legend: *bbl* = barrel; *SCF* = *standard cubic feet* (ft^3 std.); *STB* = *stock tank barrel* (oil barrel at standard conditions).

2. Considering the following properties for the reservoir fluid:

   $P_{sat} = P_{res} = 900$ psia
   $T_{res} = 140°F$
   Rs = 300 scf/stb
   $Bo_{res} = 1.30$

   What is the free-gas volume in a separator operating at 150 psia and 85°F, which oil capacity is 50 *bbl*?
   PVT properties at separator conditions:
   Bo = 1.05; Rs = 50 scf/stb; $d_g = 0.7$
3. One intends to check the performance of a multiphase flow meter, placed at a certain depth of a well, for which the PVT properties are: Bo = 1.17; Rs = 50.0 m^3 std/m^3 std; and Bg = 0.0525. The oil and gas flow rates, both given by the meter in situ, are 112.4 m^3/day and 100.5 m^3/day. One relies on the flow rates at the separator: 98.5 m^3/day (oil) and 458.3 m^3/day (gas). PVT properties at separator conditions are: Bo = 1.04; Bg = 0.0732; Rs = 8 m^3 std/m^3 std. Calculate the errors of the rates given by the meter.
4. A surface sample was analyzed in the laboratory, showing 40°API. In the formation test, the well produced a cumulated volume of 125 m^3 in 6 h. Stabilized production GOR is 100 m^3 std/m^3 std. Surface gas gravity is 0.82 and reservoir-fluid Bo is 1.250. Obtain the reservoir-fluid density.
5. A mass balance in a dry-gas reservoir assumes that the produced moles (N_p) are equal to the initial amount (N_i) minus the current one (N_c), that is:

$$N_p = N_i - N_a$$

Let G_p be the produced gas volume (measured at standard conditions) between the initial (P_i) and current pressure (P). If G is the total volume of the reservoir gas (also measured at standard conditions), derive the working equation of the dry-gas reservoir:

$$\frac{P}{z} = \frac{P_i}{z_i} - \frac{P_i}{z_i}\frac{G_p}{G}$$

The correlations of Sutton (see Danesh, 1998) for the pseudocritical properties allow calculate the compressibility factor by the graph below:

$$P_{pc} = 708.75 - 57.5d_g$$

$$T_{pc} = 169.0 + 314.0d_g$$

in which d_g is the gas gravity, referred to the air.

Using these information, solve Problems 1.6 and 1.7:

6. A dry-gas reservoir is made up by:

Component	Mole Fraction	Pc (psia)	Tc (R)
Methane	0.75	673.1	343.2
Ethane	0.20	708.3	504.8
n-Hexane	0.05	440.1	914.2

Initial pressure and temperature were 4200 psia and 180°F, respectively. The reservoir produced for a long time and pressure registers informed:

P/z (psia)	Gp (10^9 scf)
4600	0
3700	1
2800	2

What will be the cumulated production of gas when the average pressure reaches 2000 psia?

7. A recently discovered dry-gas reservoir $\left(d_g = 0.85\right)$ has the following data:

$V_{\text{bulk}} = 1.776 \times 10^{10}\ \text{ft}^3$;
$P = 4290$ psia;
$T = 660$ R;
Porosity $= 0.19$.

Connate-water saturation (fraction of the available porous volume occupied by immobile water) = 0.20.

The parties signed a selling contract of 80 MM scf/day with a certain delivery pressure, which requires a minimum of 1200 psia in the reservoir.

How long will this demand be attended? After the end of the contract, the field will produce gas to be consumed internally in the production facilities. If the abandon pressure is 400 psia, obtain the total gas volume that will be produced in this period.

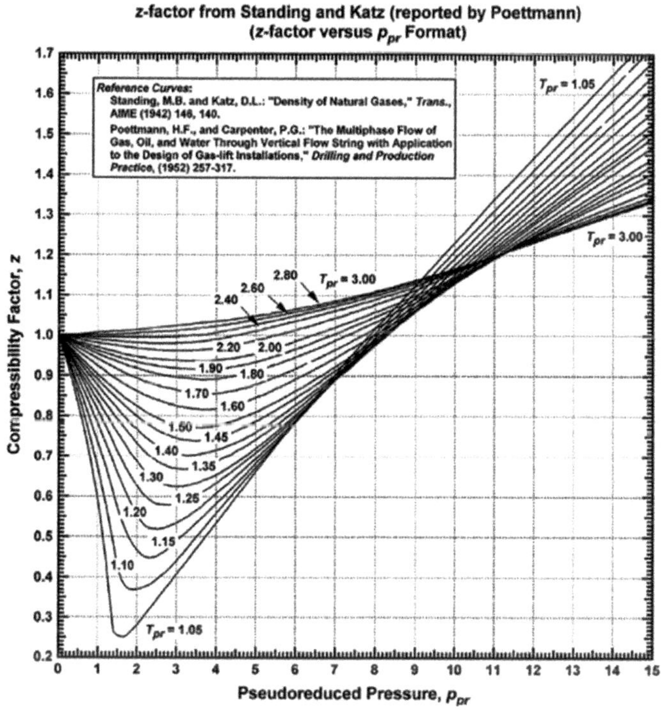

(This Image is copyright of American Institute of Mining, Metallurgical, and Petroleum Engineers, Inc. (AIME). Used with permission.)

REFERENCES

Danesh, A., 1998. PVT and Phase Behavior of Petroleum Reservoir Fluids. Elsevier Science, Amsterdam (Netherlands).

Ekundayo, J.M., Ghedan, S.G., Minimum miscibility pressure with slim tube apparatus – How unique is the value? SPE 165966-MS, Reservoir Characterization and Simulation Conference and Exhibition, Abu Dhabi, Setembro, 2013.

Gonzalez, D.L., Vargas, F.M., Hirasaki, G.J., Chapman, W.G., 2008. Modeling study of CO_2-induced asphaltene precipitation, Energy Fuels 22, 757–762.

Hammami, A., Phelps, C.H., Monger-McClure, T., 1999. Asphaltene precipitation from live oils: an experimental investigation of onset conditions and reversibility. Energy Fuels 1, 14–18.

McCain Jr., W.D., 1990. The Properties of Petroleum Fluids. Pennwell Publishing Company, Tulsa.

Pedersen, K.S., Christensen, P.L., Shaikh, J.A., 2014. Phase Behavior of Petroleum Reservoir Fluids. CRC Press, Boca Raton.

Rosa, A.J., Carvalho, R.S., Xavier, J.A.D., *Engenharia de reservatórios de petróleo*, Ed. Interciência, 2005.

Whitson, C.H., 1998. Manual PVT Analysis. Norsk Hydro, Field Development Technology, Reservoir Technology, Trondheim (Norway).

Chapter 2

Phase Equilibrium Thermodynamics

In majority of cases, the modeling of vapor–liquid equilibrium involving hydrocarbons (water being treated as an independent component) using cubic equations of state introduced in the 1970s is enough to ensure the robustness of equipments' project and the development of production systems. More complex mixtures involving macromolecules like asphaltenes, paraffins, and hydrates can lead the system to showing other phases as well as rock–fluid interactions. Such interactions can introduce chemical reactions that change permeability and porosity of the rock and require more elaborated thermodynamic models that are beyond the scope of this book. The asphaltene precipitation, which may be treated as a second ultra-viscous liquid phase that can be modeled using a more complex type equation of state (EoS), which may consider the molecular association effects in their polar-heteroatom sites, The Cubic Plus Association EoS (Kontogeorgis et al., 1996) can, for example, be employed for this task and will be discussed in detail in Chapter 7.

Here we will proceed to reproduce the Pressure, Volume and Temperature (PVT) analysis using traditional cubic equations and flash algorithms that represent the successive steps of differential liberation within the PVT cell as being closed-system equilibrium states. Parameters of these models will be estimated to fit the volumetric experimental data in order to guide the reader to achieve sensitivity about the influence of such parameters on properties prediction.

2.1 THE PHASE EQUILIBRIUM PROBLEM

It is worth to emphasize that each step of differential liberation consists of a thermodynamic equilibrium state at cell's temperature and pressure. These equilibrium stages can be understood as a succession of flashes that feed composition is the one of the saturated liquids remaining in the cell from the previous step. Therefore, one expects that the determination of the conditions for thermodynamic equilibrium in closed systems, as well as conventional flash and saturation pressure algorithms, be useful for the prediction of

Compositional Grading in Oil and Gas Reservoirs. DOI: http://dx.doi.org/10.1016/B978-0-12-812452-9.00002-2

composition and properties of the phases in equilibrium. So, let us make a brief review of such basic concepts before applying them to reproduce the phase behavior of reservoir fluids.

Starting from the so-called Fundamental Relation of Thermodynamics for internal energy, one has

$$dU = TdS - PdV + \sum_{i=1}^{nc} \mu_i dn_i \tag{2.1}$$

where U is the internal energy; T is the absolute temperature; S is the entropy; P is the pressure; V is the volume; and μ_i and n_i are, respectively, the chemical potential and number of moles of component i.

Let us consider a closed system with rigid walls, thermally insulated and without chemical reactions, in which there are nc components and nf phases, as sketched in Fig. 2.1. This system, when evolves toward equilibrium state, produces more entropy, until it reaches a maximum.

From Eq. (2.1), the entropy of each phase j is given by

$$dS_j = \frac{1}{T_j} dU_j + \frac{P_j}{T_j} dV_j - \sum_{i=1}^{nc} \frac{\mu_{ij}}{T_j} dn_{ij}, \quad \text{for } j = 1, \ldots, nf \tag{2.2}$$

So, the total entropy can be calculated by the sum of the entropies of the phases, that is

$$dS_T = \sum_{j=1}^{nf} \frac{1}{T_j} dU_j + \sum_{j=1}^{nf} \frac{P_j}{T_j} dV_j - \sum_{j=1}^{nf} \sum_{i=1}^{nc} \frac{\mu_{ij}}{T_j} dn_{ij} \tag{2.3}$$

For determining the equilibrium state, we need to maximize the system's total entropy, that is, to null its differential dS_T, subject to the restrictions of constant internal energy, volume, and mole numbers:

$$U = U_1 + U_2 + \cdots + U_{nf} \tag{2.4a}$$

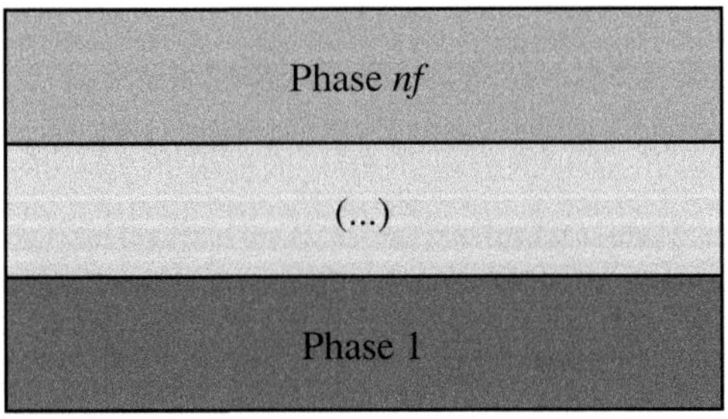

FIGURE 2.1 Phase equilibrium representation in a closed system without chemical reactions.

$$V = V_1 + V_2 + \cdots + V_{nf} \tag{2.4b}$$

$$n_i = n_{i,1} + n_{i,2} + \cdots + n_{i,nf}, \quad \text{for } i = 1, 2, \ldots, nc \tag{2.4c}$$

As one can notice, variables U_j, V_j, and n_{ij} in Eq. (2.3) are not independent due to relations expressed in Eqs. (2.4a)–(2.4c).

Differentiating these restrictions, knowing that U, V, and n_i are constant, we have

$$0 = dU_1 + dU_2 + \cdots + dU_{nf} \tag{2.5a}$$

$$0 = dV_1 + dV_2 + \cdots + dV_{nf} \tag{2.5b}$$

$$0 = dn_{i1} + dn_{i2} + \cdots + dn_{i,nf}, \quad \text{for } i = 1, 2, \ldots, nc \tag{2.5c}$$

So the properties of any phase (e.g., an arbitrary phase J) can be written in terms of other phase properties by difference, that is,

$$dU_J = -\sum_{\substack{j=1 \\ j \neq J}}^{nf} dU_j \tag{2.6a}$$

$$dV_J = -\sum_{\substack{j=1 \\ j \neq J}}^{nf} dV_j \tag{2.6b}$$

$$dn_{iJ} = -\sum_{\substack{j=1 \\ j \neq J}}^{nf} dn_{ij}, \quad \text{for } i = 1, 2, \ldots, nc \tag{2.6c}$$

Next step is to isolate the dependent phase J from the summations of Eq. (2.3):

$$\begin{aligned} dS_T = {} & \frac{1}{T_J} dU_J + \sum_{\substack{j=1 \\ j \neq J}}^{nf} \frac{1}{T_j} dU_j + \frac{P_J}{T_J} dV_J + \sum_{\substack{j=1 \\ j \neq J}}^{nf} \frac{P_j}{T_j} dV_j \\ & - \sum_{i=1}^{nc} \frac{\mu_{iJ}}{T_J} dn_{iJ} - \sum_{i=1}^{nc} \sum_{\substack{j=1 \\ j \neq J}}^{nf} \frac{\mu_{ij}}{T_j} dn_{ij} \end{aligned} \tag{2.7}$$

Substituting restrictions expressed in Eqs. (2.6a)–(2.6c) into Eq. (2.7), one reaches

$$dS_T = \frac{1}{T_J}\left(-\sum_{\substack{j=1\\ j\neq J}}^{nf} dU_j\right) + \sum_{\substack{j=1\\ j\neq J}}^{nf} \frac{1}{T_j}dU_j + \frac{P_J}{T_J}\left(-\sum_{\substack{j=1\\ j\neq J}}^{nf} dV_j\right) + \sum_{\substack{j=1\\ j\neq J}}^{nf} \frac{P_j}{T_j}dV_j - \sum_{i=1}^{nc}\frac{\mu_{iJ}}{T_J}\left(-\sum_{\substack{j=1\\ j\neq J}}^{nf} dn_{iJ}\right) - \sum_{i=1}^{nc}\sum_{\substack{j=1\\ j\neq J}}^{nf}\frac{\mu_{ij}}{T_j}dn_{ij} \tag{2.8}$$

Note that temperature, pressure, and chemical potentials in phase J that premultiply summations of Eq. (2.8) are independent of the internal terms, so they can be included in these summations as mere constants. After some rearranging in Eq. (2.8), one has

$$dS_T = \sum_{\substack{j=1\\ j\neq J}}^{nf}\left(\frac{1}{T_j} - \frac{1}{T_J}\right)dU_j + \sum_{\substack{j=1\\ j\neq J}}^{nf}\left(\frac{P_j}{T_j} - \frac{P_J}{T_J}\right)dV_j + \sum_{i=1}^{nc}\sum_{\substack{j=1\\ j\neq J}}^{nf}\left(\frac{\mu_{ij}}{T_j} - \frac{\mu_{iJ}}{T_J}\right)dn_{ij} \tag{2.9}$$

Now, all variables U_j, V_j, and n_{ij} are independent, U_J, V_J, and n_{iJ} being calculated by difference and therefore excluded from the summing index. So, the maximization of the entropy by nullifying its first derivative can only happen when every single term of the summations is simultaneously null.

Then, we can conclude that

$$\frac{1}{T_j} - \frac{1}{T_J} = \frac{P_j}{T_j} - \frac{P_J}{T_J} = \frac{\mu_{ij}}{T_j} - \frac{\mu_{iJ}}{T_J} = 0, \text{ for } i = 1, 2, \ldots, nc \text{ and } j = 1, 2, \ldots, nf \tag{2.10}$$

The equality of phase temperatures (all of them must be equal to T_J) implies that there is no heat flux between them, avoiding the increase in dS by thermal generation. Furthermore, there must not be pressure unbalance, which would dissipate mechanical energy, generating entropy; and, finally, chemical potential equality avoids mass transfer among phases, which would lead to entropy transport by convection, diffusion, or simple physical-state changing.

It is worth to emphasize that one would reach the same conclusion if one did this analysis on Gibbs or Helmholtz energies, minimizing these thermodynamic potentials with respect to their respective natural variables $(T, P, \underline{N})$ or $(T, V, \underline{N})$, where the underscore means vector magnitude, that is, $\underline{N} = [n_1, n_2, \ldots, n_{nc}]$.

The setting of temperature and pressure for determining the equilibrium state of a simple system is intuitive. Actually, the chemical potential equality provides the main equations for phase equilibrium. Obtaining "constitutive equations" for chemical potential of each component is, therefore, crucial for the accuracy of the calculations of their distribution among the phases in equilibrium. The most commonly used models in petroleum industry are the equations of state, which definition we will now consider.

The mathematical relation between the partial derivatives of the fundamental relation and its natural variables is called EoS. In the case of internal energy—Eq. (2.1)—natural variables are *S*, *V*, and *N*. One can notice, for example, that the temperature is the partial derivative of internal energy with entropy, keeping volume and mass constant:

$$T = f(S, V, \underline{N}) = \left(\frac{\partial U}{\partial S}\right)_{V,\underline{N}} \quad (2.11)$$

Analogously,

$$P = f(S, V, \underline{N}) = -\left(\frac{\partial U}{\partial V}\right)_{S,\underline{N}} \quad (2.12)$$

The definitions of other thermodynamic potentials give rise to other equations of state. The enthalpy (*H*) is the Legendre transformation of the internal energy $H = H(S, P, \underline{N})$ (Callen, 1985):

$$H = U - \left(\frac{\partial U}{\partial V}\right)_{S,\underline{N}} V = U + PV \quad (2.13)$$

$$dH = dU + PdV + VdP \quad (2.14)$$

$$dH = TdS + VdP + \sum_{i=1}^{nc} \mu_i dn_i \quad (2.15)$$

Therefore, the natural variables of enthalpy are *S*, *P*, and $\underline{N}$.

The Helmholtz energy (*A*) is the Legendre transformation of *U* defined as

$$A = U - \left(\frac{\partial U}{\partial S}\right)_{V,\underline{N}} S = U - TS \quad (2.16)$$

$$dA = dU - TdS - SdT \quad (2.17)$$

$$dA = -SdT - PdV + \sum_{i=1}^{nc} \mu_i dn_i \quad (2.18)$$

The natural variables of Helmholtz energy are *T*, *V*, and $\underline{N}$. The pressure (*P*), for example, can be written as a function of these natural variables, providing an EoS much more useful in practice, since *T*, *V*, and $\underline{N}$ are easily measurable.

$$P = f\left(T, V, \underline{N}\right) = -\left(\frac{\partial A}{\partial V}\right)_{T,\underline{N}} \tag{2.19}$$

Finally, the Gibbs energy (G) is given by

$$G = U + PV - TS = H - TS \tag{2.20}$$

$$dG = dU + PdV + VdP - TdS - SdT = dH - TdS - SdT \tag{2.21}$$

$$dG = -SdT + VdP + \sum_{i=1}^{nc} \mu_i dn_i \tag{2.22}$$

Once one sets T, P, and $\underline{N}$ (as in the majority of the phase equilibrium algorithms applied in the industry), the Gibbs energy is the thermodynamic potential to be minimized at the equilibrium condition.

Note that the chemical potential can be calculated by many ways starting from different fundamental relations:

$$\mu_i = \left(\frac{\partial U}{\partial n_i}\right)_{S,V,n_{j\neq i}} = \left(\frac{\partial H}{\partial n_i}\right)_{S,P,n_{j\neq i}} = \left(\frac{\partial A}{\partial n_i}\right)_{T,V,n_{j\neq i}} = \left(\frac{\partial G}{\partial n_i}\right)_{T,P,n_{j\neq i}}, \quad \text{for } i = 1,\ldots,nc \tag{2.23}$$

Among them all, the Gibbs energy formula is the most convenient one to obtain the chemical potential regarding the simplifications inherent to the fact that easily measurable T, P, and $\underline{N}$ are the natural variables.

This observation gets clearer when one considers the concept of partial molar property, from which one can derive both Euler's Theorem and Gibbs–Duhem equation, which we will discuss next.

Let $M(= U, H, S,\ A,\ or\ G)$ be some extensive property. By definition, partial molar property of component i ($\overline{M}_i$) is given by

$$\overline{M}_i = \left(\frac{\partial M}{\partial n_i}\right)_{T,P,n_{j\neq i}}, \quad \text{for } i = 1,\ldots,nc \tag{2.24}$$

As the thermodynamic properties are additive in mass basis, one has

$$n\overline{M} = M(T, P, \underline{N}) \tag{2.25}$$

where the molar property $\overline{M}$ is written as $\overline{M} = \overline{M}(T, P, \underline{x})$. Differentiating M, one has

$$dM = d\left(n\overline{M}\right) = \left[\frac{\partial\left(n\overline{M}\right)}{\partial T}\right]_{P,\underline{N}} dT + \left[\frac{\partial\left(n\overline{M}\right)}{\partial P}\right]_{T,\underline{N}} dP + \sum_{i=1}^{nc}\left[\frac{\partial\left(n\overline{M}\right)}{\partial n_i}\right]_{T,P,n_{j\neq i}} dn_i \tag{2.26}$$

As vector $\underline{N}$ is constant, it implies that all number of moles of the components (n_i, for i=1,...,nc) are also constant, resulting in

$$\left[\frac{\partial(n\overline{M})}{\partial T}\right]_{P,\underline{N}} = \left[n\frac{\partial \overline{M}}{\partial T}\right]_{P,\underline{N}} \tag{2.27}$$

since $n = \sum_{i=1}^{nc} n_i$.

So, Eq. (2.26) can be written as

$$dM = n\left(\frac{\partial \overline{M}}{\partial T}\right)_{P,\underline{N}} dT + n\left(\frac{\partial \overline{M}}{\partial P}\right)_{T,\underline{N}} + \sum_{i=1}^{nc} \overline{M}_i dn_i \tag{2.28}$$

The last term of Eq. (2.28) might be written in terms of mole fractions $x_i = n_i/n$:

$$dn_i = ndx_i + x_i dn, \quad \text{for } i = 1, \ldots, nc \tag{2.29}$$

Replacing Eq. (2.29) into Eq. (2.28), we get

$$dM = nd\overline{M} + \overline{M}dn = n\left(\frac{\partial \overline{M}}{\partial T}\right)_{P,\underline{N}} dT + n\left(\frac{\partial \overline{M}}{\partial P}\right)_{T,\underline{N}} + n\sum_{i=1}^{nc} \overline{M}_i dx_i + dn\sum_{i=1}^{nc} \overline{M}_i x_i \tag{2.30}$$

Grouping the terms in n (related to the system size) and dn (related to the variation on the system size), one has

$$n\left[\left(\frac{\partial \overline{M}}{\partial T}\right)_{P,\underline{N}} dT + \left(\frac{\partial \overline{M}}{\partial P}\right)_{T,\underline{N}} + \sum_{i=1}^{nc} \overline{M}_i dx_i - d\overline{M}\right] + dn\left[\sum_{i=1}^{nc} \overline{M}_i x_i - \overline{M}\right] = 0 \tag{2.31}$$

Eq. (2.31) is general and valid for any condition and any extensive thermodynamic property as well $M(U, S, H, G, A, \ldots)$. It is also necessarily valid for any size n and size variation dn. Note that, if the terms in brackets are invariant with n and dn, there is only one way to make sure that the left-hand side will be zero:

$$\overline{M} = \sum_{i=1}^{nc} \overline{M}_i x_i \text{ or } M = \sum_{i=1}^{nc} \overline{M}_i n_i \tag{2.32}$$

Eq. (2.32) characterizes the Euler's theorem. In the same way, we have

$$\left(\frac{\partial \overline{M}}{\partial T}\right)_{P,\underline{N}} dT + \left(\frac{\partial \overline{M}}{\partial P}\right)_{T,\underline{N}} + \sum_{i=1}^{nc} \overline{M}_i dx_i - d\overline{M} = 0 \tag{2.33}$$

Applying the Euler's Theorem, that is, differentiating Eq. (2.32) and substituting into Eq. (2.33), we have

$$d\overline{M} = \sum_{i=1}^{nc} \overline{M}_i dx_i + \sum_{i=1}^{nc} x_i d\overline{M}_i \tag{2.34}$$

$$\left(\frac{\partial \overline{M}}{\partial T}\right)_{P,\underline{N}} dT + \left(\frac{\partial \overline{M}}{\partial P}\right)_{T,\underline{N}} dP - \sum_{i=1}^{nc} x_i d\overline{M}_i = 0 \tag{2.35}$$

Eq. (2.35) is the Gibbs–Duhem equation. Setting $\overline{M} = \overline{G}$ and comparing Eqs. (2.22) and (2.35), one can say then that the chemical potential of component i is the partial molar Gibbs energy of this component. So, one can write

$$d\mu_i = -\overline{S}_i dT + \overline{V}_i dP, \quad \text{for } i = 1, \ldots, nc \tag{2.36}$$

Once one has an EoS available, keeping both temperature and composition constant, the chemical potential can be calculated by integrating Eq. (2.36):

$$d\mu_i = \overline{V}_i dP(T, \underline{x} = \text{constants}) \rightarrow \mu_i = \mu_i^0 + \int_0^P \overline{V}_i dP, \quad \text{for } i = 1, \ldots, nc \tag{2.37}$$

where μ_i^0 is the chemical potential of component i in an arbitrary reference state; for example, ideal gas at the zero-pressure limit.

2.2 THE DEFINITION OF FUGACITY

At the arbitrary reference state, the chemical potential μ_i^0 is known and does not need an EoS to be calculated. For the rest of its scale, in a convenience matter, Professor Gilbert N. Lewis[1] defined, in 1905, the fugacity, a parameter that responds for the influence of both pressure and composition in the integration of the molar volume of component i.

$$d\mu_i \equiv RTd(ln\hat{f}_i), \text{ at constants } T \text{ and } \underline{x}, \quad \text{for } i = 1, \ldots, nc \tag{2.38}$$

where superscript "^" refers to the fugacity of component i in a mixture.

Eq. (2.38) is a general and unrestricted definition, which tends necessarily to the same limit as Eq. (2.37) when pressure tends to zero, that is,

$$\lim_{P \to 0} d\mu_i = \overline{V}_i dP = \frac{RT}{P} dP = RTdlnP, \quad \text{for } i = 1, \ldots, nc \tag{2.39}$$

1. Gilbert Newton Lewis (1875-1946) was an American physical chemist. In addition to his contributions to thermodynamics, he is also known for the definition of a covalent bond and for his theory of acid and bases.

$$\lim_{P\to 0}\hat{f}_i = x_iP \rightarrow \lim_{P\to 0} RTd\ln\hat{f}_i = RTd\ln(x_iP) = RTd\ln P, \quad \text{at constant } \underline{x}, \quad \text{for } i = 1,\ldots,nc \tag{2.40}$$

For a pure component, this limit corresponds to the pressure of the system. Fugacity has, therefore, units of pressure. According to the definition of Eq. (2.38), the chemical potential at any state of interest may be calculated by

$$\mu_i = \mu_i^0 + RT\ln\left(\frac{\hat{f}_i}{f_i^0}\right), \quad \text{for } i = 1,\ldots,nc \tag{2.41}$$

Therefore, in a vapor–liquid equilibrium, for example, equating the chemical potential of component i in both phases also means to equate its fugacities, since μ_i^0 and f_i^0 are both defined for an arbitrary reference state that does not depend on the phase considered, having the same values in both liquid and vapor phases, which leads to automatic canceling out. So, we can write

$$\mu_i^V = \mu_i^L, \quad \text{for } i = 1,\ldots,nc \tag{2.42}$$

or

$$\hat{f}_i^V = \hat{f}_i^L, \quad \text{for } i = 1,\ldots,nc \tag{2.43}$$

The fugacity coefficient ($\hat{\varnothing}_i$) is the relation between the fugacity of component i in the mixture and its fugacity in an ideal gas mixture at the same pressure and temperature:

$$\hat{\varnothing}_i(T,P,\underline{x}) = \frac{\hat{f}_i(T,P,\underline{x})}{\hat{f}_i^{ig}(T,P,\underline{x})} = \frac{\hat{f}_i(T,P,\underline{x})}{x_iP}, \quad \text{for } i = 1,\ldots,nc \tag{2.44}$$

where $\underline{x} = (x_1, x_2, \ldots, x_{nc})$ is the vector of mole fractions.

Starting again from Eq. (2.38) and integrating it from the zero-pressure limit to the system actual pressure, keeping both T and $\underline{x}$ constants, we have

$$\mu_i = \mu_i^0 + RT\ln\frac{\hat{f}_i(T,P,x)}{f_i^0} = \mu_i^0 + \int_0^P \overline{V}_i dP, \quad \text{for } i = 1,\ldots,nc \tag{2.45}$$

Repeating the integration, this time considering the mixture as an ideal gas, we have

$$\mu_i^{ig} = \mu_i^0 + RT\ln\frac{\hat{f}_i^{ig}(T,P,\underline{x})}{f_i^0} = \mu_i^0 + \int_0^P \overline{V}_i^{ig} dP, \quad \text{for } i = 1,\ldots,nc \tag{2.46}$$

Subtracting Eq. (2.46) of Eq. (2.45), one gets

$$ln\hat{\varnothing}_i(T,P,\underline{x}) = \frac{1}{RT}\int_0^P \left(\overline{V}_i - \frac{RT}{P}\right)dP, \quad \text{for } i = 1,\ldots,nc \tag{2.47}$$

One then concludes that the limit of the fugacity coefficient of component i in a mixture that behaves like an ideal gas that is equal to unity.

With an EoS, it is possible to integrate Eq. (2.47) and obtain an expression for the fugacity coefficient. Replacing then Eq.(2.44) into Eq. (2.43), the chemical potential equality condition for phase equilibrium reduces to

$$\hat{\varnothing}_i^V(T,P,\underline{y})y_i = \hat{\varnothing}_i^L(T,P,\underline{x})\,x_i, \quad \text{for } i = 1,\ldots,nc \tag{2.48}$$

where y_i and x_i are, respectively, the vapor- and liquid-phase mole fractions. Now, let us give some insight about the thermodynamic model that provides the constitutive equations for chemical potentials, fugacities, and other properties as well: the EoS.

2.3 PENG–ROBINSON EQUATION OF STATE

The most widely used EoS in petroleum industry is the one introduced by Peng and Robinson (1976). Using Helmholtz energy's natural variables, it states P as a function of T, V, and $\underline{N}$, applying an empirical cubic form derived from the pioneer van der Walls' Nobel-Prize commended equation:

$$P = \frac{RT}{\overline{V} - b} - \frac{a(T)}{\overline{V}(\overline{V} + b) + b(\overline{V} - b)} \tag{2.49}$$

in which $\overline{V} = V/n$ is the molar volume, and the parameters $a(T)$ and b can be obtained by classic mixing rules as follows:

$$a(T) = \sum_{i=1}^{nc}\sum_{j=1}^{nc} y_i y_j (a_i a_j)^{1/2}(1 - k_{ij}) \tag{2.50}$$

$$b = \sum_{i=1}^{nc} y_i b_i \tag{2.51}$$

where pure component parameters are given by

$$a_i(T) = 0.45724\frac{R^2T_{c,i}^2}{P_{c,i}}\alpha_i(T), \quad \text{for } i = 1,\ldots,nc \tag{2.52}$$

$$b_i = 0.07780\frac{RT_{c,i}}{P_{c,i}}, \quad \text{for } i = 1,\ldots,nc \tag{2.53}$$

$$\alpha_i = \left[1+\kappa_i\left(1-\sqrt{\frac{T}{T_{c,i}}}\right)\right]^2, \quad \text{for } i=1,\ldots,nc \tag{2.54}$$

$$\kappa_i = 0.37464 + 1.5422\omega_i - 0.26992\omega_i^2, \quad \text{for } i=1,\ldots,nc \tag{2.55}$$

In Eqs. (2.52)–(2.55), $T_{c,i}$, $P_{c,i}$, and ω_i are, respectively, the critical temperature, critical pressure, and acentric factor of component i. k_{ij} is the binary interaction parameter, which responds for the attraction energies (pairwise) between molecules i and j. It is the main adjustable parameter to fit the PVT experimental data. In the following, we show the expression of the fugacity coefficient, obtained by integrating Eq. (2.47), using Peng–Robinson EoS (Eq. (2.49)):

$$\ln\hat{\varnothing}_i^V = \frac{b_i}{b^V}\left(Z^V-1\right) - \ln\left(Z^V - \frac{Pb^V}{RT}\right) - \frac{a^V}{2\sqrt{2}b^V RT}\left(\frac{2\sum_{j=1}^{nc} y_j a_{ij}}{a^V} - \frac{b_i}{b^V}\right)\ln\left(\frac{Z^V + (\sqrt{2}+1)\frac{Pb^V}{RT}}{Z^V - (\sqrt{2}-1)\frac{Pb^V}{RT}}\right), \text{ for } i=1,\ldots,nc \tag{2.56}$$

$$\ln\hat{\varnothing}_i^L = \frac{b_i}{b^L}\left(Z^L-1\right) - \ln\left(Z^L - \frac{Pb^L}{RT}\right) - \frac{a^L}{2\sqrt{2}b^L RT}\left(\frac{2\sum_{j=1}^{nc} y_j a_{ij}}{a^L} - \frac{b_i}{b^L}\right)\ln\left(\frac{Z^L + (\sqrt{2}+1)\frac{Pb^L}{RT}}{Z^L - (\sqrt{2}-1)\frac{Pb^L}{RT}}\right), \text{ for } i=1,\ldots,nc \tag{2.57}$$

where Z^V and Z^L are, respectively, the compressibility factors of vapor and liquid phases. Given T, P, and $\underline{N}$ of each phase, the solution of the EoS for the molar volume provides automatically the value of Z by the relation $Z = P\overline{V}/(RT)$.

2.4 VAPOR–LIQUID EQUILIBRIUM CALCULATIONS

In this section, one introduces the equations that model the vapor–liquid equilibrium at each step of differential liberation presented in Chapter 1. In order that the thermodynamic behavior of the reservoir fluid is well described, it is necessary that its PVT properties are reasonably predicted by the EoS adopted. This one, in turn, must satisfy as much accurately as possible the chemical potential equality at each equilibrium stage. *Flash* is the unit operation through which one separates components by volatility difference, in a single stage. This equilibrium stage is reached when a liquid mixture is heated up and/or

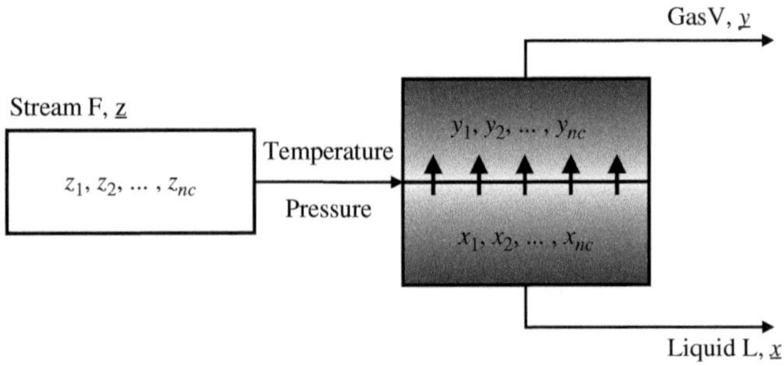

FIGURE 2.2 Single-equilibrium stage characterizing the Flash: global composition $\underline{z}$ (not to be confused with compressibility factor, Z) of stream F distributes through phases V (of composition $\underline{y}$) and L (of composition $\underline{x}$).

depressurized to stimulate the appearance of a vapor phase. Or, analogously, when a gaseous mixture is cooled down or pressurized to cause a condensation. Or, still, when a retrograde gas mixture is depressurized. Regardless the situation, once temperature and pressure are set for each stage, the original-fluid composition distributes by the two phases in equilibrium and this mass balance must be respected. Fig. 2.2 shows a general sketch of this unit operation on a stream of arbitrary molar rate F.

In molar basis, the material balance for component i provides the following equation:

$$Fz_i = Vy_i + Lx_i, \quad \text{for } i = 1, \ldots, nc \tag{2.58}$$

Defining the vaporized fraction $\beta = V/F$, one can write

$$z_i = \beta y_i + (1-\beta)x_i, \quad \text{for } i = 1, \ldots, nc \tag{2.59}$$

Besides, the chemical potentials of each component are the same for both liquid and vapor phases, resulting in the relation expressed in Eq. (2.48). Eqs. (2.59) and (2.48) make up a system of $(nc-1) + nc = 2nc - 1$ equations. Variables are x_i, y_i, and β; that is $(nc-1) + (nc-1) + 1 = 2nc - 1$ unknowns. Therefore, the system presents no degree of freedom, being perfectly determined.

Some simulators solve material balance equations in terms of the extensive mole number variables, which, differently from mole fractions, are all independent, that is,

$$N_i = n_i^L + n_i^V, \quad \text{for } i = 1, \ldots, nc \tag{2.60}$$

for which, $z_i = \frac{N_i}{\sum_{k=1}^{nc} N_k}$, $x_i = \frac{n_i^L}{\sum_{k=1}^{nc} n_k^L}$, and $y_i = \frac{n_i^V}{\sum_{k=1}^{nc} n_k^V}$.

In this case, the system turns to have $2nc$ equations for $2nc$ unknowns, which are n_i^L and n_i^V (for i=1,. . .,nc), in which the vaporized fraction is given by $\beta = \sum_{i=1}^{nc} n_i^V / \sum_{i=1}^{nc} N_i$ and obtained afterwards.

Naturally, one expects the results of both formulations to be equivalent. Prioritizing the former formulation (with intensive variables), there is a procedure granted in the literature, proposed by Rachford and Rice (1952) to generate good initial guesses for further applying of a robust numerical method that solves Eqs. (2.48) and (2.59) simultaneously. These authors opted by decoupling them, making convergence easier when one is far from solution.

Defining the so-called equilibrium constant, equilibrium ratio, or distribution coefficient of component i:

$$K_i \equiv \frac{y_i}{x_i} = \frac{\hat{\varnothing}_i^L(T, P, \underline{x})}{\hat{\varnothing}_i^V(T, P, \underline{y})}, \quad \text{for } i = 1, \ldots, nc \tag{2.61}$$

And replacing Eq. (2.61) into Eq. (2.59), we have

$$z_i = \beta K_i x_i + (1 - \beta) x_i, \quad \text{for } i = 1, \ldots, nc \tag{2.62}$$

Isolating x_i on the left-hand side of Eq. (2.62),

$$x_i = \frac{z_i}{1 + \beta(K_i - 1)}, \quad \text{for } i = 1, \ldots, nc \tag{2.63}$$

Applying again the definition of equilibrium constant to Eq.(2.63), we have

$$y_i = \frac{K_i z_i}{1 + \beta(K_i - 1)}, \quad \text{for } i = 1, \ldots, nc \tag{2.64}$$

Knowing that $\sum_{i=1}^{nc} y_i = 1$ and that $\sum_{i=1}^{nc} x_i = 1$, one can sum up Eqs. (2.63) and (2.64) over index i (i.e., over all components) and subtract them to obtain

$$\sum_{i=1}^{nc} y_i - \sum_{i=1}^{nc} x_i = \sum_{i=1}^{nc} \frac{z_i(K_i - 1)}{1 + \beta(K_i - 1)} = f(\beta) = 0 \tag{2.65}$$

With a good guess for the equilibrium constants, Eq. (2.65) turns to be a function of β (which is necessarily between 0 and 1) only. It can be solved then by Newton Raphson's method, for example. The correlation proposed by Wilson (1969), which does not depend on composition, has shown to be a good option for hydrocarbons:

$$\ln K_i = 5.37(1 + \omega_i)\left(1 - \frac{T_{ci}}{T}\right) + \ln\left(\frac{P_{ci}}{P}\right), \quad \text{for } i = 1, \ldots, nc \tag{2.66}$$

Once determined the vaporized fraction β, the mole fractions x_i and y_i might be estimated, respectively, by Eqs. (2.63) and (2.64). So, the updated compositions might be inserted in the fugacity coefficient expressions (Eq. (2.61)) to renew the values of the equilibrium constants, then getting back to Eq. (2.65) to solve it for β once more till one reaches the desired convergence.

Eq. (2.65) is the so-called Rachford–Rice equation and the method proposed to its solution is named *successive substitution*. In some simulators, as already mentioned, the method of successive substitution is applied until one reaches a convergence tight enough for a robust method (e.g., Newton) to be successfully employed to solve Eqs. (2.59) and (2.48) simultaneously, starting form good initial guesses for β, x_i, and y_i. Fig. 2.3 shows the algorithm of successive substitution method.

Differential liberation is a succession of flashes, which feed composition at each stage is the composition of the liquid remaining in the cell from the previous step. Given the volume occupied by the residual oil in a *flash* at 60°F, all PVT properties might be calculated and compared with the experimental data. Then, some parameters of the EoS are adjusted to minimize the

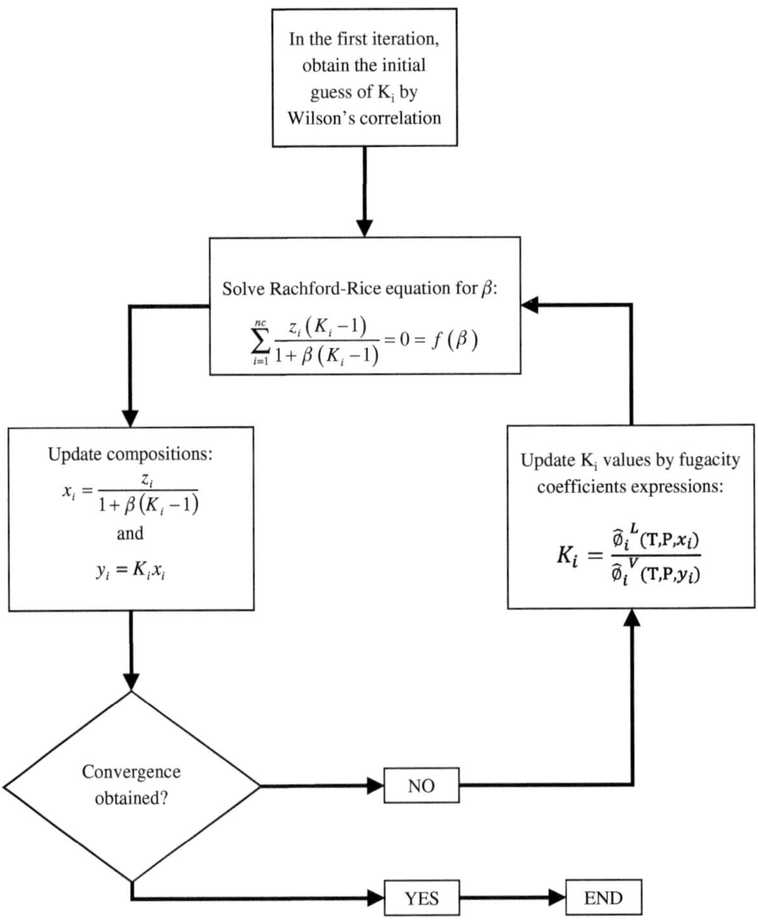

FIGURE 2.3 *Flash* algorithm scheme by successive substitution method, setting $\underline{z}$, T, and P.

objective function, that is, the sum of the square errors of each experimental point of the liberation (experimental errors not considered; see Appendix E for a more detailed discussion).

2.5 EQUATION OF STATE FITTING BASED ON PVT ANALYSIS

Any EoS-fitting procedure starts from a suitable compositional modeling of the reservoir fluid. As already mentioned, the flash liberation assay provides this composition through a material balance between the chromatographic compositions of both gas and dead-oil, pondered by gas-oil ratio (GOR). It would be impossible to feed the high-pressure reservoir–fluid sample to a chromatograph, thence the importance of the indirect measurement of its composition from information of the streams generated at atmospheric pressure.

Eq. (2.59) is used to determine mole fractions z_i, vaporized fraction β being calculated from GOR. The relation between them both is in the transformation from volume to molar basis, using their respective definitions:

$$\mathrm{GOR} = \frac{\text{Volume of gas (std)}}{\text{Volume of oil (std)}} \tag{2.67}$$

$$\beta = \frac{n_{\mathrm{gas}}}{n_{\mathrm{gas}} + n_{\mathrm{oil}}} \tag{2.68}$$

Let us now define the so-called molar GOR:

$$\mathrm{GOR}_{\mathrm{molar}} = \frac{n_{\mathrm{gas}}}{n_{\mathrm{oil}}} \tag{2.69}$$

Comparing Eq. (2.69) to Eq. (2.68), one finds

$$\beta = \frac{\mathrm{GOR}_{\mathrm{molar}}}{\mathrm{GOR}_{\mathrm{molar}} + 1} \tag{2.70}$$

Considering the flash gas as ideal at standard conditions, to obtain the molar GOR from volumetric GOR, the numerator of the latter must be multiplied by $P_{\mathrm{std}}/(RT)$. As for the volume of oil to be transformed into mole numbers, it must be multiplied by its density (easily obtained from °API)[2] and divided by its molar mass (experimentally measured by cryoscopy). Therefore, one has

$$\mathrm{GOR}_{\mathrm{molar}} = \mathrm{GOR}\frac{P_{\mathrm{std}}/(RT)}{(\rho_{o,\mathrm{std}}/M_o)} \tag{2.71}$$

---

2. The American Petroleum Institute gravity, or API gravity, is a measure of how heavy or light a petroleum liquid is compared to water.

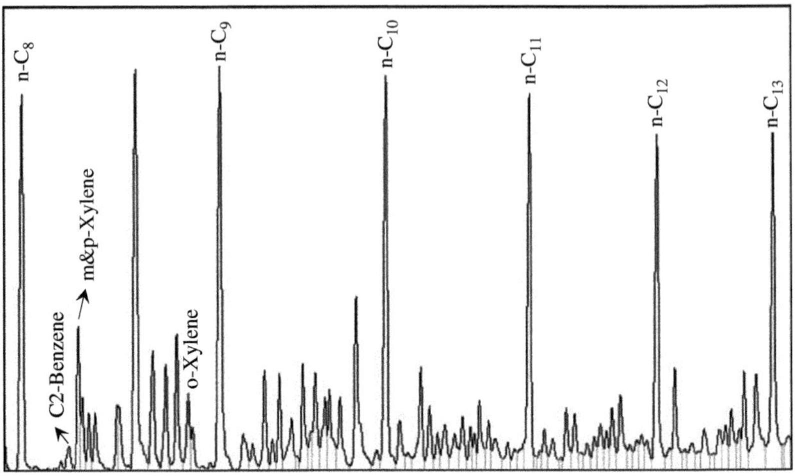

FIGURE 2.4 Section of a gas condensate chromatogram, locating the normal-paraffins peaks by retention time. Intermediate peaks will make up the pseudocomponents of Table 2.1. *Adapted from Folsta, K.C.B.M., Camargo, G.M. and Espósito, R.O., Gas condensate characterization from chromatogram areas and retention times, Fluid Phase Equilib., 292, 1–2, pp. 87–95, 2010.*

where P_{std} and $\rho_{o,\text{std}}$ are, respectively, the pressure and oil density, both at standard conditions. M_o is the molar mass of the oil.

Oil chromatography is analogous to its distillation curve. More volatile and less polar hydrocarbons leave the column faster due to their lower affinity with its polar matrix. The methodology to interpret the areas of the peaks is very simple: Normal paraffins have their retention times tabled and calibrated in the equipment, and then their peaks are easily identified. So are the lightest components (up to n-C_5). From C_6 on, the number of isomers starts increasing considerably, and all the peaks between two normal paraffins are lumped into the pseudocomponent of the same carbon number of the last one. Fig. 2.4 shows a section of a chromatogram used in a EoS-fitting procedure proposed by Folsta et al. (2010). For example, the mass fraction of pseudocomponent C_9 is the sum of the areas of all peaks between paraffin n-C_8 (exclusive) and paraffin n-C_9 (inclusive), divided by the total area of the chromatogram, which is also calibrated by a so-called external chromatographic standard. This is a mixture (e.g., a distillation cut), the composition of which is known a priori and also injected in the chromatograph before the sample. According to this method, C_{20+} (i.e., components with more than 20 carbon atoms grouped all together) does not even leave the column, its mass fraction being calculated by the difference between sample and external standard amounts. Details of the laboratory experimental procedure are beyond the scope of this book. See, for example, McCain (1990), Danesh (1998), and Folsta et al. (2010).

Pseudocomponents up to C_{19} have their average properties presented in Table 2.1, elaborated initially by Katz and Firoozabadi (1978) based on

TABLE 2.1 Basic Properties of Pseudocomponents, After Katz and Firoozabadi (1978)

Component	*M* (g/mol)	Density at 60°F (g/cm^3)	Average Boiling Point (°F)
C_6	84	0.685	146.93
C_7	96	0.727	199.13
C_8	107	0.749	242.33
C_9	121	0.768	289.13
C_{10}	134	0.782	330.53
C_{11}	147	0.793	370.13
C_{12}	161	0.804	407.93
C_{13}	175	0.815	442.13
C_{14}	190	0.826	476.33
C_{15}	206	0.836	510.53
C_{16}	222	0.843	542.93
C_{17}	237	0.851	571.73
C_{18}	251	0.856	595.13
C_{19}	263	0.861	616.73
C_{20}	275	0.866	641.93
C_{21}	291	0.871	663.53
C_{22}	300	0.876	686.93
C_{23}	312	0.881	706.73
C_{24}	324	0.885	726.53
C_{25}	337	0.888	748.13
C_{26}	349	0.892	766.13
C_{27}	360	0.896	784.13
C_{28}	372	0.899	802.13
C_{29}	382	0.902	816.53
C_{30}	394	0.905	834.53
C_{31}	404	0.909	850.73
C_{32}	415	0.912	866.93
C_{33}	426	0.915	881.33

(Continued)

TABLE 2.1 (Continued)

Component	*M* (g/mol)	Density at 60°F (g/cm^3)	Average Boiling Point (°F)
C_{34}	437	0.917	895.73
C_{35}	445	0.920	908.33
C_{36}	456	0.922	922.73
C_{37}	464	0.925	933.53
C_{38}	475	0.927	947.93
C_{39}	484	0.929	958.73
C_{40}	495	0.931	973.13
C_{41}	502	0.933	982.13
C_{42}	512	0.934	992.93
C_{43}	521	0.936	1003.73
C_{44}	531	0.938	1018.13
C_{45}	539	0.940	1027.13

Reproduced from Firoozabadi, A., 1999. Thermodynamics of Hydrocarbon Reservoirs. McGraw Hill, New York.

TABLE 2.2 Global Composition of a Reservoir–Fluid After Component Lumping from Chromatography of Flash Liberation

Components	Composition (mole%)
CO_2	0.2
C_1	43.0
C_2	2.2
C_3–C_4	2.8
C_5–C_7	5.9
C_8–C_{12}	12.7
C_{13}–C_{19}	12.0
C_{20+}	21.1
C_{20+} molar mass	573
C_{20+} specific gravity	0.9709
Flash GOR	62.71 m^{3}std /m^3 std
API	22.87

various dead-oil hydrocarbon distillation cuts. Table 2.2 presents molar-based compositions of a reservoir fluid lumped into its definite pseudocomponents, according to the user's convenience. Before this lumping, the molar mass of C_{20+} is calculated by difference between the dead-oil's one and the sum of all other average values up to C_{19}, the latter obtained in Table 2.1. The same thought holds for the density of the heavy fraction, also calculated by difference, given the experimental dead-oil API and the gravities of all other fractions till C_{19}, again obtained in Table 2.1. Reservoir–fluid composition is obtained straightforwardly by Eq. (2.62), after turning GOR into β. After lumping, the basic properties of the pseudocomponents might be calculated by the individual mole fractions of their internal components starting from the same information present in Table 2.1.

Two of the three basic properties shown in Table 2.1—molar mass, density, and average boiling point—are enough to obtain critical properties and acentric factors necessary to the components' characterization for the EoS, for, in practice, one observes that these three variables are correlated, for example, by Riazi and Daubert's (1987) equation:

$$M = 42.965\left(\frac{T_B}{1.8}\right)^{1.26007}(d_{60/60})^{4.98308} \times \exp\left[2.097\times10^{-4}\left(\frac{T_B}{1.8}\right) - 7.78712 d_{60/60} + 2.08476\times10^{-3}\left(\frac{T_B}{1.8}\right)d_{60/60}\right] \tag{2.72}$$

where T_B is given in Rankine and $d_{60/60}$ is the specific gravity at 60°F.

Among all empirical correlations used in the literature for the calculation of critical properties from basic ones, the correlations proposed by Cavett (1962) outstand:

$$T_c = 768.071 + 1.7134T_B - 0.0010834T_B^2 + 0.3889\times10^{-6}T_B^3 - 0.0089213T_B\text{API} + 0.53095\times10^{-5}T_B^2\text{API} + 0.32712\times10^{-7}T_B^2\text{API}^2 \tag{2.73}$$

$$\log_{10}P_c = 2.829 + 0.9412\times10^{-3}T_B - 0.30475\times10^{-5}T_B^2 + 0.15141\times10^{-8}T_B^3 - 0.20876\times10^{-4}T_B\text{API} + 0.11048\times10^{-7}T_B^2\text{API} + 0.1395\times10^{-9}T_B^2\text{API}^2 - 0.4827\times10^{-7}T_B\text{API}^2 \tag{2.74}$$

where T_B is in °F, T_c is in Rankine and P_c is in psia.

Recalling Eq. (1.1), the relation between specific gravity at 60°F ($d_{60/60}$) and API is given by

$$\text{API} = \frac{141.5}{d_{60/60}} - 131.5 \tag{2.75}$$

With both critical temperature and pressure, the correlation of Kesler and Lee (1976), for example, can be used to obtain the acentric factor, which wraps up the minimum characterization required by a cubic EoS:

$$\omega = \frac{\ln P_{Br} - 5.92714 + \dfrac{6.09649}{T_{Br}} + 1.28862\ln T_{Br} - 0.169347T_{Br}^{6}}{15.2518 - \dfrac{15.6875}{T_{Br}} - 13.4721\ln T_{Br} + 0.43577T_{Br}^{6}}, \text{ for } T_{Br} < 0.8 \tag{2.76}$$

$$\omega = -7.904 + 0.1352K_W - 0.007465K_W^2 + 8.359T_{Br} + \frac{1.408 - 0.01063K_w}{T_{Br}}, \quad \text{for } T_{Br} > 0.8 \tag{2.77}$$

where $T_{Br} = T_b/T_c$, $P_{Br} = P_{\text{atm}}/P_c$, and K_W is the Watson characterization factor, given by $K_W = T_B^{1/3}/d_{60/60}$, with T_B in Rankine.

In the case of C_{20+}, which properties are not tabled, density, and molar mass are calculated by difference between dead-oil properties (experimental API and molar mass) and the sum of them up to C_{19}. The average boiling temperature T_B might be calculated numerically solving Eq. (2.72). Good initial guesses for T_c, P_c, and ω might be calculated by Eqs. (2.73)–(2.77).

The mean values in Table 2.1 are not obviously exact and fixed for all petroleums. But it is interesting to note that the higher the aromaticity of the fraction, the higher its density and the lower its molar mass, relative to the analogous normal paraffin. For example, benzene has a density greater than the one of *n*-hexane, although its molar mass is lower. Fig. 2.5 shows clearly this behavior for density.

As all basic properties of C_{20+} (including its mole fraction) are calculated by difference, one expects the highest uncertainties—responsible by the errors in the EoS predictions—to be within the characterization of this lumped heavy fraction, which does not even come out of the chromatographic column. Following a thumb rule, they should follow the tendency of increasing (T_c and ω) and decreasing (P_c) with molar mass and must also be adjusted from this initial estimate to reproduce the experimental data of PVT analysis. Intuitively, we note that the critical properties, volume translations (discussed later in this text) and binary interaction coefficients of C_{20+} might be adjustable parameters of the EoS on its mission of reproducing the PVT properties by simulation of successive flashes.

Figs. 2.6 and 2.7 show, respectively, the tendencies of T_c and P_c with molar mass, according to Cavett's correlations, using basic properties of Table 2.1. Naji (2010) does a good review of many other correlations used in the literature to obtain critical properties and acentric factors from basic inputs. Firoozabadi (1999) plots some experimental data along with other correlations of the literature.

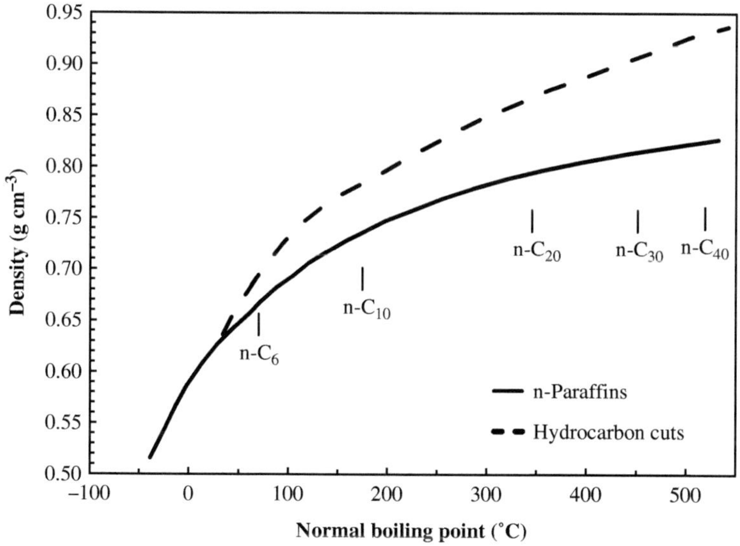

FIGURE 2.5 Densities at 60°F of hydrocarbon cuts (pseudocomponents) obtained by distillation, compared to those of normal paraffins. *Adapted from Firoozabadi, A., 1999. Thermodynamics of Hydrocarbon Reservoirs. McGraw Hill, New York.*

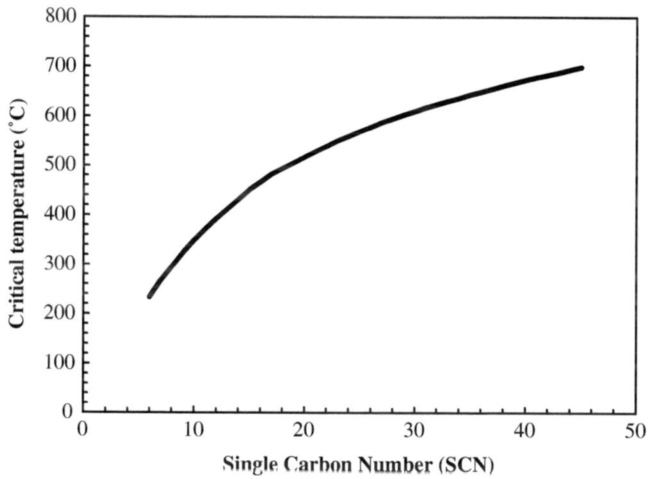

FIGURE 2.6 Critical temperature as a function of single carbon number by Cavett's correlation with basic inputs from Table 2.1.

Beside these, there might be other adjustable parameters in the EoS. For example, the binary interaction coefficients of methane with other hydrocarbons should increase with the latters' molar mass, due to increasingly repulsion resulting from the higher molar mass related polarity, even if this is still

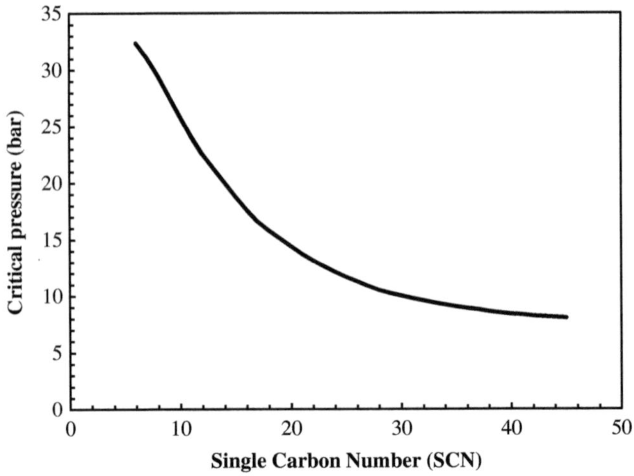

FIGURE 2.7 Critical pressure as a function of single carbon number by Cavett's correlation with basic inputs from Table 2.1.

relatively low. The values of $k_{C_1-C_n}$ must be coherent with this rule. Generally, one can propose a linear relation with molar mass:

$$k_{C_1-C_n} = A + B\,M_{C_n} \tag{2.78}$$

where M_{C_n} is the molar mass of C_n.

The constants of Eq. (2.78) could be re-estimated for any regional context. The correlation of Arbabi and Firoozabadi (1995) is a good initial guess for Peng–Robinson EoS:

$$k_{C_1-C_n} = 0.0289 + 0.0001633\,M_{C_n} \tag{2.79}$$

In reservoirs containing CO_2, its binary interaction coefficient with methane ($k_{CO_2-C_1}$) varies around 0.15 (Firoozabadi, 1999), decreasing with hydrocarbons' molar mass. A good initial estimate might be given by

$$k_{CO_2-C_n} = 0.1515 - 0.0002\,M_{C_n} \tag{2.80}$$

Other very interesting and influencing parameter for EoS fitting is the so-called volume shift (or volume translation), proposed initially by Peneloux et al. (1982). These authors noticed that cubic EoSs presented systematic errors in calculating densities of pure hydrocarbons in liquid state at room temperature. Then they proposed a summing constant responsible for such a difference, which tended to increase with hydrocarbons' molar mass.

$$v_{i,\text{cor}} = v_{i,\text{EoS}} + c_i, \quad \text{for } i = 1, \ldots, nc \tag{2.81}$$

where $v_{i,\text{EoS}}$ is the molar volume originally calculated by the EoS, and c_i is the volume translation of pure component i, simply a constant to be added to the original value to eliminate its error. The task is relatively simple for pure hydrocarbons up to C_6, since their experimental densities at ambient conditions are known. For mixtures, the thought is extended using a linear simple rule:

$$c = \sum_{i=1}^{nc} x_i c_i \tag{2.82}$$

Several correlations for c_i as a function of molar mass from C_6 have been proposed in literature (Jhaveri and Youngren, 1984; Pedersen et al., 2014). Its parameters have been adjusted for some practical cases. The original values of volume translations reported in commercial softwares might be correlated as a function of the logarithm of the molar mass of each component:

$$\frac{c_i}{b_i} = 0.0887 \ln M_i - 0.4668, \quad \text{for } i = 1, \ldots, nc \tag{2.83}$$

where b_i has already been defined in Eq. (2.53) for Peng–Robinson EoS.

It is important to emphasize that the Peneloux's volume translation does not change saturation pressure or the compositions of the phases in equilibrium, but only the volumes filled by them (Peneloux et al., 1982; Zabaloy and Brignole, 1997). It is an important parameter to obtain an accurate prediction of, for example, fluid saturations within the rock pores or even surface-equipment design.

Depending on concentrations of contaminants like CO_2, N_2, or H_2S, their volume shifts or binary interaction coefficients with hydrocarbons may become important. For CO_2 at low pressures (up to 250 bar), a volume shift of 0.2569 provides good results for its pure density estimation. For higher pressures, the value of -0.0718 gives better results.

In some cases, splitting procedures of C_{20+} fraction or other pseudocomponents' lumping may also be necessary, changing the number of variables in regression. In the development of oil fields with water injection, for which one does not change the global composition above the bubble point, the adjustment of critical properties and binary interaction coefficients of C_{20+} from good initial guesses is enough to obtain a good prediction of thermodynamic properties along reservoir simulation. In gas-injection projects, however, the resulting mixture can approach the critical point or even cross lines toward retrograde or wet gas. Mainly in the vicinity of gas-injector wells, heavies traces can influence some properties like dew points and condensates' densities. In these cases, a *swelling test*, using a gas composition close to the one that will be effectively injected should also be performed in the lab, its experimental data also being part of the objective function in the numerical regression. This increases its complexity, mainly with respect to critical transition evidenced, in this assay, by the change in the characteristic of the saturation pressure of the swollen fluid (bubble → dew point). In other words, in the simulation, some components with molar mass just above

C_{20}'s, which would remain longer in gaseous phase, might drop out in the retrograde liquid much above the real dew point, because they could be transferred to the liquid phase only because they are still grouped to heavier components, also overestimating the condensate volume. So, an enhanced detailing via splitting of C_{20+} fraction, with further lumping into heavier pseudocomponents, becomes necessary. There are many techniques of splitting and lumping in the literature, all applied with similar efficiency (see, e.g., Whitson, 1983; Shibata et al., 1987; Pedersen et al., 2014). In the following, we present two examples of real case studies: a water- and a gas-injection project. For the latter, we will detail the splitting/lumping procedure introduced by Shibata et al. (1987), optimized by Rochocz et al. (1997), and used by Espósito et al. (2000).

2.6 EXAMPLE 2.1: THE DIFFERENTIAL LIBERATION FITTING FOR THE RESERVOIR FLUID OF TABLE 2.2

Once obtained the basic properties of the reservoir fluid of Table 2.2, we can adopt, for example, the correlations already presented in Eqs. (2.73)–(2.77)—or any other one reviewed by Naji (2010)—for critical properties and acentric factors up to C_{19} in Table 2.2. As to initial estimates for C_{20+}, let us get its molar mass and gravity $d_{60/60}$ in Table 2.2: 573 g/mol and 0.9709, respectively. Solving Eq. (2.72) for T_B, we obtain the value of 1481.9 $R = 549.9$ K. Eqs. (2.73)–(2.77) may then be used to obtain T_c, P_c, and ω.

As the heavy fraction concentrates all the uncertainty coming from the more aromatic or paraffinic characteristic of the pseudocomponents relative to the average values of Table 2.1, one suggests a numerical regression of the critical properties of C_{20+} using, as initial estimates, the values obtained by the calculations described above. Simultaneously, we can adjust the binary interaction coefficients of methane with the other pseudocomponents starting from the values calculated, for example, by the correlation of Arbabi and Firoozabadi (1995). Volume translations estimated by Eq. (2.83) can also be optimized together with the other parameters cited above.

Figs. 2.8–2.10 show the comparison between results before and after regression for some PVT properties. Viscosity is not a volumetric property simulated by EoS and therefore its adjustment is easily carried out by fitting the parameters of some suitable empirical correlation after EoS volumetric properties are done. Pedersen et al. (2014) review two main correlations for viscosity: Lohrenz–Bray–Clark (LBC) and Corresponding States Principle (CSP). Their functional forms are large, the former showing a higher number of adjustable parameters (critical volumes inclusive) in relation to the latter. That is why LBC correlation, described in Eqs. (2.84)–(2.87), despite its more empirical character, is chosen to represent the fluid's viscosity, although CSP correlation—resumed in Eq. (2.88)—has a more consistent

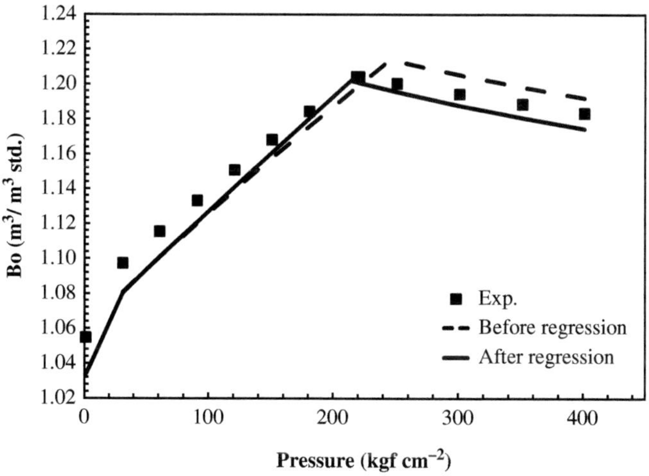

FIGURE 2.8 Comparison between B_o results before and after regression for Example 2.1.

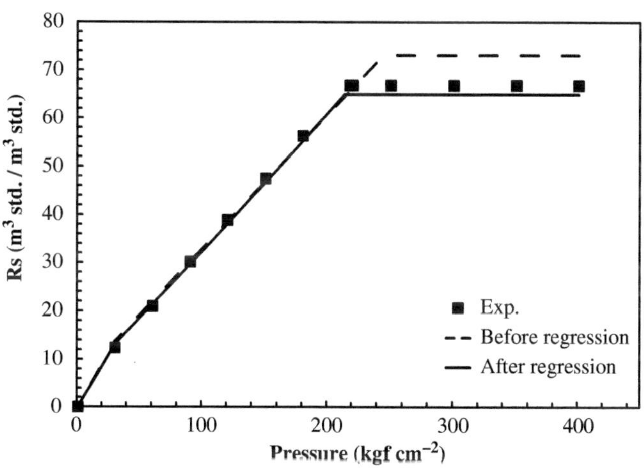

FIGURE 2.9 Comparison between R_s results before and after regression for Example 2.1.

theoretical basis. As viscosity is important to compute Darcy's law velocity (as will be shown in Chapter 4), we suggest the reader to get through Section 10.1 of Pedersen et al. (2014) for computer programming. We strongly recommend that viscosity adjustment is performed after EoS fitting, since critical properties influence both reviewed correlations. Results for CSP parameter estimation for this fluid in Example 2.1 are shown in Fig. 2.11.

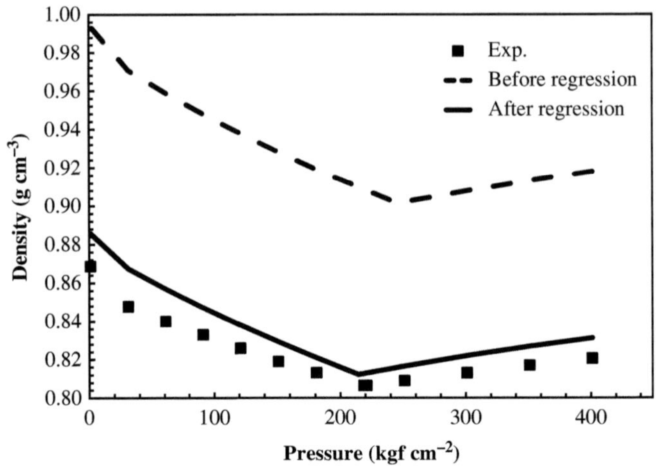

FIGURE 2.10 Comparison between density results before and after regression for Example 2.1.

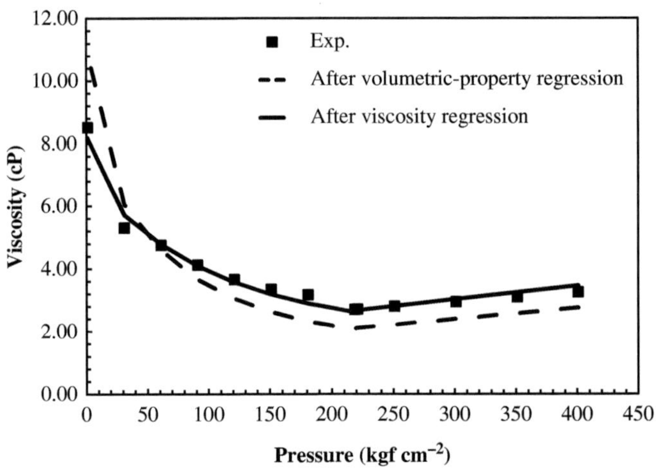

FIGURE 2.11 Comparison of viscosity results before and after CSP parameter regression, which was done after volumetric-property optimization.

$$\left[(\mu-\mu^0)\xi+10^{-4}\right]^{0.25}=a_0+a_1\rho_r+a_2\rho_r^2+a_3\rho_r^3+a_4\rho_r^4 \tag{2.84}$$

$$\rho_r=\frac{\rho}{\rho_c}=\frac{v_c}{v},\quad \text{where } v_c=\left(\sum_{i=1}^{nc}x_iv_{c,i}^{\alpha}\right)^{1/\alpha} \tag{2.85}$$

$$\xi=\frac{T_c^{1/6}}{M^{1/2}P_c^{2/3}},\quad \mu^0=\frac{34\times10^{-8}T_r^{0.94}}{\xi},\quad \text{for } T_r<1.5 \tag{2.86}$$

$$\mu^0 = \frac{17.78 \times 10^{-8}(4.58T_r - 1.67)^{5/8}}{\xi}, \quad \text{for } T_r > 1.5 \tag{2.87}$$

$$\mu^{\text{mix}}(T,P) = \left(\frac{T_c^{\text{mix}}}{T_{c0}}\right)^{-1/6}\left(\frac{P_c^{\text{mix}}}{P_{c0}}\right)^{2/3}\left(\frac{M^{\text{mix}}}{M_0}\right)^{1/2}\left(\frac{\alpha^{\text{mix}}}{\alpha_0}\right)\mu_0(T_0,P_0) \tag{2.88}$$

where the superscript *mix* refers to the mixture and the other parameters are detailed in Pedersen et al. (2014).

2.7 EXAMPLE 2.2: PVT ANALYSIS WITH SWELLING TEST

Reinjection of miscible gas in the reservoir during the production period implies necessarily in reaching the critical point of the resulting mixture in some point of the simulation domain. A gas miscible at the first contact is the one whose final mixture with the oil presents necessarily a single phase for any proportion between injected gas and reservoir fluid. A gas miscible in multiple contacts is the one for which the final mixture around the injector well might be in a two-phase state due to the high proportion injected gas/live oil in this region. However, due to mass-transfer processes along the trajectory of the injection bank toward the producer well, at some point in the reservoir one necessarily reaches the critical point of the resulting mixture, so there will not be distinction between liquid and vapor phases anymore. The interfacial tension between these phases tends to zero, that is why oil-recovery values are theoretically higher. In practice, sweeping inefficiencies, numerical errors around the critical point, and inaccuracies in relative-permeability curves can influence simulation's results. Therefore, a good modeling of the swelling test's experimental data is necessary, because this assay shows the volumetric behavior of the injected gas-reservoir fluid mixture at different proportions within the PVT cell. The swelling test, as well as Constant Composition Expansion and Differential Liberation, does not reproduce exactly the flow in the porous media, since, in three dimensions, this global composition varies differently from what happens within the cell where the fluid sample is confined. But the EoS fitting to reproduce these closed-system experimental data will certainly provide a good representation of the volumetric behavior at the varying pressure, temperature, and composition conditions along the simulation domain.

Table 2.3 shows the composition range in the samples of a reservoir–fluid classified as a volatile oil, most likely due to high CO_2 content in situ. The cubic-EOS fitting for the bubble-to-dew transition in a swelling test is impaired by the proximity of the critical point. Most likely, in any regression involving gases, the splitting of C_{20+} fraction becomes necessary to increase representativeness of the components that effectively condense when one reaches the retrograde dew point. The method presented here considers that,

TABLE 2.3 Approximated Composition of Both Reservoir and Injection Gas in the Swelling Test of this Example. Molar Mass and Density of C_{20+}: 559 g/mol e 0.9512 g/cm^3, Respectively. Splitting Was Carried Out Using the Technique of Shibata et al. (1987) with Two Quadrature Points

Pseudocomponent	M (g/mol)	Composition Range (mole %)	
		Reservoir Fluid	Injection Gas
CO_2	44	35–39	43–46
N_2–CH_4	16	34–38	42–45
C_2–nC_5	45	10–13	9–11
C_6–C_{12}	121	4–8	0
C_{13}–C_{19}	215	2–4	0
$QC_{28.9}$	490	2–4	0
$QC_{62.7}$	780	1–2	0

GOR and API in flash: 442 m^3 std./m^3 std. and 27, respectively.
The fractional carbon numbers (28.9 and 62.7) refer to the pseudocomponents obtained after the expansion of C_{20+} fraction and will be detailed later in this text.

from C_{20} on, the mole fraction of the hydrocarbons obey an exponential decay, as stated by Shibata et al. (1987):

$$F(I) = Ce^{-DI} \tag{2.89}$$

where I is the variable of the distribution, that is, the single carbon number (considered continuous in this approach), C is a normalization constant, and D is another constant that defines the decay intensity, respecting the distribution mean, which, in turn, is related to the average molar mass, known a priori for the C_{20+} fraction.

According to the distribution definition, the mole fraction of a species with a single carbon number equal to I is given by

$$X(I) = \frac{Ce^{-DI}dI}{\int_{\eta}^{\phi} Ce^{-DI}dI} \tag{2.90}$$

where ϕ is the maximum single carbon number of the distribution (which value depends on the analyst's interpretation and will be discussed later in this example), and η is the minimum single carbon number, that is, the origin of the distribution, which means 20 in this case.

Determination of the constants C and D is done, respectively, by the normalization of the distribution and by the calculation of its mean, that is, the average single carbon number, which is related to the molar mass (in this case, 559 g/mol) by the following expression (Katz and Firoozabadi, 1978; Espósito et al., 2000):

$$\bar{I} = \frac{\overline{M} + 4}{14} \tag{2.91}$$

So, one has

$$\int_{\eta}^{\phi} Ce^{-DI} dI = 1 \tag{2.92}$$

and also

$$\int_{\eta}^{\phi} ICe^{-DI} dI = \bar{I} \tag{2.93}$$

For some ultra-heavy oils, the asphaltenic associated fractions can reach 200 carbon atoms (Pedersen et al., 2014). If not associated, asphaltene molecules may have less carbon atoms. If asphaltenes are not present and heavy normal alkanes predominate, they can go to, say, 100. Some authors get even to use infinity as the upper integration limit. There is a drawback in this assumption. Obviously, at some stage of their modeling, it will be necessary to discretize the distribution, since it is impossible to work with infinite components. Some pseudocomponents generated by such discretization process will have physical properties far beyond the spam recommended by the main empirical correlations already cited in this chapter (Shibata et al., 1987). But the main advantage would be the possibility of using the Gauss–Laguerre quadrature, which decreasing-exponential weight function facilitates the discretization process, since the quadrature points and weights are already tabled.

Shibata et al. (1987) state that, statistically, components with more than 70 carbon atoms are uncommon in the majority of the oils. Pedersen et al. (2014) propose the upper limit of 80. With a finite upper limit, the discretization process of the distribution function proposed by Shibata et al. (1987) was named "Generalized Gauss–Laguerre Quadrature." Direct integration of Eq. (2.92) gives

$$C = \frac{D}{e^{-D\eta} - e^{-D\cdot\phi}} \tag{2.94}$$

To obtain the constant D, we define new variables (z and Δ) for the resolution of Eq. (2.93) (Espósito et al., 2000):

$$z = D(I - 20) \tag{2.95}$$

$$\Delta = D(\phi - 20) \tag{2.96}$$

Resulting in (see Exercise 2.6.4):

$$\frac{\bar{I} - \eta}{\phi - \eta} = \frac{1}{\Delta} - \frac{e^{-\Delta}}{1 - e^{-\Delta}} \tag{2.97}$$

Given the upper and lower integration limits of the distribution, as well as the molar mass of C_{20+}—from which one can easily obtain $\bar{I}$ by Eqs. (2.91)–(2.97) can be solved for Δ, and we can finally obtain the constant D by Eq. (2.96). From now on, the discretization of the function for the generation of the new pseudocomponents can be done through the technique of Gaussian-quadrature points and weights.

The result of the integral of any function $g(z)$ pondered by its weight function (in this case, e^{-z}) can be approximated by a sum of g-values evaluated at the so-called quadrature points of the function (Carnahan et al., 1969; Shibata et al., 1987):

$$\int_0^{\Delta} e^{-z} g(z) dz = \sum_{i=1}^{n} w_i g(z_i) \tag{2.98}$$

where n is the number of quadrature points, that is, the values of z_i in which the function will be evaluated, and w_i are the quadrature weights. Once n is an integer number chosen arbitrarily by the user, the determination of w_i- and z_i-values is based on the fact that the Gaussian-quadrature integration is exact when function g is a polynomial of degree up to $2n-1$ (Carnahan et al., 1969; Shibata et al., 1987).

Therefore, once n is chosen, the analytical result of the integral for polynomials of the type $g(z) = z^j$, for $j = 0, 1, \ldots, 2n-1$, can be used on the left-hand side of Eq. (2.98) for the numerical determination of w_i and z_i. For example, choosing two quadrature points, we have a system of 4 equations to be solved for w_1, w_2, z_1, and z_2:

$$\int_0^{\Delta} e^{-z} z^0 dz = 1 - e^{-\Delta} = \sum_{i=1}^{2} w_i z_i^0 = w_1 + w_2 \tag{2.99}$$

$$\int_0^{\Delta} e^{-z} z^1 dz = 1 - (\Delta + 1)e^{-\Delta} = \sum_{i=1}^{2} w_i z_i^1 = w_1 z_1 + w_2 z_2 \tag{2.100}$$

$$\int_0^{\Delta} e^{-z} z^2 dz = 2 - [\Delta^2 + 2(\Delta + 1)]e^{-\Delta} = \sum_{i=1}^{2} w_i z_i^2 = w_1 z_1^2 + w_2 z_2^2 \tag{2.101}$$

$$\int_0^{\Delta} e^{-z} z^3 dz = 6 - [\Delta^3 + 3\Delta^2 + 6(\Delta + 1)]e^{-\Delta} = \sum_{i=1}^{2} w_i z_i^3 = w_1 z_1^3 + w_2 z_2^3 \tag{2.102}$$

Shibata et al. (1987) present the values of w_1, w_2, z_1, and z_2 obtained for a wide range of Δ. They also show values of w_1, w_2, w_3 z_1, z_2, and z_3 in a completely analogous thought for three quadrature points. Tables 2.4 and 2.5 show these results, over which the user will have to interpolate, given the chosen number of quadrature points and the calculated Δ-value for the case distribution.

TABLE 2.4 Solution of the System of Eqs. (2.99)–(2.102), for Two Quadrature Points

Δ	z_1	z_2	w_1	w_2
0.3	0.0615	0.2347	0.5324	0.4676
0.4	0.0795	0.3101	0.5353	0.4647
0.5	0.0977	0.3857	0.5431	0.4569
0.6	0.1155	0.4607	0.5518	0.4482
0.7	0.1326	0.5347	0.5601	0.4399
0.8	0.1492	0.6082	0.5685	0.4315
0.9	0.1652	0.6807	0.5767	0.4233
1.0	0.1808	0.7524	0.5849	0.4151
1.1	0.1959	0.8233	0.5932	0.4068
1.2	0.2104	0.8933	0.6011	0.3989
1.3	0.2245	0.9625	0.6091	0.3909
1.4	0.2381	1.0307	0.6169	0.3831
1.5	0.2512	1.0980	0.6245	0.3755
1.6	0.2639	1.1644	0.6321	0.3679
1.7	0.2763	1.2299	0.6395	0.3605
1.8	0.2881	1.2944	0.6468	0.3532
1.9	0.2996	1.3579	0.6539	0.3461
2.0	0.3107	1.4204	0.6610	0.3390
2.1	0.3215	1.4819	0.6678	0.3322
2.2	0.3318	1.5424	0.6745	0.3255
2.3	0.3418	1.6018	0.6810	0.3190
2.4	0.3515	1.6602	0.6874	0.3126
2.5	0.3608	1.7175	0.6937	0.3063
2.6	0.3699	1.7738	0.6997	0.3003
2.7	0.3786	1.8289	0.7056	0.2944
2.8	0.3870	1.8830	0.7114	0.2886
2.9	0.3951	1.9360	0.7170	0.2830
3.0	0.4029	1.9878	0.7224	0.2776

(Continued)

TABLE 2.4 (Continued)

Δ	z_1	z_2	w_1	w_2
3.1	0.4104	2.0386	0.7277	0.2723
3.2	0.4177	2.0882	0.7328	0.2672
3.3	0.4247	2.1367	0.7378	0.2622
3.4	0.4315	2.1840	0.7426	0.2574
3.5	0.4380	2.2303	0.7472	0.2528
3.6	0.4443	2.2754	0.7517	0.2483
3.7	0.4504	2.3193	0.7561	0.2439
3.8	0.4562	2.3621	0.7603	0.2397
3.9	0.4618	2.4038	0.7644	0.2356
4.0	0.4672	2.4444	0.7683	0.2317
4.1	0.4724	2.4838	0.7721	0.2279
4.2	0.4775	2.5221	0.7757	0.2243
4.3	0.4823	2.5593	0.7792	0.2208
4.4	0.4869	2.5954	0.7826	0.2174
4.5	0.4914	2.6304	0.7858	0.2142
4.6	0.4957	2.6643	0.7890	0.2110
4.7	0.4998	2.6971	0.7920	0.2080
4.8	0.5038	2.7289	0.7949	0.2051
4.9	0.5076	2.7596	0.7977	0.2023
5.0	0.5112	2.7893	0.8003	0.1997
5.1	0.5148	2.8179	0.8029	0.1971
5.2	0.5181	2.8456	0.8054	0.1946
5.3	0.5214	2.8722	0.8077	0.1923
5.4	0.5245	2.8979	0.8100	0.1900
5.5	0.5274	2.9226	0.8121	0.1879
5.6	0.5303	2.9464	0.8142	0.1858
5.7	0.5330	2.9693	0.8162	0.1838
5.8	0.5356	2.9913	0.8181	0.1819
5.9	0.5381	3.0124	0.8199	0.1801
6.0	0.5405	3.0327	0.8216	0.1784

(Continued)

TABLE 2.4 (Continued)

Δ	z_1	z_2	w_1	w_2
6.2	0.5450	3.0707	0.8248	0.1752
6.4	0.5491	3.1056	0.8278	0.1722
6.6	0.5528	3.1375	0.8305	0.1695
6.8	0.5562	3.1666	0.8329	0.1671
7.0	0.5593	3.1930	0.8351	0.1649
7.4	0.5621	3.2170	0.8371	0.1629
7.7	0.5646	3.2388	0.8389	0.1611
8.1	0.5680	3.2674	0.8413	0.1587
8.5	0.5717	3.2992	0.8439	0.1561
9.0	0.5748	3.3247	0.8460	0.1540
10.0	0.5777	3.3494	0.8480	0.1520
11.0	0.5816	3.3811	0.8507	0.1493
12.0	0.5836	3.3978	0.8521	0.1479
14.0	0.5847	3.4063	0.8529	0.1471
16.0	0.5856	3.4125	0.8534	0.1466
18.0	0.5857	3.4139	0.8535	0.1465
20.0	0.5858	3.4141	0.8536	0.1464
25.0	0.5858	3.4142	0.8536	0.1464
30.0	0.5858	3.4142	0.8536	0.1464
40.0	0.5858	3.4142	0.8536	0.1464
60.0	0.5858	3.4142	0.8536	0.1464
100.0	0.5858	3.4142	0.8536	0.1464
∞	0.5858	3.4142	0.8536	0.1464

Source: Shibata, S.K., Sandler, S.I. and Behrens, R.A., Phase equilibrium calculations for continuous and semicontinuous mixtures, Chem. Eng. Sci., 42, 8, pp.1977–1988, 1987.

The advantage of using two quadrature points, that is, two pseudocomponents to represent the C_{20+} fraction, is in the ease of calculating their mole fractions in other samples of the same field. Suppose that another sample collected somewhere else also contributes for the EoS parameter regression, or even has its PVT analysis validate the optimum parameters after

TABLE 2.5 System of Equations of Shibata et al. (1987) for Three Quadrature Points

Δ	z_1	z_2	z_3	w_1	w_2	w_3
1.0000	0.0982	0.4613	0.8706	0.4	0.4401	0.2072
1.1000	0.1096	0.5113	0.9593	0.4	0.4365	0.1956
1.2000	0.1186	0.5551	1.0453	0.4	0.4348	0.1882
1.3000	0.1264	0.5948	1.1290	0.4	0.4334	0.1831
1.4000	0.1338	0.6337	1.2124	0.4	0.4324	0.1778
1.5000	0.1404	0.6697	1.2942	0.4	0.4319	0.1735
1.6000	0.1486	0.7109	1.3787	0.4	0.4297	0.1666
1.7000	0.1557	0.7484	1.4611	0.4	0.4282	0.1612
1.8000	0.1628	0.7855	1.5432	0.4	0.4265	0.1558
1.9000	0.1699	0.8229	1.6254	0.4	0.4245	0.1503
2.0000	0.1766	0.8587	1.7067	0.4	0.4227	0.1451
2.1000	0.1834	0.8949	1.7881	0.4	0.4206	0.1398
2.2000	0.1898	0.9297	1.8685	0.4	0.4186	0.1348
2.3000	0.1962	0.9641	1.9487	0.5	0.4165	0.1299
2.4000	0.2025	0.9983	2.0286	0.5	0.4142	0.1251
2.5000	0.2085	1.0313	2.1076	0.5	0.4121	0.1205
2.6000	0.2144	1.0638	2.1863	0.5	0.4098	0.1161
2.7000	0.2200	1.0954	2.2642	0.5	0.4076	0.1119
2.8000	0.2257	1.1270	2.3420	0.5	0.4053	0.1076
2.9000	0.2312	1.1576	2.4190	0.5	0.4029	0.1036
3.0000	0.2365	1.1878	2.4955	0.5	0.4006	0.0997
3.1000	0.2416	1.2171	2.5713	0.5	0.3983	0.0959
3.2000	0.2467	1.2460	2.6467	0.5	0.3959	0.0924
3.3000	0.2516	1.2743	2.7214	0.5	0.3935	0.0889
3.4000	0.2564	1.3020	2.7954	0.5	0.3911	0.0856
3.5000	0.2611	1.3293	2.8690	0.5	0.3887	0.0823
3.6000	0.2656	1.3557	2.9418	0.5	0.3863	0.0793
3.7000	0.2700	1.3818	3.0140	0.5	0.3840	0.0762
3.8000	0.2744	1.4074	3.0856	0.5	0.3815	0.0734

(Continued)

TABLE 2.5 (Continued)

Δ	z_1	z_2	z_3	w_1	w_2	w_3
3.9000	0.2786	1.4323	3.1565	0.6	0.3792	0.0706
4.0000	0.2827	1.4566	3.2266	0.6	0.3768	0.0681
4.1000	0.2867	1.4805	3.2961	0.6	0.3745	0.0655
4.2000	0.2906	1.5038	3.3649	0.6	0.3722	0.0630
4.3000	0.2944	1.5266	3.4329	0.6	0.3699	0.0607
4.4000	0.2981	1.5490	3.5003	0.6	0.3676	0.0585
4.5000	0.3017	1.5707	3.5668	0.6	0.3653	0.0564
4.6000	0.3052	1.5920	3.6326	0.6	0.3631	0.0544
4.7000	0.3086	1.6128	3.6976	0.6	0.3609	0.0524
4.8000	0.3119	1.6332	3.7619	0.6	0.3587	0.0505
4.9000	0.3152	1.6530	3.8254	0.6	0.3566	0.0487
5.0000	0.3183	1.6724	3.8880	0.6	0.3545	0.0470
5.1000	0.3214	1.6914	3.9499	0.6	0.3524	0.0453
5.2000	0.3244	1.7099	4.0110	0.6	0.3503	0.0438
5.3000	0.3273	1.7279	4.0712	0.6	0.3483	0.0423
5.4000	0.3301	1.7455	4.1305	0.6	0.3463	0.0409
5.5000	0.3329	1.7627	4.1890	0.6	0.3444	0.0394
5.6000	0.3355	1.7795	4.2467	0.6	0.3425	0.0381
5.7000	0.3382	1.7959	4.3035	0.6	0.3406	0.0369
5.8000	0.3407	1.8118	4.3594	0.6	0.3387	0.0357
5.9000	0.3432	1.8273	4.4143	0.6	0.3369	0.0346
6.0000	0.3456	1.8425	4.4685	0.6	0.3351	0.0335
6.2000	0.3502	1.8717	4.5740	0.6	0.3317	0.0314
6.4000	0.3545	1.8993	4.6758	0.6	0.3284	0.0295
6.6000	0.3586	1.9255	4.7740	0.6	0.3252	0.0278
6.8000	0.3625	1.9504	4.8685	0.7	0.3222	0.0263
7.0000	0.3662	1.9739	4.9591	0.7	0.3194	0.0248
7.2000	0.3697	1.9961	5.0460	0.7	0.3166	0.0236
7.4000	0.3729	2.0171	5.1292	0.7	0.3141	0.0223
7.7000	0.3775	2.0464	5.2469	0.7	0.3104	0.0208

(Continued)

TABLE 2.5 (Continued)

Δ	z_1	z_2	z_3	w_1	w_2	w_3
8.1000	0.3829	2.0816	5.3907	0.7	0.3060	0.0190
8.5000	0.3877	2.1126	5.5200	0.7	0.3021	0.0174
9.0000	0.3928	2.1462	5.6619	0.7	0.2978	0.0160
10.0000	0.4009	2.1982	5.8853	0.7	0.2912	0.0137
11.0000	0.4064	2.2341	6.0400	0.7	0.2865	0.0124
12.0000	0.4100	2.2578	6.1415	0.7	0.2834	0.0116
14.0000	0.4138	2.2822	6.2429	0.7	0.2802	0.0107
16.0000	0.4152	2.2907	6.2767	0.7	0.2790	0.0105
18.0000	0.4156	2.2933	6.2865	0.7	0.2787	0.0104
20.0000	0.4157	2.2940	6.2891	0.7	0.2786	0.0103
25.0000	0.4158	2.2943	6.2899	0.7	0.2785	0.0104
30.0000	0.4158	2.2943	6.2899	0.7	0.2785	0.0104
40.0000	0.4158	2.2943	6.2899	0.7	0.2785	0.0104
60.0000	0.4158	2.2943	6.2899	0.7	0.2785	0.0104
100.0000	0.4158	2.2943	6.2899	0.7	0.2785	0.0104
∞	0.4158	2.2943	6.2899	0.7	0.2785	0.0104

regression. The experimental molar mass of C_{20+} in this new sample requires mole fractions of both pseudocomponents to be calculated by the following material balance restriction:

$$x_1 M_1 + (1 - x_1) M_2 = M_{C_{20+}} \tag{2.103}$$

where x_1 and $x_2 = 1 - x_1$ are, respectively, the mole fractions of the first and second quadrature points, M_1 and M_2 being their respective and previously determined molar masses.

Obviously, the restriction imposed by Eq. (2.103) gets automatically fulfilled for the original sample from which composition the C_{20+}'s pseudocomponents were created by the generalized Gaussian-quadrature.

For a molar mass of 559 g/mol and lower and upper distribution function integration limits of 20 and 80, respectively, resolution of Eq. (2.97) gives Δ=2.055. The linear interpolation in Table 2.4 (two quadrature points) gives $z_1 = 0.3167$ and $z_2 = 1.4546$, with weights (mole fractions) $w_1 = 0.6648$ and $w_2 = 0.3352$. Constants C and D are, respectively, 0.0745 and 0.0337. As global mole percent of C_{20+} is equal to 4.74%, we have both pseudocomponents' single carbon number given by $I_1 = \frac{0.3167}{0.0337} + 20 = 28.90$ and

$I_2 = \frac{1.4546}{0.0337} + 20 = 62.70$; with global mole fractions equal to $0.6648 \cdot 4.74 = 3.15\%$ and $0.3352 \cdot 4.74 = 1.59\%$. The density distribution function, according to the format proposed by Pedersen et al. (2014), is given by $\rho = 0.4871 + 0.1256\ln(I)$, which constants were obtained by the continuity condition with lighter fractions—$\rho(19) = 0.857$ – honoring also the average density of C_{20+}, that is, $\rho\left(\bar{I} = \frac{559+4}{14} = 40.2\right) = 0.9512$, where $\bar{I} = 40.2$ is the average single carbon number of the distribution $F(I) = 0.0745e^{-0.0337I}$. With molar mass and density of both heavy pseudocomponents, Eq. (2.72) can be solved for T_B, so that, finally, the Cavett correlations (Eqs. (2.73) and (2.74)) can be used to calculate T_c and P_c, respectively. Acentric factors can be calculated, for example, by Lee–Kesler equations (Eqs. (2.76) and (2.77)).

In order to reduce the number of pseudocomponents for reservoir simulation, a lumping procedure might also be applied to the other fractions (up to C_{19}) and individual components (up to C_5). The choice of the lumping procedure is subjective, also depending on the enhanced oil recovery (EOR) process. Due to its importance in EOR, CO_2 must remain single, that is, not lumped with any other component. Depending on the N_2 content, likely at low concentrations, this component might be incorporated to methane, since it is not utilized for EOR. As for ethane, since it is incondensable at standard conditions, it shows single very often. To represent some refining fractions, C_3 and C_4 could be lumped together to simulate LPG, whilst C_5 to C_{12} could make up the so-called direct-distillation naphta, one of the streams of gasoline pool. C_{13} to C_{19} would make up kerosene, being the "Diesel+" fractions, including residues, thoroughly represented by the (in this case, two) pseudocomponents of C_{20+} distribution. Such criterion, as any other, is approximated, since each of the distillated fractions also contains in its composition, due to thermodynamic-equilibrium phase distribution in the theoretical plates, more than traces of the other components from the different lumps. Some users, after isolating methane (or mixing it up with N_2) tend to group all the predefined components in a single lump (C_2 to C_5). As each individual component has its critical properties already tabled, the resulting values of T_c and P_c for the lump might be obtained by some mixing rules, the empirical correlations (like Cavett's) being used from C_6.

After obtaining these initial estimates, as well as the ones for CO_2 and methane's binary interaction coefficients (the only nonzero), a nonlinear regression based on the experimental data of both PVT analysis and swelling test can be done. Pseudocomponents molar masses, obtained by the procedure of Shibata et al. (1987), can also be used as initial estimates in the regression process, together with the critical properties estimated by Cavett correlation and the volume translations given, for example, by Eq. (2.83). The definitive values of the molar masses, combined with their respective mole fractions in Eq. (2.103), can shift the global molar mass of C_{20+} by 5% to 10% relative to the original value (559 g/mol, according to Table 2.3). This information is coherent with the experimental error from dead-oil's molar mass measurement by cryoscopy.

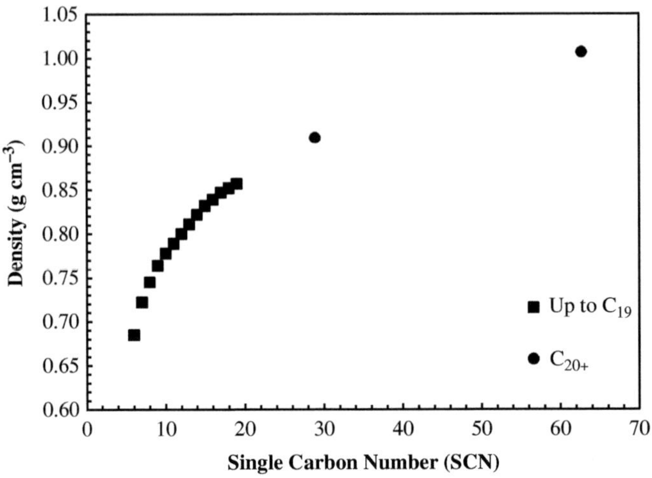

FIGURE 2.12 Densities of C_{20+} pseudocomponents compared to the other ones up to C_{19}.

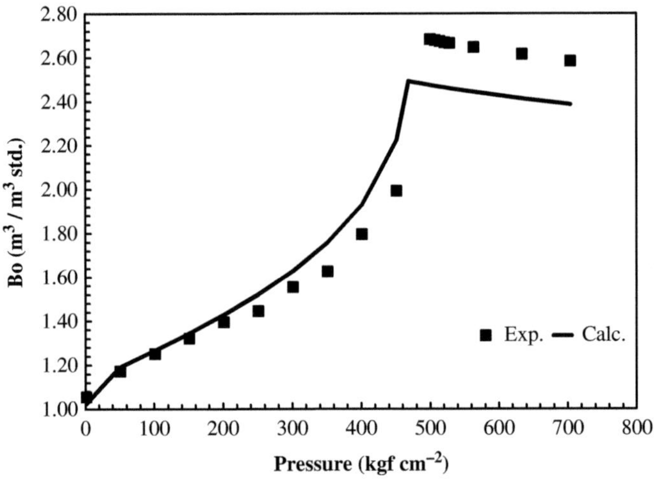

FIGURE 2.13 Oil formation volume factor (Bo) in differential liberation at 90°C.

Fig. 2.12 shows the densities of the two pseudocomponents of C_{20+} obtained by quadrature, compared to those stated by Katz and Firoozabadi (1978) for the other components up to C_{19}. Continuity is very reasonable, suggesting that the sample has its paraffin and aromatic contents following the mean-values of Table 2.1, therefore providing good inputs for Cavett's correlation.

Figs. 2.13–2.16 show, respectively, the comparison between experimental and simulated values of Bo, Rs, density, and viscosity for the oil in the differential liberation at 90°C. Although there are relatively high systematic

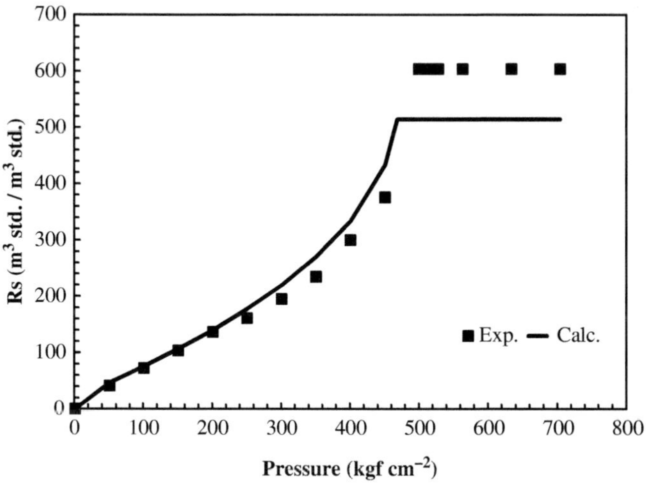

FIGURE 2.14 Solubility ratio (Rs) in differential liberation at 90°C.

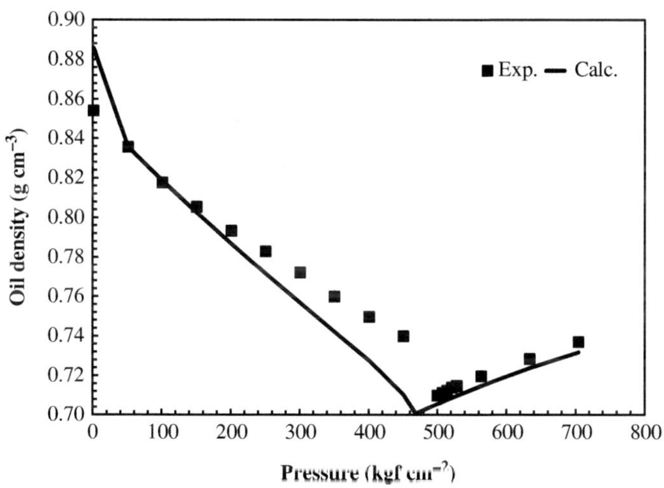

FIGURE 2.15 Oil density in the differential liberation at 90°C.

errors for both Bo and Rs above P_{sat} (possibly due to limitations in both thermodynamic model and internal optimization algorithms), global tendencies are coherent with experimental data.

The introduction of a temperature dependence in the volume shift, mainly for the lighter components, as well as the utilization of a greater number of quadrature points in C_{20+} distribution function, might reduce this systematic error but can also increase complexity and computational effort, also impairing convergence in regression. Everything depends on the tolerance required by the user. It is worth to emphasize that the set of parameters obtained in

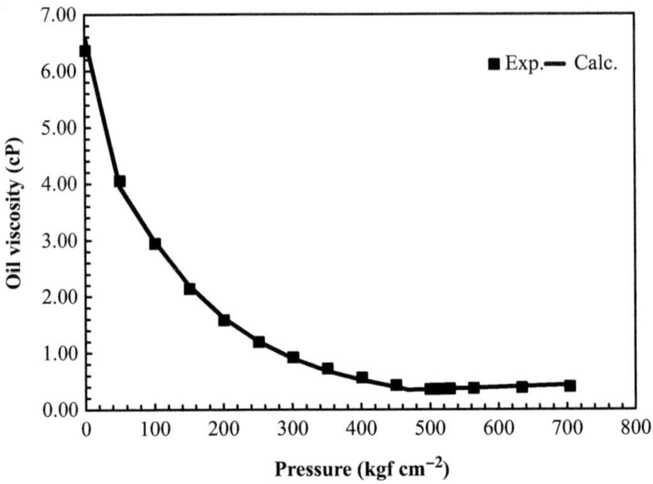

FIGURE 2.16 Oil viscosity in the differential liberation at 90°C. Parameters of Jossi–Stiel–Thodos correlation (Pedersen et al., 2014) fitted after the adjustment of volumetric properties using PR-EoS.

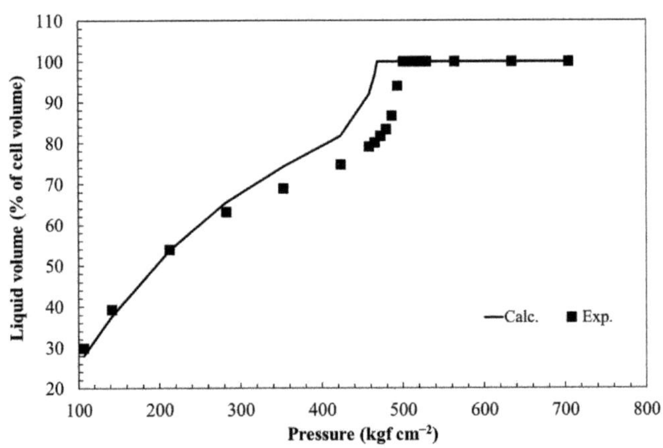

FIGURE 2.17 Liquid volume in the constant composition expansion at 90°C, no injected gas.

the regression (not shown) represents, with reasonable approximation, the swelling test and its critical transition (i.e., transition of saturation pressure from bubble to dew point, Figs. 2.17–2.21). This is a phenomenon of difficult reproduction when one uses the traditional cubic equations of state. In other words, we most likely have reached the performance limit in the parameter-adjustment process of Peng–Robinson equation, which, even with systematic errors of the order of 7% for some properties, represents very well the principal phenomena and tendencies involved in this fluid's modeling.

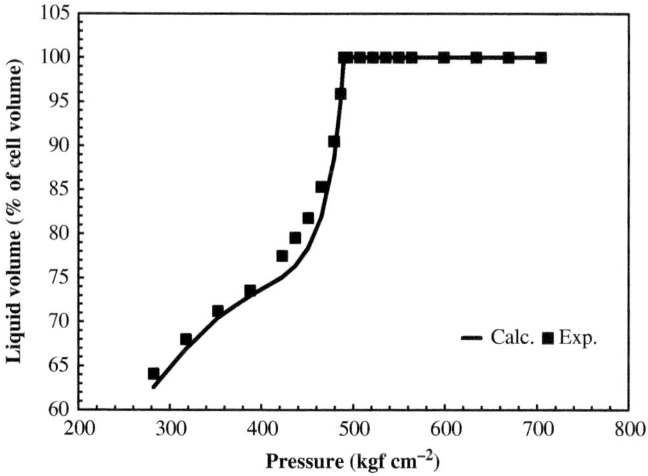

FIGURE 2.18 Liquid volume in the constant composition expansion at 90°C with 5% of gas injected.

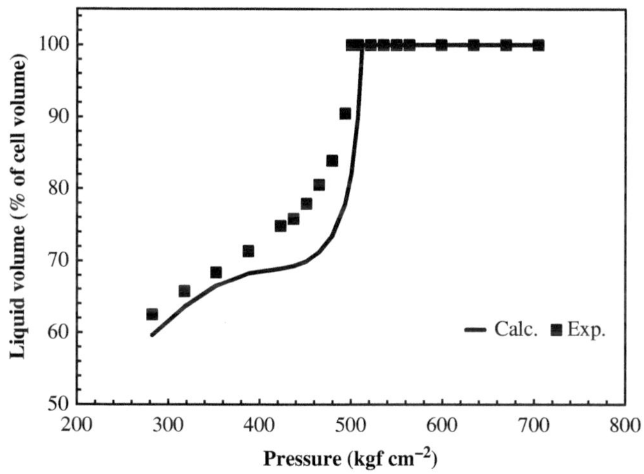

FIGURE 2.19 Liquid volume in the 90°C with 10% injected gas.

Figs. 2.22 and 2.23 also show results for gas compressibility factor, liquid volume percentage, and cumulative gas production in constant volume depletion at 90°C, all showing reasonable agreement with experimental data.

Finally, it is worth to emphasize that the flash liberation (done in a single stage at 40°C and 1 atm) of this fluid, as shown in Table 2.3, produced a GOR of 442 m^3 std/m^3 std, considerably lower than differential liberation's maximum Rs (603 m^3 std/m^3 std). This can be explained by the relatively higher temperature of this latter assay (90°C), which leads the fluid toward the critical region, in which the liberation of gas is much more pronounced,

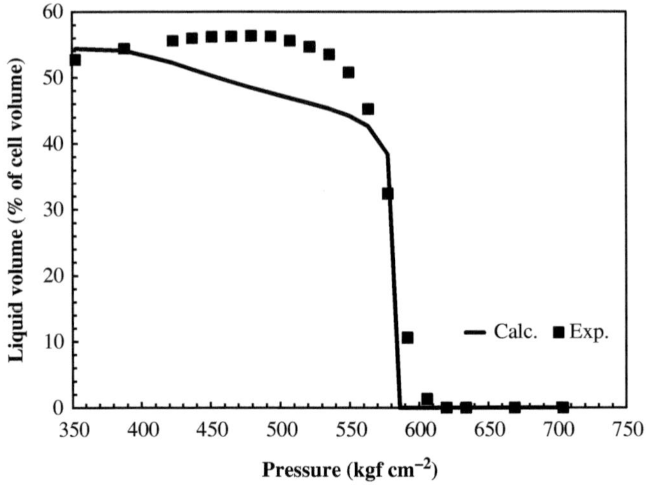

FIGURE 2.20 Liquid dropout in the constant composition expansion at 90°C, with 25% of gas injected.

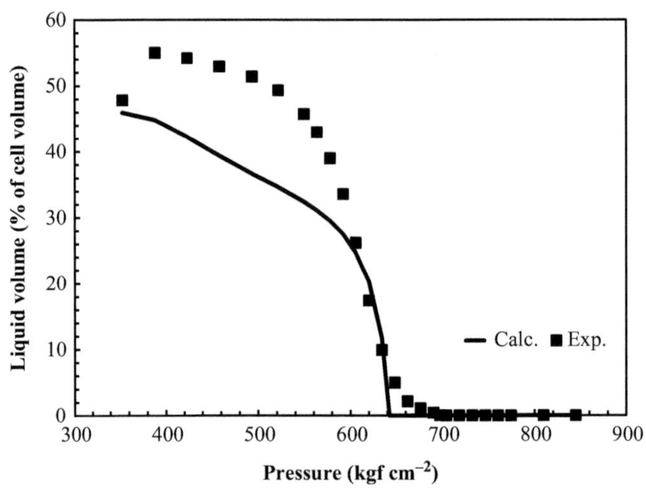

FIGURE 2.21 Liquid volume dropout in the constant composition expansion at 90°C, with 35% of injected gas.

even with pressure reduction carried out through several stages. Fig. 2.24 presents the phase (*P*—*T*) diagram of the adjusted fluid. Although the critical point is around 300°C, the critical region is broad, with quality lines already approaching each other around *T*=100°C (the vertical dashed line is the temperature of the PVT analysis, 90°C). Again, this modeling showed a 7% error for flash GOR, but matched the 27 API sharply, suggesting adequacy of heavies' volume shifts.

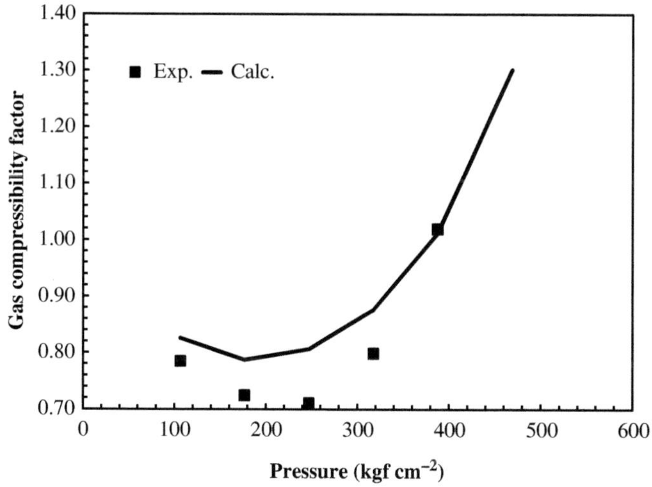

FIGURE 2.22 Gas compressibility factor in constant volume depletion at 90°C.

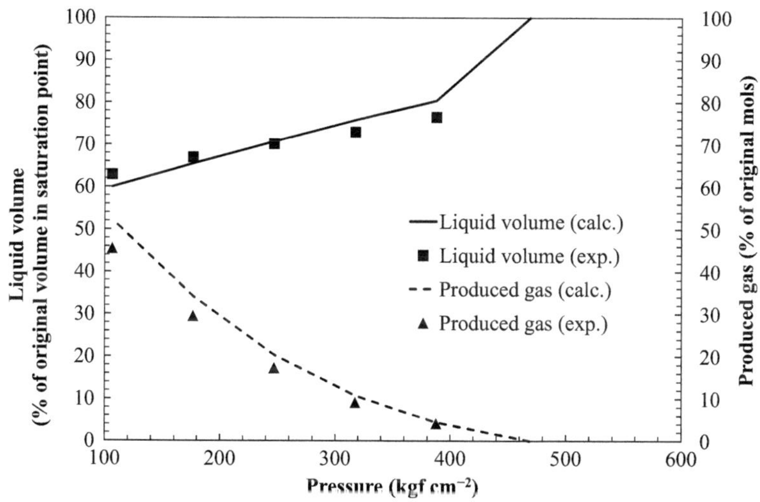

FIGURE 2.23 Liquid volume and cumulative gas production in constant volume depletion at 90°C.

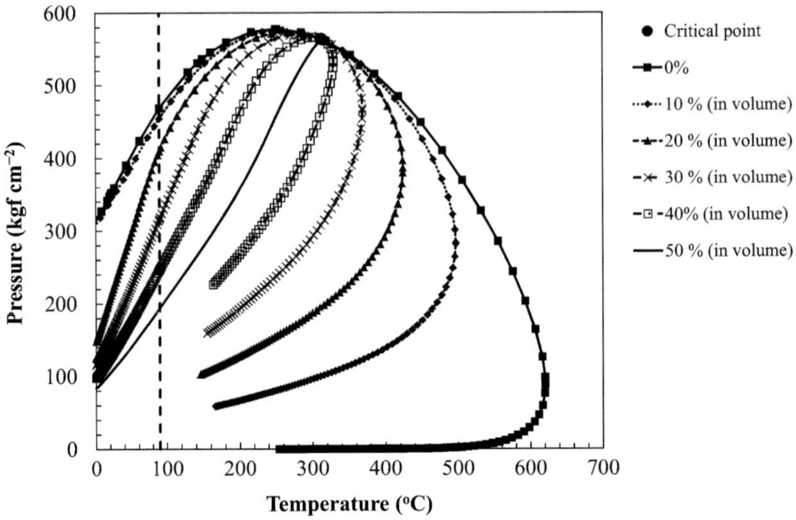

FIGURE 2.24 Simulated phase envelope of the adjusted fluid in Example 2.2. Dashed line is 90°C, the temperature of PVT analysis.

2.8 EXERCISES

1. Starting from Eq. (2.47) and using the definition of partial molar volume, obtain the expression of the fugacity coefficient for the Peng–Robinson EoS (Eq. (2.56)).
2. The critical point of a pure component is a metastable equilibrium state, that is, not only the first, but also the second and third derivatives of the Helmholtz energy with volume are zero. Show that this is equivalent to writing:

$$\left(\frac{\partial P}{\partial V}\right)_{T_c} = \left(\frac{\partial^2 P}{\partial V^2}\right)_{T_c} = 0$$

   And perform these derivatives to obtain the expressions of a_i and b_i for Peng–Robinson EoS. Also, show that these equations predict a Z_c-value equal to 0.303 for any component.
3. Consult the paper of Zabaloy and Brignole (1997) to prove that the *volume shift* changes only the volumetric properties and not the compositions of the phases in equilibrium. Use, for example, the Peng–Robinson EoS.
4. Derive Eq. (2.97), using the variable changes of Eq. (2.96) on the integral of Eq. (2.93).

REFERENCES

Arbabi, S., Firoozabadi, A., 1995. Near-critical phase behavior of reservoir fluids using equations of state. SPE Adv. Technol. Ser. 3, 1.

Callen, H.B., 1985. Thermodynamics and an Introduction to Thermostatistics, second ed John Wiley and sons.

Carnahan, B., Luther, H.A., Wilkes, J.O., 1969. *Applied Numerical Methods*. New York. John Wiley and Sons.

Cavett, R.H., 1962. Physical data for distillation calculations-vapor-liquid equilibrium. Proc. 27th Meeting. API, San Francisco, pp. 351–366.

Danesh, A., 1998. PVT and Phase Behavior of Petroleum Reservoir Fluids. Amsterdam, Elsevier Science & Technology Books.

Espósito, R.O., Castier, M., Tavares, F.W., 2000. Phase equilibrium calculations for semicontinuous mixtures subject do gravitational fields. Ind. Eng. Chem. Res. 39, 4415–4421.

Firoozabadi, A., 1999. Thermodynamics of Hydrocarbon Reservoirs. McGraw Hill, New York.

Folsta, K.C.B.M., Camargo, G.M., Espósito, R.O., 2010. Gas condensate characterization from chromatogram areas and retention times. Fluid Phase Equilib. 292 (1–2), 87–95.

Jhaveri, B.S., and Youngren, G.K., Three-parameter modification of the Peng-Robinson equation of state to improve volumetric predictions. SPE paper 13118, presented at the 59th Annual Technical Conference and Exhibition, Houston, Texas, 1984.

Katz, D.L. and Firoozabadi, A.: "Predicting phase behavior of condensate/crude-oil systems using methane interaction coefficients," 1978 J. Pet. Technol. 30, 11.

Kesler, M.G., Lee., B.I., 1976. Improved prediction of enthalpy of fractions. Hydrocarbon Process.153–158, March.

Kontogeorgis, G.M., Voutsas, E.C., Yacoumis, I.V., Tassios, D.P., 1996. An equation of state for associating fluids. Ind. Eng. Chem. Res. 35, 4310–4318.

McCain Jr., W.D., 1990. The Properties of Petroleum Fluids, 2nd edition, Tulsa, Oklahoma. Pennwell Publishing Company.

Naji, H.S., 2010. Characterizing pure and undefined petroleum components. Int. J. Eng. Tech. (IJET-IJENS) 10 (2), 28–48.

Pedersen, K.S., Christensen, P.L., Shaikh, J.A., 2014. Phase Behavior of Petroleum Reservoir Fluids, 2nd edition, Boca Raton, FL. CRC Press.

Peneloux, A., Rauzy, E., Freze, R., 1982. A consistent correction for Redlich–Kwong–Soave volumes. Fluid Phase Equilib. 8, 7–23.

Peng, D.Y., Robinson, D.B., 1976. A simple two-constant equation of state. Ind. Eng. Chem. Fundam. 15 (1), 59–64.

Rachford, H.H., Rice, J.D., 1952. Procedure for use of electronic digital computers in calculating flash vaporization hydrocarbon equilibrium. J. Pet. Technol. 4 (10), section 1, p.19, October.

Riazi, M.R., Daubert, T.E., 1987. Characterizing parameters for petroleum fractions. Ind. Eng. Chem. Res. 26 (24), 755–759.

Rochocz, G.L., Castier, M., Sandler, S.I., 1997. Critical point calculations for semicontinuous mixtures. Fluid Phase Equilib. 139, 137.

Shibata, S.K., Sandler, S.I., Behrens, R.A., 1987. Phase equilibrium calculations for continuous and semicontinuous mixtures. Chem. Eng. Sci. 42 (8), 1977–1988.

Whitson, C.H., 1983. Characterizing hydrocarbon plus fractions. SPE 12233-PA, SPE J. 23 (4), 683.

Wilson, G.M., A modified Redlich-Kwong equation of state, application to general physical data calculation. *Paper No. 15C Presented at the 1969 AIChE 65th National Meeting*, Cleveland, OH, 1969.

Zabaloy, M.S., Brignole, E.A., 1997. On volume translations in equations of state. Fluid Phase Equilib. 140, 87–95.

Chapter 3

Phase Equilibrium Under the Influence of the Gravitational Field

In this chapter, we study the influence of the gravitational field on the conditions necessary for the establishment of the thermodynamic equilibrium in systems with relatively considerable height. The derivation of the maximum-entropy condition (or minimum energy) remains on the specification of total volume and mass, but the total energy now contains a potential-energy term to be added to the internal-energy one. We will see that, although the temperature-equality condition remains the same, the hydrostatic charge generated by the gravitational field leads to equating the so-called piezometric pressure at each height level, introducing also an additional term in the condition of both chemical potential and fugacity equality. After a brief literature review, we discuss the application of this approach in the study of compositional grading in oil reservoirs considered as isothermal, revisiting some classical examples of the literature.

3.1 THE NEW EQUILIBRIUM CONDITION

In systems under the influence of external fields, particularly the gravitational one, the equilibrium condition related to the minimum of free energy or to the entropy maximum (for a isolated system) remains the same. One must add to the internal energy (U) a potential-energy term (E_{pot}) relative to an arbitrary reference level. Therefore, still neglecting the kinetic energy of a static closed system, one can define the total energy as

$$E = U + E_{pot} \tag{3.1}$$

So, reconsidering Fig. 2.1 (see Fig. 3.1), now with the phases ordered by their densities (increasing from top to bottom), one can relate the center of mass of each one of them to a discrete height level z_j.

Due to the action of the gravitational field, one expects that, in a isothermal system, the heaviest components segregate to the bottom of the recipient. Generalizing the behavior suggested in Fig. 3.1, composition can change with height inside a single phase due to the action of gravity. Theoretically,

Compositional Grading in Oil and Gas Reservoirs. DOI: http://dx.doi.org/10.1016/B978-0-12-812452-9.00003-4

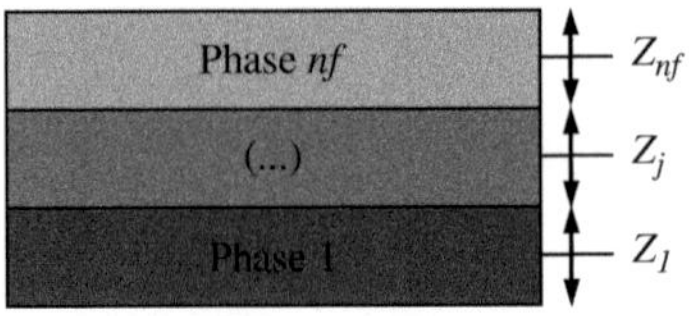

FIGURE 3.1 Closed system in equilibrium under the action of gravity, phases being ordered by density from bottom to top.

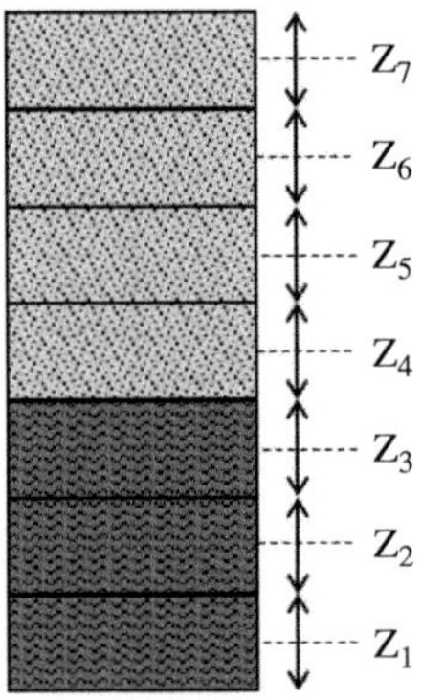

FIGURE 3.2 Vapor–liquid equilibrium under the influence of the gravitational field (liquid at the bottom, vapor on top). Total height arbitrarily discretized in seven levels, each one with a composition distinct from everyone else, so being considered as a pseudophase of the system. In this way, the figure shows three liquid phases (darker tones) and four vapor phases (lighter tones).

this changes the usual concept of "phase." According to it, *phase* is a homogeneous part of the system, with distinct properties from everyone else. If composition changes along height, each point of the system can be considered a new *pseudophase* because its composition (and therefore its properties) differs from those located at other discrete depths. So, at the continuous limit, one can have infinite phases in the system, which would not be convenient for the equilibrium modeling. Redrawing Fig. 3.1, assuming, for example, a liquid–vapor equilibrium under the influence of gravity. Fig. 3.2 shows such a system, arbitrarily discretized in, say, seven height levels, each one presenting a distinct composition, independent of the phase to which it belongs, which would not happen if not by the gravitational field.

Knowing that potential energy is given by

$$E_{\text{pot}} = \sum_{i=1}^{nc} m_i g z \tag{3.2}$$

where m_i is the mass of component i; g is the gravity acceleration, and z is the height relative to an arbitrary reference level $z_0 = 0$, one can differentiate Eq. (3.1) and use the fundamental relation of internal energy to obtain

$$dE = TdS - PdV + \sum_{i=1}^{nc} \mu_i dn_i + \sum_{i=1}^{nc} m_i g dz + \sum_{i=1}^{nc} gz\, dm_i \tag{3.3}$$

In a unidimensional system, it is easy to note the relation between the differential elements of volume and height:

$$dV = Adz \tag{3.4}$$

where A is the cross-sectional area to the z axis.

Replacing Eq. (3.4) in the first term of potential-energy differential in Eq. (3.3), and isolating dS on the left-hand side, one has

$$dS = \frac{1}{T}dU + \left(\frac{P}{T} + \frac{\sum_{i=1}^{nc} m_i g}{AT}\right)dV - \sum_{i=1}^{nc} \frac{\mu_i + M_i gz}{T} dn_i \tag{3.5}$$

where M_i is the molar mass of component i, related to its number of moles (n_i) by

$$dm_i = M_i dn_i, \quad \text{for } i = 1, \ldots, nc \tag{3.6}$$

The total internal energy of the system, as well as volume and number of moles of each component, is additive, in this case, over the *np pseudophases*. So, one can write

$$dS = \sum_{j=1}^{np} \frac{dU_j}{T_j} + \sum_{j=1}^{np} \left(\frac{P_j}{T_j} + \frac{\sum_{i=1}^{nc} m_{ij} g z_j}{A_j T_j z_j}\right) dV_j - \sum_{j=1}^{np} \sum_{i=1}^{nc} \frac{\mu_{ij} + M_i g z_j}{T} dn_{ij} \tag{3.7}$$

Note that, in Eq. (3.7), we multiply both numerator and denominator of the second term of volume differential by height z_j. This step will have been important in our derivation to state that piezometric pressure (and not the conventional pressure anymore) is constant along the system in equilibrium, as will be shown further.

Continuing our derivation, one can see that not all of the differential variables on the right-hand side of Eq. (3.7) are independent. Mass- and energy-balance restrictions for a rigid-wall closed system relate U_j, V_j, and n_{ij} in the following way:

$$U = U_1 + U_2 + \cdots + U_{np} = \text{constant} \tag{3.8}$$

$$dU = dU_1 + dU_2 + \cdots + dU_{np} = 0 \tag{3.9}$$

$$V = V_1 + V_2 + \cdots + V_{np} = \text{constant} \tag{3.10}$$

$$dV = dV_1 + dV_2 + \ldots + dV_{np} = 0 \tag{3.11}$$

$$n_i = n_{i1} + n_{i2} + \cdots + n_{i,np} = \text{constant}, \quad \text{for } i = 1, \ldots, nc \tag{3.12}$$

$$dn_i = dn_{i1} + dn_{i2} + \cdots + dn_{i,np} = 0, \quad \text{for } i = 1, \ldots, nc \tag{3.13}$$

From Eqs. (3.9), (3.11), and (3.13), one can note that any pseudophase (arbitrarily chosen) may be written by difference with respect to the global properties, that is,

$$dU_J = -\sum_{\substack{j=1 \\ j \neq J}}^{np} dU_j \tag{3.14}$$

$$dV_J = -\sum_{\substack{j=1 \\ j \neq J}}^{np} dV_j \tag{3.15}$$

$$dn_{iJ} = -\sum_{\substack{j=1 \\ j \neq J}}^{np} dn_{ij} \tag{3.16}$$

where J is the index of the arbitrary pseudophase, in which properties will be written by difference. Rewriting Eq. (3.7), taking the terms relative to the index J off the sums, one has

$$dS = \sum_{\substack{j=1 \\ j \neq J}}^{np} \frac{dU_j}{T_j} + \frac{dU_J}{T_J} + \sum_{\substack{j=1 \\ j \neq J}}^{np} \left(\frac{P_j}{T_j} + \frac{\sum_{i=1}^{nc} m_{ij} g z_j}{A_j T_j z_j} \right) dV_j + \left(\frac{P_J}{T_J} + \frac{\sum_{i=1}^{nc} m_{iJ} g z_J}{A_J T_J z_J} \right) dV_J - \sum_{\substack{j=1 \\ j \neq J}}^{np} \sum_{i=1}^{nc} \frac{\mu_{ij} + M_i g z_j}{T_j} dn_{ij} - \frac{\mu_{iJ} + M_i g z_J}{T_J} dn_{iJ} \tag{3.17}$$

Replacing Eqs. (3.14)–(3.16) into Eq. (3.17), we have

$$dS = \sum_{\substack{j=1 \\ j \neq J}}^{np} \left(\frac{1}{T_j} - \frac{1}{T_J} \right) dU_j + \sum_{\substack{j=1 \\ j \neq J}}^{np} \left[\left(\frac{P_j}{T_j} + \frac{\sum_{i=1}^{nc} m_{ij} g z_j}{A_j T_j z_j} \right) - \left(\frac{P_J}{T_J} + \frac{\sum_{i=1}^{nc} m_{iJ} g z_J}{A_J T_J z_J} \right) \right] dV_j - \sum_{\substack{j=1 \\ j \neq J}}^{np} \sum_{i=1}^{nc} \left(\frac{\mu_{ij} + M_i g z_j}{T_j} - \frac{\mu_{iJ} + M_i g z_J}{T_J} \right) dn_{ij} \tag{3.18}$$

Now, all the differential variables of Eq. (3.18) are independent and, by hypothesis, cannot be zero simultaneously. Therefore, the only way of nullifying dS and assuring that the system reaches the equilibrium is to nullify the terms within the parenthesis, for all the differentials and all their indexes. So, one can write

$$T_j = T_J, \quad \text{for } j = 1, \ldots, np \tag{3.19}$$

where Eq. (3.19) describes the thermal equilibrium, which comes from internal-energy differential. This condition shows that the temperature is a uniform field at equilibrium.

Knowing that the density of pseudophase j can be given by $\rho_j = \frac{\sum_{i=1}^{nc} m_{ij}}{A_j z_j}$, we have, from the condition of mechanical equilibrium in piezometric pressure,

$$P_j + \rho_j g z_j = P_J + \rho_J g z_J, \quad \text{for } j = 1, \ldots, np \tag{3.20}$$

And finally, to eliminate the mass transfer among the pseudophases (and neglecting possible chemical reactions),

$$\mu_{ij} + M_i g z_j = \mu_{iJ} + M_i g z_J, \quad \text{for } i = 1, \ldots, nc \text{ and } j = 1, \ldots, np \tag{3.21}$$

3.2 LITERATURE REVIEW

In flash algorithms for closed systems under the influence of gravitational field like the ones in Figs. 3.1 and 3.2, it is not convenient to specify the pressure together with temperature and global composition, as the density-dependent hydrostatic charge makes P vary continuously with height. Therefore, differently from the traditional flash algorithms of Gibbs free energy minimization for flash problem with T, P, and $\underline{N}$ specified, the presence of the gravitational field suggests the minimization of Helmholtz free energy, specifying total volume V of the system, together with T and $\underline{N}$. Espósito et al. (2000a, 2000b) formulated this flash problem at T, V, and $\underline{N}$ specified, defining both gravity-modified Helmholtz free energy (A^*) and chemical potential (μ_i^*) as:

$$A^* = A + E_{\text{pot}} = A + \sum_{i=1}^{nc} m_i g z \tag{3.22}$$

$$\mu_i^* = \mu_i + M_i g z, \quad \text{for } i = 1, \ldots, nc \tag{3.23}$$

Generally speaking, the methodology employed by Espósito et al. (2000a, 2000b) to minimize the Helmholtz free energy in Eq. (3.22) consists in obtaining the optimum number of moles of each component in each pseudophase j (n_{ij}), also indicating the precise location of the interfaces between the properly old-defined *phases* and the static-pressure profile that attends the restriction imposed by Eq. (3.23):

$$F_{\text{OBJ}} = A^*(\underline{\underline{n}}) \tag{3.24}$$

where $\underline{\underline{n}} = \begin{pmatrix} n_{11} & \cdots & n_{1np} \\ \vdots & \ddots & \vdots \\ n_{nc1} & \cdots & n_{ncnp} \end{pmatrix}$ is a matrix, in which elements are the number of moles (n_{ij}) of each component i in each pseudophase j.

An outer optimization loop updates the height relative to the interface(s) between the phases (or *contacts*, as one says in petroleum industry), whereas an inner loop minimizes the modified Helmholtz energy, recalculating the number of moles in each pseudophase (or level). Details of this fascinating modified-flash problem are beyond the scope of this book and can be found in Espósito (1999) and Espósito et al. (2000a).

New techniques for solution of problems at T, V, and $\underline{N}$ specified in simulators, both process and reservoir ones, have been developed since then (see, e.g., Nichita et al., 2002; Mikyska and Firoozabadi, 2012; Castier, 2014). The great advantage of setting volume instead of pressure is the straight resolution of the cubic equation of state (EoS), which is likely written as $P = P(T, V, \underline{N})$, saving computational effort at each step of the equilibrium calculation.

For the calculation of compositional grading, that is, the determination of composition and pressure along depth in a petroleum reservoir, this type of algorithm based on the minimization of Helmholtz free energy is not convenient because one does not have how to determine a priori the global composition of such a huge system. Only a few sampling points have their local composition known. The formulation of the problem must, therefore, be altered in order that, considering the temperature uniform and with pressure and composition measured at a determined height level, one can determine P and $\underline{x}$ at any other depth by equilibrium condition of Eq. (3.23). Differentiating this equation, one can write

$$d\mu_i = M_i g dz, \quad \text{for } i = 1, \ldots, nc \tag{3.25}$$

Using the relation between chemical potential and fugacity:

$$d\mu_i = RTd \ln \hat{f}_i, \quad \text{for } i = 1, \ldots, nc \tag{3.26}$$

And replacing Eq. (3.26) into Eq. (3.25), after taking the exponential, we get

$$\hat{f}_i^z(T, P^z, \underline{x}^z) = \hat{f}_i^{z_{\text{ref}}} \exp\left[\frac{M_i g(z - z_{\text{ref}})}{RT}\right], \quad \text{for } i = 1, \ldots, nc \tag{3.27}$$

Once we know pressure and composition measured at the reference level z_{ref}, the system with nc equations in Eq. (3.27) has, as unknowns, the $(nc - 1)$-independent mole fractions and the pressure at the level of interest z. Note that this problem is analogous (but not the same) to that employed for calculations of saturation pressure, in which only the fugacity equality conditions, and not the material balance equations, are required. Note also that, as fugacities are obtained by the resolution of the EoS for compositions at each level, the densities so obtained for each *pseudophase* satisfy automatically the piezometric pressure equality condition of Eq. (3.20).

The calculations are done starting from the reference level to obtain P^z and $\underline{x}^z$ at the level just above (or just below), independent of the other ones. It becomes evident that, for a liquid–vapor equilibrium problem as the one shown in Fig. 3.2, a stability test will have to be performed depending on the refinement degree of the discrete mesh close to the interface. For the latter to be precisely determined between the last liquid pseudophase and the first vapor one, the thickness of the discrete levels, at least within that region, will have to be reset via mesh refinement sometime in the solution algorithm of Eq. (3.27).

Studies relative to the influence of the gravitational field in phase equilibrium date from the pioneer works of Gibbs (1906). Muskat (1930), and Sage and Lacey (1939) simplified Eq. (3.25) for gases and ideal mixtures, presenting the first results of gravity segregation for relatively simple systems. In the petroleum industry, compositional grading in reservoirs started being quantitatively studied from the work of Schulte (1980), who, for the first time, used the cubic EoS to calculate the fugacities in Eq. (3.27). Schulte (1980) presented several simulations, solving Eq. (3.27) from a certain arbitrary height level for hydrocarbon mixtures similar to typical reservoir fluids, like Brent petroleum, for example. The author already stated that results would be sensitive to EoS parameters, like the binary interaction coefficients, without, however, showing such adjustments, neither field experimental data.

A more general contribution for the modeling of thermodynamic systems under the influence of various external contributions, like, for example, electric and magnetic fields, surface phenomena and solid–fluid interactions, not only applicable to petroleum industry, might be found in the text book of Sychev (1981). According to this author, the gravitational effects on the state of a thermodynamic system manifest themselves first of all through the variation of pressure with height. For systems with moderate heights, this variation is generally negligible compared to the absolute pressure within the recipient. For this reason, the gravity effect is neglected in the majority of industrial processes. However, in some situations for which the compressibility of a substance is high, even small variations of pressure with height lead to noticeable variations of density and other thermodynamic properties of the substance. This is a characteristic of systems close to the critical point, at which the isothermal compressibility of a component is infinite:

$$\lim_{\substack{T \to T_c \\ P \to P_c}} \frac{1}{v}\left(\frac{\partial v}{\partial P}\right)_T = \infty \tag{3.28}$$

In other words, the proximity of the critical point increases the gravitational segregation due to the significant increase in compressibility.

Chang et al. (1983) calculated density and composition profiles for systems very close to the critical point within 1-mm sized cells, using the EoS of Leung and Griffiths (1973). According to these authors, in the case of a pure component at its critical temperature, if the cell is filled with a global density close to the critical one (ρ_c), the later would be actually obtained at a certain height level within the cell. It is at this level that the meniscus (liquid–vapor interface) will first form when the cell is cooled down to a value infinitesimally below the critical temperature. It was observed that the density profile, in the case of a pure component, is antisymmetric relative to this level. For mixtures, as composition also varies along height, the meniscus will form in which temperature and static pressure match, respectively, both critical temperature and pressure of the composition at that level.

Back to petroleum industry, Creek and Schrader (1985) managed to reproduce the compositional grading occurred in the East Painter reservoir, in Wyoming (USA), fitting a cubic EoS based in PVT analysis of recombined samples. These authors did not show the parameters of their modeling, but this work became a classic reference in the subject, for being one of the first to present a really quantitative analysis, including comparisons with field data.

Wheaton (1991) also presented a more complete procedure for the calculation of compositional grading, including the determination of gas–oil and water–oil contacts, as well as their respective transition zones for which capillary effects are important. Consideration of such effects is also beyond the scope of this book.

Lira-Galeana et al. (1994) also used the thermodynamics of semicontinuous mixtures to try to reproduce the results of Creek and Schrader (1985). Firoozabadi (1999), another classic reference in the subject, also did a full literary review that can be useful to the interested reader.

The works of Espósito (1999), Espósito et al. (2000a, 2000b) are also inserted in this context. The latter is also based on the thermodynamics of semicontinuous mixtures and generalizes the method of Shibata et al. (1987) for numbers of quadrature points greater than 3, also presenting examples of compositional grading in hydrocarbon mixtures.

In the following, we will proceed to the solution of some examples (classic and semiquantitative) introduced by some already cited references. The methodology adopted in these examples serves as a basis for the reader to understand the real case studies to be discussed in the next chapters.

3.3 GRAVITY SEGREGATION OF PURE CO_2 CLOSE TO THE CRITICAL POINT

As CO_2 is an important agent in the petroleum industry due to its high potential for enhanced recovery, this example treats its phase behavior and gravity segregation in the vicinity of its critical point (304.2K and 73.805 bar) within a hypothetical 1-cm high recipient. Using, specifically for this case, a volume shift of 0.455, the critical density of CO_2 is around 468.7 kg/m^3 (Poling et al., 2001). According to Chang et al. (1983), a pure component at its critical temperature tends to present an antisymmetric behavior in the density distribution along height when specified global density is close to the critical one. The density referent to the antisymmetric level corresponds exactly to this value. It is at this level that the meniscus will form when, due to a differential variation in temperature, the partition of the system in two phases occur. The height correspondent to the meniscus can change with the specified global density. For the interested reader, there are two ways of formulating and solving this problem: The first one consists of, given the constant temperature of 304.2K, set the critical pressure of CO_2 in a certain reference level and solve Eq. (3.27) for the other levels. The second one consists of specifying a global density around 468 kg/m^3 at the same temperature and proceed to the Helmholtz free energy minimization modified by gravity (Eq. (3.22)), using, for example, the algorithm proposed by Espósito (1999) and also described by Espósito et al. (2000a). Results should be equivalent. Fig. 3.3 shows the simulated density profiles for global densities of 468.0; 468.6, and 470.2 kg/m^3. Note that the antisymmetric level shows increasingly close to the top of the system as global density increases. This result is reasonable, for, with increasing density, the amount of liquid existing in the system when partition takes place is higher. Fig. 3.4 shows the pressure profile along the depth of the recipient. Due to its low dimensions, pressure variation is minimum around the critical pressure, but it can cause considerable variations in density.

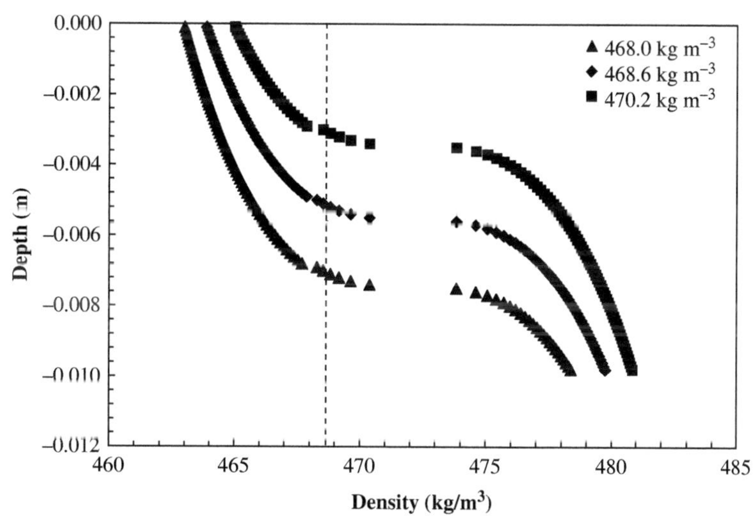

FIGURE 3.3 Density profile along depth for pure CO_2 at 304.2K, setting three overall densities close to the critical one. The dotted vertical line indicates this value of $\rho_c = 468.7\ \mathrm{kg}/m^3$.

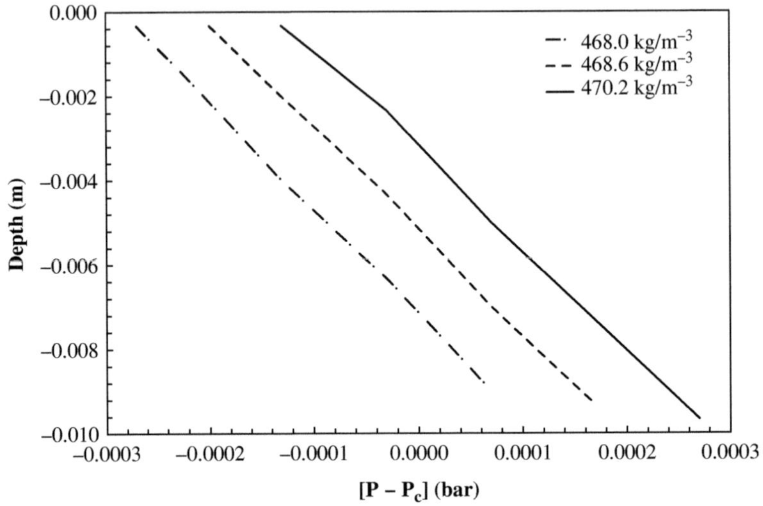

FIGURE 3.4 Deviation in relation to critical pressure along height for near-critical pure CO_2, at 304.2K, setting three overall densities close to the critical one.

3.4 RESERVOIR AT VAPOR–LIQUID EQUILIBRIUM

Schulte (1980) was one of the pioneers in considering the gravitational field in reservoir–fluid studies. This author presented calculations done with Peng–Robinson EoS in hypothetical hydrocarbon mixtures, solving Eq. (3.27) to check not only the gravity segregation of these components with depth but also to detect the interfaces between phases in equilibrium or the critical transitions like those observed for CO_2 in Section 3.3.

Let us see the example of Table 3 of Schulte (1980), in which fluid presents, at the reference level, the composition shown in Table 3.1.

Schulte (1980) did not present the parameters of the thermodynamic model used in his work. Here, we expand the C_{7+} fraction, following the methodology of Shibata et al. (1987) for two quadrature points, which represent pseudocomponents with 11.08 and 33.10 atoms of carbon, respectively. Densities were calculated on the basis of Fig. 2.4 and Table 2.1, following the corresponding hydrocarbon cuts. Numerically, one can say that density varies with the logarithm of single carbon number (SCN), that is,

$$\rho = A + B\ \ln(\mathrm{SCN}) \tag{3.29}$$

where A and B are adjustable parameters.

To ensure continuity of C_{7+} fraction with lighter hydrocarbons, the density of C_6 (0.685 g/cm^3, according to Table 2.1) must be honored by Eq. (3.29). The average density of the fraction (not informed) can also be obtained by interpolation in molar mass (197.50 g/mol) or in average single carbon number given by the methodology of Shibata et al. (1987): $\bar{I} = 14.39 \rightarrow d_{60/60} = 0.8267$. So, one calculates constants A and B of

TABLE 3.1 Composition at the Reference Level for the Fluid Shown in Table 3 of Schulte (1980). Static Pressure at the Reference Level: 293.02 bar. Saturation Pressure at the Reference Level: 290.14 bar. $T = 361$K

Component	Composition at the Reference Level (mol%)	Molar Mass (g/mol)
CO_2	0.12	44.01
N_2	0.13	28.00
C_1	54.98	16.00
C_2	5.44	30.00
C_3	4.88	44.00
iC_4	1.88	58.00
nC_4	2.48	58.00
iC_5	1.45	72.00
nC_5	1.30	72.00
nC_6	1.91	86.00
C_{7+}	25.43	271.49
Global	**100.00**	**69.04**

TABLE 3.2 Basic Properties of the Pseudocomponents Represented by Quadrature Points with 11.08 and 33.10 atoms of Carbon, after Shibata et al. (1987)

Pseudocomponent	Composition at the Reference Level (mol%)	M (g/mol)	$d_{60/60}$
$QC_{11.08}$	21.53	151.16	0.7840
$QC_{33.10}$	3.90	459.34	0.9606

Eq. (3.29), which can now be applied to the pseudocomponents individually. Basic properties obtained for both pseudocomponents are shown in Table 3.2. Using Cavett's (1962) correlation—Eqs. (2.73) and (2.74)—for critical properties, Kesler and Lee's (1976)—Eqs. (2.76) and (2.77)—for acentric factors, and also Eq. (2.79) for methane's binary interaction coefficients (the only nonzero), a fine tuning in these coefficients for the heavies (and also in their mole fractions) was performed to reproduce precisely the saturation pressure at the reference level. The complete modeling is shown in Tables 3.3 and 3.4.

TABLE 3.3 Properties of Peng–Robinson Equation of State for the Fluid of Table 3 of Schulte (1980)

Component	Composition at the Reference Level (mol%)	P_c (bar)	T_c (K)	ω	M (g/mol)	Volume Shift
CO_2	0.12	73.76	304.20	0.2250	44.01	−0.0718
N_2	0.13	33.94	126.20	0.0400	28.01	0.0000
C_1	54.98	46.00	190.60	0.0080	16.04	−0.2209
C_2	5.44	48.84	305.40	0.0980	30.07	−0.1649
C_3	4.88	42.46	369.80	0.1520	44.10	−0.1309
iC_4	1.88	36.48	408.10	0.1760	58.12	−0.1064
nC_4	2.48	38.00	425.20	0.1930	58.12	−0.1064
iC_5	1.45	33.84	460.40	0.2270	72.15	−0.0873
nC_5	1.30	33.74	469.60	0.2510	72.15	−0.0873
nC_6	1.91	29.69	507.40	0.2960	86.18	−0.0715
$QC_{11.08}$	21.80	23.43	642.51	0.4882	151.16	−0.0217
$QC_{33.10}$	3.63	9.90	929.34	1.0854	459.34	0.0769

TABLE 3.4 Binary Interaction Coefficients (k_{C1-Cn}) for methane in Peng–Robinson Equation of State

Component	k_{C1-Cn}
C_2	0.0256
C_3	0.0270
iC_4	0.0284
nC_4	0.0284
iC_5	0.0298
nC_5	0.0298
nC_6	0.0312
$QC_{11.08}$	0.0486
$QC_{33.10}$	0.0791

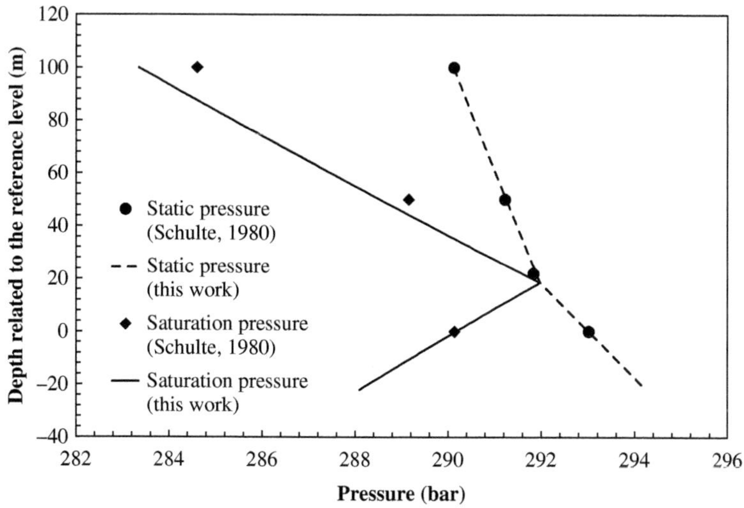

FIGURE 3.5 Static and saturation pressures along depth for the fluid of Table 3 of Schulte (1980). Gas–oil contact simulated at 18.75 m above reference level.

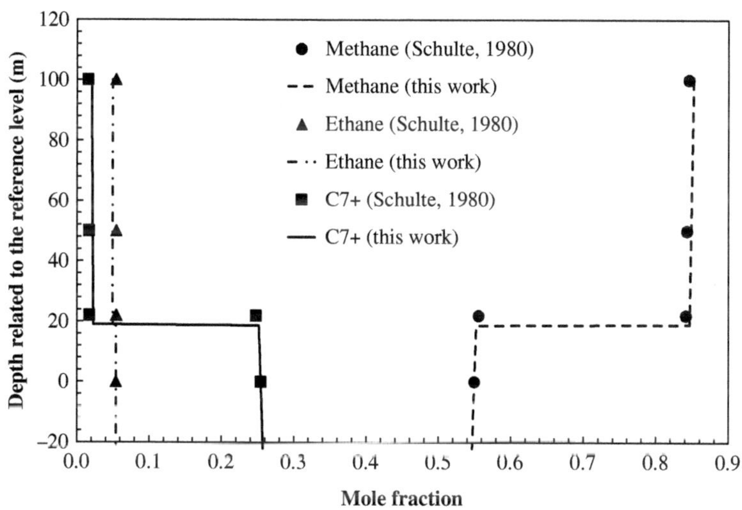

FIGURE 3.6 Mole fractions of the main components along depth for the fluid of Table 3 of Schulte (1980).

The results of the compositional grading calculations are compared with those presented by Schulte (1980) in Figs. 3.5–3.7. The agreement of our methodology with the estimation performed (but not shown) by Schulte (1980) is very reasonable. We emphasize that, except for the saturation pressure at the reference level, no other data were available to make up the

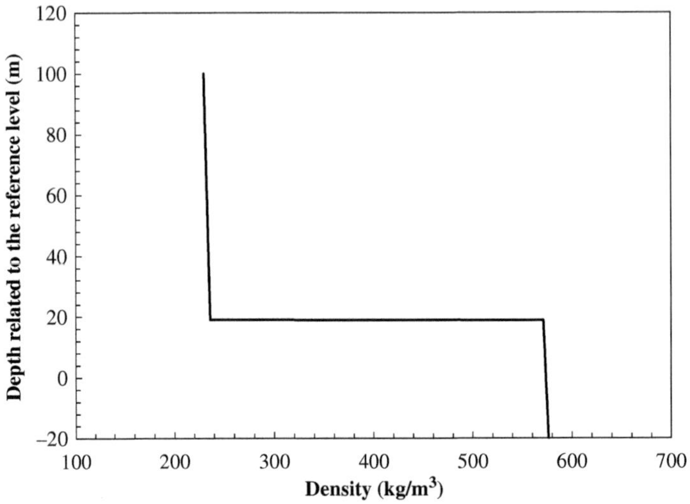

FIGURE 3.7 Density as a function of depth for the fluid of Table 3 of Schulte (1980). The author presented no data. Coherently with the two previous figures, the system shows a liquid–vapor interface at the relative height of 18.75 m.

basis for regression. This pioneer example of Schulte (1980) must be considered as semiquantitative, although it is very convenient for explaining purposes.

3.5 RESERVOIR WITH CRITICAL TRANSITION

In this second example, we will study a mixture of six normal paraffins also presented by Schulte (1980). Its composition at the reference level is presented in Table 3.5. Critical properties, acentric factors, and molar masses are also shown. Such properties, regarding normal paraffins, are all tabled and therefore are not adjusted. Not informed by Schulte (1980), methane's binary interaction coefficients (the only nonzero and shown in Table 3.6) and volume shifts were obtained using Eqs. (2.79) and (2.83), respectively.

Temperature was fixed at 275K, and the pressure at the reference level was 275 bar. The results of the solution of the system represented by Eq. (3.27) are presented in Figs. 3.8–3.11. Mole fractions of methane (Fig. 3.8), ethane, and *n*-decane (Fig. 3.9) are in excellent agreement with the ones (also simulated) reported by Schulte (1980). Fig. 3.10 presents the variation of density with depth, although, once more, Schulte (1980) has not presented any data or result related to this property.

The critical transition is confirmed in Fig. 3.11: The saturation pressure plot does not touch the static-pressure one, passing through a maximum very close to the reference level, which composition is the closest to its critical

TABLE 3.5 Properties of the Mixture of *N*-Paraffins of Table 4 of Schulte (1980), using Peng–Robinson EoS

Component	Composition at the Reference Level (mol%)	P_c (bar)	T_c (K)	ω	M (g/mol[1])	Volume Shift
C_1	80.97	46.00	190.6	0.008	16.04	−0.2209
C_2	5.66	48.84	305.4	0.098	30.07	−0.1649
C_3	3.06	42.46	369.8	0.152	44.10	−0.1309
nC_5	4.57	33.74	469.6	0.251	72.15	−0.0873
nC_7	3.30	27.36	540.2	0.351	100.21	−0.0581
nC_{10}	2.44	21.08	617.6	0.490	142.29	−0.0270

TABLE 3.6 Methane's Binary Interaction Coefficients (Peng–Robinson EoS) for the Mixture of Normal Paraffins of Table 4 of Schulte (1980)

Component	$k_{C1\ Cn}$
C_2	0.0256
C_3	0.0270
nC_5	0.0298
nC_7	0.0326
nC_{10}	0.0368

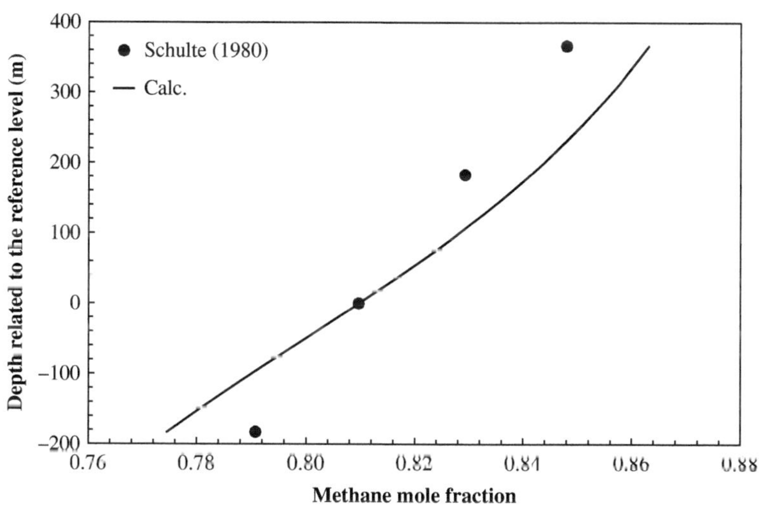

FIGURE 3.8 Methane's distribution along height for the mixture of *n*-paraffins of Table 4 of Schulte (1980).

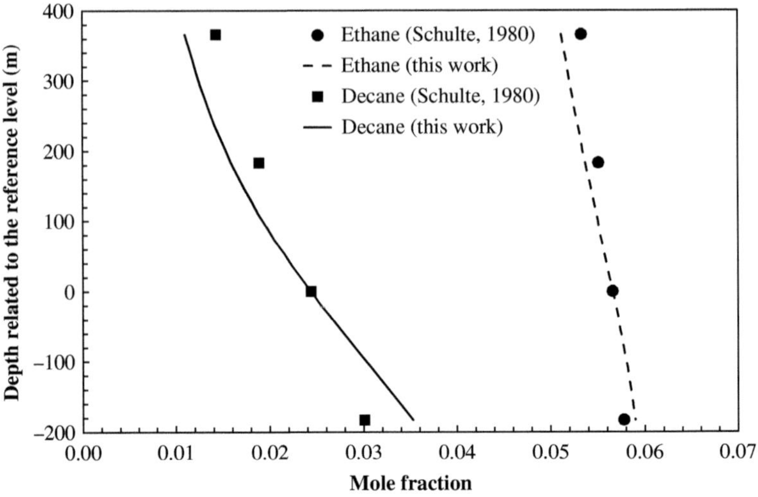

FIGURE 3.9 Distributions of ethane and *n*-decane along depth for the mixture of normal paraffins in Table 4 of Schulte (1980).

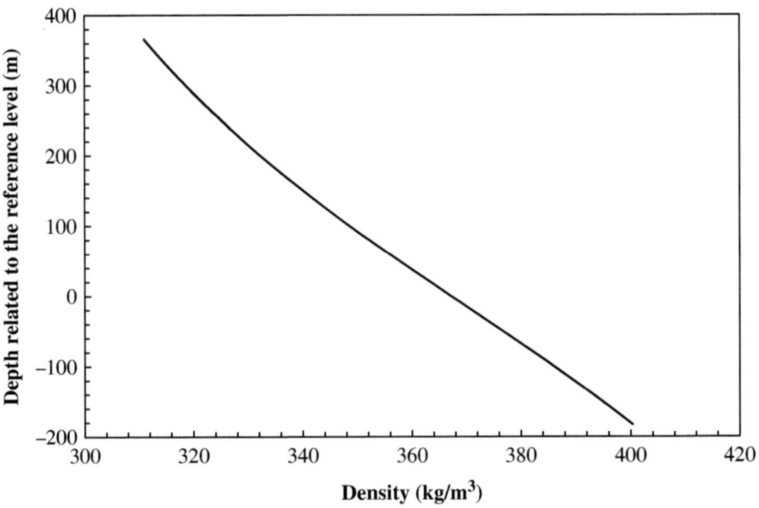

FIGURE 3.10 Density as a function of depth for the fluid of Table 4 of Schulte (1980).

point in all the system. It is noteworthy that, for mixtures, as compositions varies with height, the meniscus will form only if, for a certain level, the temperature of the system is the same as the critical temperature of that composition. In other words, although the density profile remains antisymmetrical, the critical density will not be reproduced necessarily at the antisymmetric level.

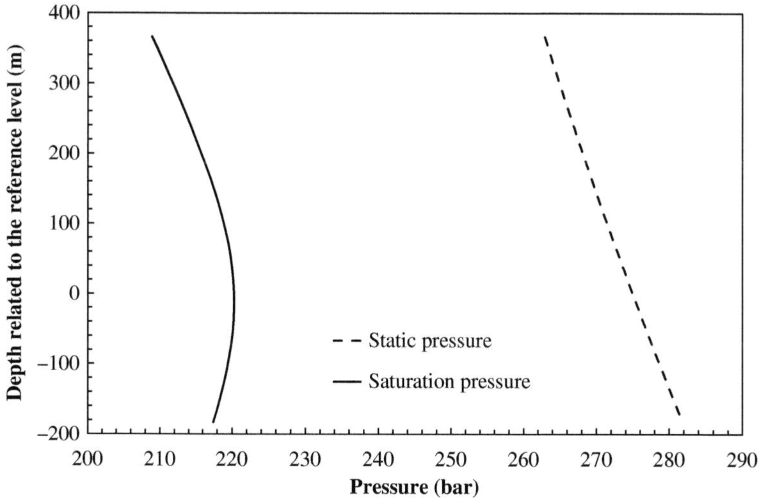

FIGURE 3.11 Static and saturation pressure profiles as a function of depth for the fluid of Table 4 of Schulte (1980).

3.6 EAST PAINTER RESERVOIR

Creek and Schrader (1985) applied the technique of EoS fitting based on PVT analysis data for some reservoirs of volatile oil and gas condensate of the so-called Wyoming Overthrust Belt, more precisely, the *East Painter* reservoir. Specifically for the case of this reservoir, from various samples obtained by recombination, these authors selected the medium point of the perforated interval in well 42-7 A (−4255 m) as being the reference level for (with Peng−Robinson EoS fitted) solving Eq. (3.27) and simulate the composition and some properties measured along depth. The work of Creek and Schrader (1985) can be considered the first one to reproduce quantitatively this kind of compositional grading. The authors did not present the parameters of their fitting. So, with the molar mass (158 g/mol) and specific gravity (0.796) of C_{7+} fraction, we proceeded, as in the previous examples, to the exponential-decay splitting and further two-quadrature-point lumping, following the methodology of Shibata et al. (1987). The same correlations used in Section 3.5 were hereby used for the calculation of critical properties, acentric factors, volume translations, and binary interaction coefficients. A fine tuning of some of these parameters was done to reproduce saturation pressure and GOR informed at the reference level. The final set of parameters is presented in Tables 3.7 and 3.8. Figs. 3.12−3.14 show the comparison of some simulated and experimental data reported by Creek and Schrader (1985): methane and C_{7+} profiles and variation of the latter's molar mass along depth. The agreement of the simulated gravity segregation

TABLE 3.7 Properties of Peng–Robinson Equation of State for the Fluid of Well 42-7 A, Perforated Interval 4179–4337 m (Creek and Schrader, 1985)

Component	P_c (bar)	T_c (K)	ω	M (g/mol)	Volume Shift
CO_2	73.76	304.20	0.2250	44.01	−0.0718
N_2	33.94	126.20	0.0400	28.01	0.0000
C_1	46.00	190.60	0.0080	16.04	−0.2206
C_2	48.84	305.40	0.0980	30.07	−0.1649
C_3	42.46	369.80	0.1520	44.10	−0.1309
iC_4	36.48	408.10	0.1760	58.12	−0.1064
nC_4	38.00	425.20	0.1930	58.12	−0.1064
iC_5	33.84	460.40	0.2270	72.15	−0.0873
nC_5	33.74	469.60	0.2510	72.15	−0.0873
C_6	32.89	507.50	0.2750	86.00	−0.0717
$QC_{9.48}$	25.57	618.61	0.4283	128.69	−0.0340
$QC_{23.85}$	12.17	863.04	0.8012	329.91	0.0437

TABLE 3.8 Nonzero Binary Interaction Coefficients for the Fluid of Well 42-7 A, Perforated Interval 4179–4337 m (Creek and Schrader, 1985)

Component	k_{C1-Cn}
C_2	0.00509
C_3	0.00649
iC_4	0.00789
nC_4	0.00789
iC_5	0.00930
nC_5	0.00930
C_6	0.01068
$QC_{9.48}$	0.01495
$QC_{23.85}$	0.03507

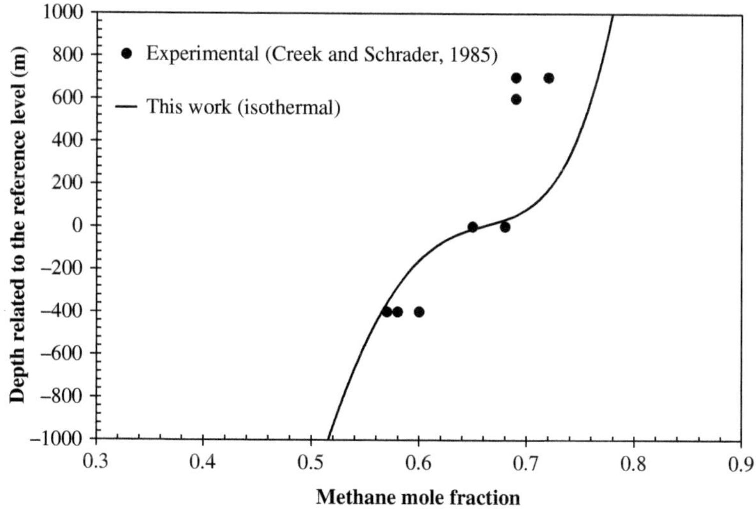

FIGURE 3.12 Methane mole fraction distribution along depth in East Painter reservoir.

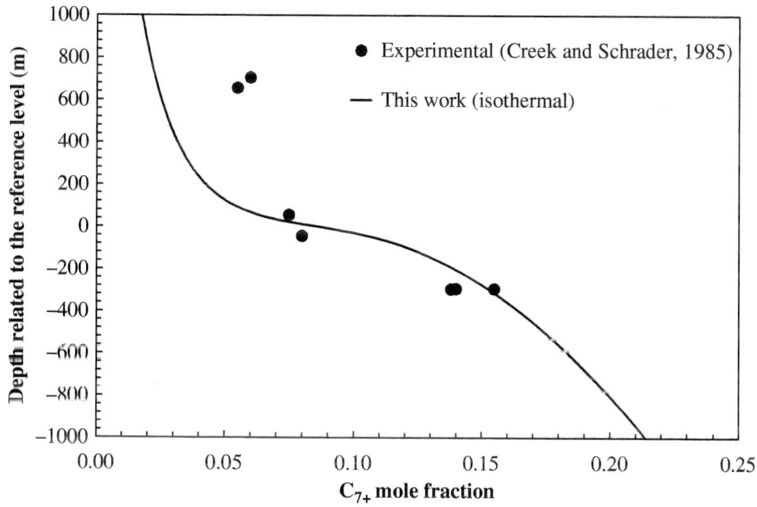

FIGURE 3.13 C_{7+} mole fraction distribution along depth in East Painter reservoir.

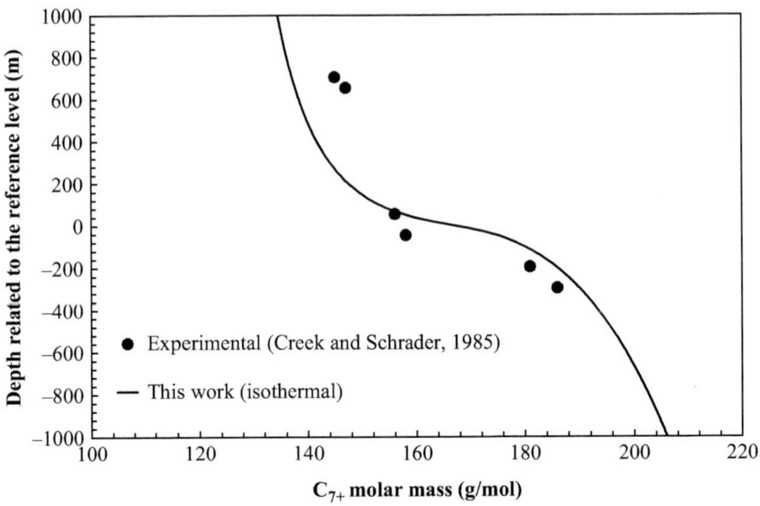

FIGURE 3.14 Variation of the molar mass of C_{7+} fraction along depth in East Painter reservoir.

with the one observed in practice is very reasonable, considering the scarce PVT data made available by Creek and Schrader (1985) and also the uncertainties related to the field measurements and recombinations of surface samples.

3.7 EXERCISES

1. Show that, in a mixture with nc components submitted to the gravitational field in an isothermal unidimensional column, the distribution of component j along height is given by

$$\sum_{j=1}^{nc-1} \frac{\partial \mu_i}{\partial x_j} \frac{dx_j}{dz} = (\rho \bar{v}_i - M_i)g, \quad \text{for } i = 1, \ldots, nc - 1$$

2. Rewrite the equation derived in Exercise 1 in matrix form and discuss two causes by which the variation of x_j with z might be very pronounced. Suggestion: consult Firoozabadi (1999) (Thermodynamics of hydrocarbon reservoirs, pp. 59 and 60, McGraw Hill, 1999).
3. Using the thermodynamic modeling proposed in Example 3.4 (Creek and Schrader, 1985), plot the determinant of the matrix $\frac{\partial \mu_i}{\partial x_j}$ along depth. Discuss its behavior.
4. The *gas–oil contact* (*GOC*) is the interface between the gas cap and the oil zone in a saturated black-oil reservoir. Starting from a sample at some point in the oil zone, would the exact location of the GOC in an isothermal reservoir, and the determination of the compositions in the gas zone, require some modification in Eq. (3.27)? Why?

REFERENCES

Castier, M., 2014. Helmholtz function-based global phase stability test and its link to the isothermal-isochoric flash problem. Fluid Phase Equilib. 379, 104–111.

Cavett, R.H., *Physical Data for Distillation Calculations—Vapor–Liquid Equilibrium*, Proc. 27th Meeting, API, San Francisco, pp. 351–366, 1962.

Chang, R.F., Levelt-Sengers, J.M.H., Doirn, T., Jones, J., 1983. Gravity-induced density and concentration profiles in binary mixtures near gas-liquid critical lines. J. Chem. Phys. 79 6, 3058–3066.

Creek, J.L., Schrader, M.L., *East Painter Reservoir: An Example of a Compositional Gradient from a Gravitational Field*, SPE 14411, 60th Annual Technical Conference and Exhibition, Las Vegas-NV, September/1985.

Espósito, R.O., Cálculo de equilíbrio termodinâmico em sistemas sob a influência de campos gravitacionais, Tese de Mestrado, Escola de Química—UFRJ, 1999.

Espósito, R.O., Castier, M., Tavares, F.W., 2000a. Calculations of thermodynamic equilibrium in systems subject to gravitational fields. Chem. Eng. Sci. 55, 3495–3504.

Espósito, R.O., Castier, M., Tavares, F.W., 2000b. Phase equilibrium calculations for semicontinuous mixtures subject do gravitational fields. Ind. Eng. Chem. Res. 39, 4415–4421.

Firoozabadi, A., 1999. Thermodynamics of Hydrocarbon Reservoirs. McGraw Hill, New York, NY.

Gibbs, J.W., The Scientific Papers, Vol. 1 Thermodynamics, Longmans, Green and Cia, 1906.

Kesler, M.G., Lee, B.I., 1976. Improved Prediction of Enthalpy of Fractions. Hydrocarbon Process.153–158.

Leung, S.S., Griffiths, R.B., 1973. Thermodynamic properties near the liquid-vapor critical line in mixtures of He3 and He4. Phys. Rev A8, pp. 2760–2683.

Lira-Galeana, C., Firoozabadi, A., Prausnitz, J.M., 1994. Computation of compositional grading in hydrocarbon reservoirs. Applications of continuous thermodynamics. Fluid Phase Equilib. 102, 143–158.

Mikyska, J., Firoozabadi, A., 2012. Investigation of mixture stability at given volume, temperature, and number of moles. Fluid Phase Equilib. 321, 1–9.

Muskat, M., 1930. Distribution of non-reacting fluids in the gravitational field. Phys. Rev. 35, 1384–1393.

Nichita, D.V., Valencia, C.A.D and Gomez, S., *Isochoric Phase Stability Analysis and Flash Calculations*, AIChE Spring Meeting, Paper 142c, New Orleans, March 10th to 14th, 2002.

Poling, B.E., Prausnitz, J.M., O'Connell, J.P., 2001. The Properties of Gases and Liquids, 5th Edition McGraw Hill, New York, NY.

Sage, B.H., Lacey, W.N., 1939. Gravitational concentration gradients in static columns of hydrocarbon fluids. Trans. AIME 132, 120–131.

Schulte, A.M., 1980. Compositional variations within a hydrocarbon column due to gravity, SPE 9235. Dallas-Texas 21–24.

Shibata, S.K., Sandler, S.I., Behrens, R.A., 1987. Phase equilibrium calculations for continuous and semicontinuous mixtures. Chem. Eng. Sci. 42 (8), 1977–1988.

Sychev, V.V., 1981. Complex Thermodynamic Systems. Mir Publishers, Moscow.

Wheaton, R., 1991. Treatment of variations of composition with depth in gas-condensate reservoirs. SPE Res. Eng. 239–244.

FURTHER READING

Sage, B.H., Lacey, W.N., 1936. Phase equilibria in hydrocarbon systems. Ind. Eng. Chem. 28 (2), 249–255.

Chapter 4

Irreversible Thermodynamics Applied to Reservoir Engineering

Rigorously, no petroleum reservoir is in thermodynamic equilibrium due to the spontaneous heat flux from the nuclei of the Earth to the surface, which, in turn, is a function of the thermal gradient from Earth's center to crust. This heat flux, which is also influenced by lithology (type of rock forming the submerse layers), by itself, already causes an uninterrupted entropy increase, invalidating the equations of chemical-potential equality for components inside a petroleum reservoir. The impact of entropy generation in a closed unidimensional system as the one in Fig. 3.2 might be noticed by the change in fluids distribution predicted by Eq. (3.27). Even if the system keeps static, that is, without flow occurrence, the components of the fluid will not have the same gravity-modified fugacities in all points of the domain. This happens because, depending on temperature, pressure, and composition conditions, some components tend to diffuse to the hotter or colder regions of the reservoir, a phenomenon known as "Soret effect," that is, a mass flux due to a temperature gradient. The objective of this chapter is to derive the equations that govern the components flow in reservoirs under not only the action of the gravitational field but also of the geothermal gradient, due to which the isofugacity equations hold no more. Concepts of Irreversible Thermodynamics are used to predict these fluxes, the simulation of compositional distribution done by the resolution of another type of nonlinear-equation system in both transient and steady state.

4.1 INTRODUCTION

Consider then a two-component unidimensional system, originally in thermodynamic equilibrium, sketched in Fig. 4.1A. The isolated action of the gravitational field makes heavier component's molecules concentrate at the bottom of the recipient, while the lighter stay preferably at the top. Soret effect may intensify (Fig. 4.1B) or attenuate (Fig. 4.1C) the effect of gravity, rearranging the components by thermal diffusion.

Compositional Grading in Oil and Gas Reservoirs. DOI: http://dx.doi.org/10.1016/B978-0-12-812452-9.00004-6

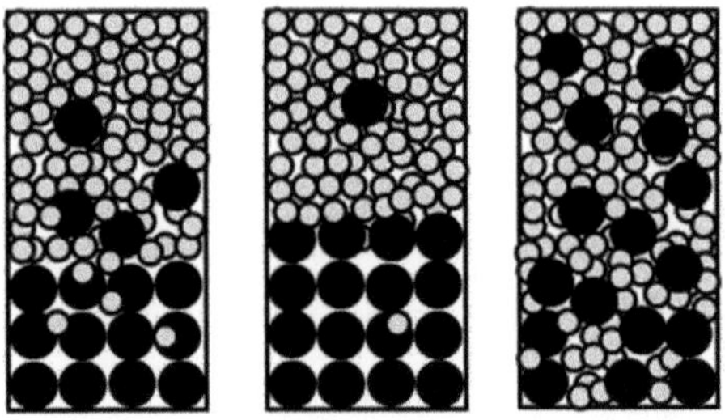

FIGURE 4.1 (A) Sketch of a binary system initially in thermodynamic equilibrium. Gravitational field's action concentrates the heavier component's molecules at the bottom of the reservoir; (B) applied thermal gradient enhances segregation, because thermal diffusive flux has the same direction as gravity; and (C) applied thermal gradient attenuates segregation because thermal diffusive flux's direction is opposite to gravity.

Consider now a mixture flowing in a porous media of arbitrary dimensions submitted simultaneously to pressure, temperature, and composition gradients.

Given the velocity of component i ($\underline{v}_i$) in relation to a fixed coordinate axis, the flux of i in mass per unit area per unit time ($\underline{N}_i$) is given by

$$\underline{N}_i = \rho_i \underline{v}_i, \quad \text{for} \quad i = 1, \ldots, nc \tag{4.1}$$

where ρ_i is the density (or "mass concentration," SI unit of which is kg/m^3) of component i.

Consequently the total mass flux ($\underline{N}_t$) of the mixture relative to the same fixed axis will be given by

$$\underline{N}_t = \sum_{i=1}^{nc} \rho_i \underline{v}_i = \rho \underline{v} \tag{4.2}$$

where ρ is the global mass concentration, and $\underline{v}$ is the velocity of the center of mass of the mixture, given by

$$\underline{v} = \sum_{i=1}^{nc} w_i \underline{v}_i \tag{4.3}$$

in which w_i is the mass fraction of component i, given by

$$w_i = \frac{\rho_i}{\rho}, \quad \text{for} \quad i = 1, \ldots, nc \tag{4.4}$$

We can also define the flux of i in relation to a moving coordinate axis, that is, which follows the velocity difference between an individual component and the mixture as a whole. One calls such a flux $\underline{j}_i$, *the diffusive flux of component i*. It is responsible for the shift in direction undergone by component i, relative to the bulk mass flow of the mixture. Therefore, one can write

$$\underline{j}_i = \rho_i(\underline{v}_i - \underline{v}), \quad \text{for} \quad i = 1, \ldots, nc \tag{4.5}$$

Summing up over all components, it is easy to note that $\sum_{i=1}^{nc} \underline{j}_i = 0$. Rearranging Eq. (4.5), we can write

$$\rho_i \underline{v}_i = \underline{j}_i + \rho_i \underline{v}, \quad \text{for} \quad i = 1, \ldots, nc \tag{4.6}$$

Replacing $\underline{v} = \frac{1}{\rho}\underline{N}_t$ in Eq. (4.6), one gets

$$\underline{N}_i = \rho_i \underline{v}_i = \underline{j}_i + w_i \underline{N}_t, \quad \text{for} \quad i = 1, \ldots, nc \tag{4.7}$$

which means that the total mass flux of component i relative to a fixed coordinate axis is the sum of two contributions. The first one relative to a moving axis (i.e., the diffusive flux $\underline{j}_i$). It responds for the relative moving of the individual component relative to the one of the mixture as a whole. The second one is responsible for the contribution of i itself for this bulk flow ($w_i \underline{N}_t$), also called *convective flux* of *i*.

An individual component mass balance in this reservoir can be written equating the accumulation rate of i in an arbitrary volume element to the net rate of i that crosses the surface of this element, adding also a source (or sink) term, which is nonzero when one has wells (injectors or producers) and/or filling or leakage by migration streams is given by

$$\int_V \frac{\partial(\phi\rho_i)}{\partial t} dV = -\int_A \phi\rho_i \underline{v}_i \cdot \underline{n} dA + \int_V q_i dV, \quad \text{for} \quad i = 1, \ldots, nc \tag{4.8}$$

where ϕ is the porosity and q_i is the rate of generation of i per unit rock volume.

Applying Gauss Theorem on the surface integral of Eq. (4.8) and grouping the integrands, we have

$$\frac{\partial(\phi\rho_i)}{\partial t} = -\nabla\cdot(\phi\rho_i \underline{v}_i) + q_i, \quad \text{for} \quad i = 1, \ldots, nc \tag{4.9}$$

Replacing Eq. (4.7) in Eq. (4.9):

$$\frac{\partial(\phi\rho_i)}{\partial t} = -\nabla\cdot\left[\phi(\underline{j}_i + w_i \rho \underline{v})\right] + q_i, \quad \text{for} \quad i = 1, \ldots, nc \tag{4.10}$$

We can also define the bulk velocity in molar basis:

$$\underline{v}_{\text{mol}} = \sum_{i=1}^{nc} x_i \underline{v}_i \tag{4.11}$$

And the molar diffusive flux of component i in relation to this bulk flow might be defined in the following way:

$$\underline{J}_i = c_i(\underline{v}_i - \underline{v}_{\text{mol}}), \quad \text{for} \quad i = 1, \ldots, nc \tag{4.12}$$

where c_i is the concentration of component i in mole per unit volume.

Analogously to the mass flux, $\sum_{i=1}^{nc} \underline{J}_i = 0$.

Abandoning the index mol, we can use a molar balance equation for component i absolutely analogous to Eq. (4.10):

$$\frac{\partial(\phi c_i)}{\partial t} = -\nabla \cdot \left[\phi(\underline{J}_i + x_i c\underline{v})\right] + q_i, \quad \text{for} \quad i = 1, \ldots, nc \tag{4.13}$$

where the units of q_i are now in mol/(volume of rock)/(time).

Eqs. (4.10) and (4.13), the way they were written, do not consider capillary effects (which relate to two-phase flow and may not be relevant here). Their main application is to obtain the production forecast in petroleum fields, that is, the continuous flow rate of oil produced at surface (computed from the source term) as a function of the properties of fluids and rock, which are included in the auxiliary expressions for diffusive and convective fluxes. In the case of phase equilibrium occurrence during simulation, the term on which one applies the divergent in the right-hand side must be written for each phase and the equations of isofugacity, plus the flash mass-balance ones, must be solved simultaneously to calculate the distribution coefficients and the produced oil, gas and water flow rates along time. Details of reservoir simulation for the development of petroleum fields and production forecasts computation are beyond the scope of this book. For more information, see Coats (1969) or Ertekin et al. (2001).

The main application discussed here is the computation of the fluid composition all around the domain, from the reference point where samples were collected. The prediction of this compositional grading will be useful for a more precise initialization in the simulation of the field to be developed, including the location of the interfaces between the existing phases (if there is any) and the source or sink points that avoid reaching steady state. These aspects influence decisions about drainage plans, i.e., the possible arrangements of producer and injector wells.

Eqs. (4.10) and (4.13), generally without the source terms, make up a system of nc differential equations, which unknowns are $nc - 1$ independent mole fractions plus pressure. The initial condition of this problem will be directly related to the composition information from the reference-point sample, while the boundary conditions will be basically of second type (Neumann), considering the impermeable limits of the domain.

With the convective flux in porous media given by Darcy's law:

$$\underline{v}^\alpha = -\frac{K \cdot k_r^\alpha}{\mu^\alpha}(\nabla P + \rho^\alpha \underline{g}) \tag{4.14}$$

where K is the absolute permeability of the medium and k_r^α, μ^α, and ρ^α are, respectively, the relative permeability, viscosity, and density of phase α. One still has to obtain a suitable expression for the diffusive flux also defined in the bulk phase.

In general, both diffusive and heat fluxes may be written as a function of their own driving forces, that is, temperature and chemical-potential gradients, as stated by Irreversible Thermodynamics (De Groot and Mazur, 1962; Fitts, 1962; Haase, 1969). In Appendix A, the reader will find the complete derivation of the following equations:

Heat flux:

$$\underline{q} = L_{00}\nabla(1/T) + \sum_{i=1}^{nc-1} L_{0i}\left[-\frac{1}{T}\nabla_T(\tilde{\mu}_i - \tilde{\mu}_{nc})\right] \tag{4.15}$$

Diffusive flux of component i:

$$\underline{j}_i = L_{i0}\nabla(1/T) + \sum_{k=1}^{nc-1} L_{ik}\left[-\frac{1}{T}\nabla_T(\tilde{\mu}_i - \tilde{\mu}_{nc})\right], \quad \text{for} \quad i = 1,\ldots,nc-1 \tag{4.16}$$

where coefficients L_{00}, L_{ik}, and L_{i0} are called *Onsager coefficients*, proportionality factors between fluxes and their driving forces. Coefficients L_{i0} represent the contribution of the Soret effect, that is, the mass flux caused by a temperature gradient, and of Dufour effect, that is, a heat flux due to a chemical-potential gradient. These two effects are called *cross phenomena*, L_{00} and L_{ik} being the proportionality factors of the *direct phenomena*. The latter represent, respectively, the heat flux due to temperature gradient and mass flux due to a chemical-potential gradient. $\tilde{\mu}_i$ is the chemical-potential of component i in mass basis; its units in SI are J/kg. The chemical potentials of the mixture components are also complex functions of temperature, pressure and composition. Appendix A also presents—according to the formalism proposed by Ghorayeb and Firoozabadi (2000)—all the algebraic manipulation necessary to be done on Eqs. (4.15) and (4.16) in order that the diffusive flux is written conveniently as a function of $\nabla\underline{w}$ or $\nabla\underline{x}$, ∇T, and ∇P and, therefore, might be used in Eqs. (4.10) or (4.13).

$$J_i = -c\left(\begin{bmatrix} \sum_{l=1}^{nc-1} D_{il}^M \frac{\partial x_l}{\partial x} \\ \sum_{l=1}^{nc-1} D_{il}^M \frac{\partial x_l}{\partial y} \\ \sum_{l=1}^{nc-1} D_{il}^M \frac{\partial x_l}{\partial z} \end{bmatrix} + D_i^{T,M}\begin{bmatrix} \frac{\partial T}{\partial x} \\ \frac{\partial T}{\partial y} \\ \frac{\partial T}{\partial z} \end{bmatrix} + D_i^{P,M}\begin{bmatrix} \frac{\partial P}{\partial x} \\ \frac{\partial P}{\partial y} \\ \frac{\partial P}{\partial z} \end{bmatrix}\right), \quad \text{for} \quad i = 1,\ldots,nc-1 \tag{4.17a}$$

or

$$\underline{j}_i = -\rho\left(\begin{bmatrix}\sum_{l=1}^{nc-1} D_{il}^m \dfrac{\partial w_l}{\partial x}\\ \sum_{l=1}^{nc-1} D_{il}^m \dfrac{\partial w_l}{\partial y}\\ \sum_{l=1}^{nc-1} D_{il}^m \dfrac{\partial w_l}{\partial z}\end{bmatrix} + D_i^{T,m}\begin{bmatrix}\dfrac{\partial T}{\partial x}\\ \dfrac{\partial T}{\partial y}\\ \dfrac{\partial T}{\partial z}\end{bmatrix} + D_i^{P,m}\begin{bmatrix}\dfrac{\partial P}{\partial x}\\ \dfrac{\partial P}{\partial y}\\ \dfrac{\partial P}{\partial z}\end{bmatrix}\right), \quad \text{for} \quad i=1,\ldots,nc-1 \tag{4.17b}$$

where D_{il}^M and D_{il}^m are, respectively the molecular diffusion coefficients (relative to molar and mass bulk velocity axis) of i for a concentration gradient of component l; D_i^T and D_i^P are, respectively, the temperature and pressure thermal diffusion coefficients of component i. The three coefficients are very complex functions of T, P, and $\underline{x}$ or $\underline{w}$. In a system kept at constant T and P, the remaining term constitutes the Fick's Law. It is possible to obtain the diffusive flux referent to the mass bulk velocity $\underline{j}_i$ from the flux referent to the molar bulk velocity $\underline{J}_i$ through the following relation (see Appendix A):

$$\underline{j}_i = \sum_{k=1}^{nc-1} \underline{J}_k M_k\left[\delta_{ik} + \frac{w_i}{w_k}\left(-w_k + w_{nc}\frac{x_k}{x_{nc}}\right)\right], \quad \text{for} \quad i=1,\ldots,nc-1 \tag{4.18}$$

where δ_{ik} is the Kronecker Delta function.

According to Bolton and Firoozabadi (2014) and Firoozabadi (2015), this is not just a unit conversion in which one simply divides the mass flux of component i by its molar mass. The so-obtained flux, i.e., $\underline{J}_i^* \equiv \frac{\underline{j}_i}{M_i}$, shows an inconsistency if it is to be used in the balance equations: it is interesting to note that $\underline{J}_i \neq \underline{J}_i^* = \frac{\underline{j}_i}{M_i}$, since J_i^*, although has SI units of kmol/m^2/s, keeps, as reference axis, the mass bulk velocity and, therefore, may not be used conceptually in Eqs. (4.13) and (4.17a). Finally, it is easy to note that $\sum_{i=1}^{nc} \underline{J}_i^* \neq 0$.

It is also possible to show (see Appendix A) that the molecular-diffusion coefficient referent to the mass bulk velocity axis D_{il}^m (for $i, l = 1,\ldots,nc-1$) is related to the molar bulk velocity one by the following relation:

$$\underline{\underline{D}}^m = \underline{\underline{K}}\,\underline{\underline{W}}\,\underline{\underline{X}}^{-1}\underline{\underline{D}}^M\underline{\underline{X}}\,\underline{\underline{W}}^{-1} \tag{4.19}$$

where

$$K_{ik} = \delta_{ik} + \frac{w_i}{w_k}\left(-w_k + w_{nc}\frac{x_k}{x_{nc}}\right), \quad \text{for} \quad i,k = 1,\ldots,nc-1 \tag{4.20}$$

and $\underline{\underline{W}}$ and $\underline{\underline{X}}$ are diagonal matrices of the mass and mole fractions, respectively:

$$\underline{\underline{W}} = \begin{bmatrix} w_1 & 0 & \dots & 0 \\ 0 & w_2 & \dots & 0 \\ & \dots & & \\ 0 & 0 & \dots & w_{nc-1} \end{bmatrix}; \quad \underline{\underline{X}} = \begin{bmatrix} x_1 & 0 & \dots & 0 \\ 0 & x_2 & \dots & 0 \\ & \dots & & \\ 0 & 0 & \dots & x_{nc-1} \end{bmatrix} \tag{4.21}$$

Now, we make a review of the efforts that have been made in the literature to compute these coefficients, be it by empirical correlations or from derivations of more complex expressions, which terms will also depend on the solution of well-fitted equations of state. We also approach the classic types of problems to be solved with the fluxes' expressions, discussing the basic hypothesis and some simplifications sometimes necessary on Eqs. (4.10) and (4.13).

4.2 LITERATURE REVIEW AND CONSTITUTIVE EQUATIONS

Shukla and Firoozabadi (1998) proposed simplified expressions for the calculation of the thermal diffusion coefficient for binary mixtures. Later, Ghorayeb and Firoozabadi (2000) generalized the idea for multi-component mixtures, proposing constitutive equations for the three diffusion coefficients using some tools of Irreversible Thermodynamics. Appendix A presents the full derivation of the equations proposed by Ghorayeb and Firoozabadi (2000). In this chapter, we will present only the final form of these flux equations, which are directly used in Eqs. (4.10) and (4.13). Defining matrix $\underline{\underline{j}}$, containing the vector components of diffusive fluxes from $i = 1, \dots, nc - 1$ in relation to the average mass velocity axis:

$$\underline{\underline{j}} = \begin{bmatrix} j_{1,x} & j_{2,x} & \dots & j_{nc-1,x} \\ j_{1,y} & j_{2,y} & \dots & j_{nc-1,y} \\ j_{1,z} & j_{2,z} & \dots & j_{nc-1,z} \end{bmatrix} \tag{4.22}$$

Ghorayeb and Firoozabadi (2000) reached to

$$\underline{\underline{j}}^t = -\rho\left[\underline{\underline{D}}\,\underline{\underline{M}}\,\underline{\underline{L}}\,\underline{\underline{W}}\,\underline{\underline{F}}(\nabla \mathrm{w}^t)^t + \frac{w_{nc}}{RT^2}\underline{\underline{D}}\,\underline{\underline{M}}\,\underline{\underline{L}}\,\widehat{\underline{Q}}^{*,m}(\nabla \mathrm{T})^t + \underline{\underline{D}}\,\underline{\underline{M}}\,\underline{\underline{L}}\,\underline{V}(\nabla \mathrm{P})^t\right] \tag{4.23}$$

where

$$\underline{\underline{D}}^m = \begin{bmatrix} D^m_{11} & D^m_{12} & \dots & D^m_{1,nc-1} \\ D^m_{21} & D^m_{22} & \dots & D^m_{2,nc-1} \\ & & \dots & \\ D^m_{nc-1,1} & D^m_{nc-1,2} & \dots & D^m_{nc-1,nc-1} \end{bmatrix} = \underline{\underline{D}}\,\underline{\underline{M}}\,\underline{\underline{L}}\,\underline{\underline{W}}\,\underline{\underline{F}} \tag{4.24}$$

$$\underline{D}^{T,m} = \begin{bmatrix} D_1^T \\ D_2^T \\ \cdot \\ \cdot \\ \cdot \\ D_{nc-1}^T \end{bmatrix} = \frac{w_{nc}}{RT^2} \underline{\underline{D}}\, \underline{\underline{M}}\, \underline{\underline{L}}\, \underline{\widehat{Q}}^{*,m} \tag{4.25}$$

$$\underline{D}^{P,m} = \begin{bmatrix} D_1^P \\ D_2^P \\ \cdot \\ \cdot \\ \cdot \\ D_{nc-1}^P \end{bmatrix} = \underline{\underline{D}}\, \underline{\underline{M}}\, \underline{\underline{L}}\, \underline{V} \tag{4.26}$$

where $\underline{\underline{D}}$, $\underline{\underline{M}}$, $\underline{\underline{W}}$, $\underline{\underline{F}}$, and $\underline{V}$ are auxiliary matrices and vector defined by the authors; $\underline{\underline{L}}$ is the matrix of Onsager coefficients, related to molecular diffusion by direct phenomena:

$$\underline{\underline{L}} = \begin{bmatrix} L_{11} & L_{12} & \dots & L_{1,nc-1} \\ L_{21} & L_{22} & \dots & L_{2,nc-1} \\ & & \dots & \\ L_{nc-1,1} & L_{nc-1,2} & \dots & L_{nc-1,nc-1,} \end{bmatrix} \tag{4.27}$$

and $\underline{\widehat{Q}}^{*,m}$ is the vector of the so-called *Net Heats of Transport* in relation to the average mass velocity axis for components $i = 1, \dots, nc - 1$:

$$\underline{\widehat{Q}}^{*,m} = \begin{bmatrix} Q_1^{*,m} - Q_{nc}^{*,m} \\ Q_2^{*,m} - Q_{nc}^{*,m} \\ \dots \\ Q_{nc-1}^{*,m} - Q_{nc}^{*,m} \end{bmatrix} \tag{4.28}$$

Eqs. (4.24)–(4.28), replaced in Eq. (4.23), provide the diffusive flux to be used in Eq. (4.10) or, if transformed the average molar velocity axis, in Eq. (4.13). Of all auxiliary matrices and vectors, $\underline{\underline{L}}$ and $\underline{\widehat{Q}}^{*,m}$ are the only ones that cannot be calculated only with terms from the equation of state and therefore need some fundamentally dynamic properties (or nonequilibrium quantities) for their computation. Now, let us resume the calculation procedure for these variables, while Appendix A shows implementation in detail.

4.2.1 The Heat of Transport

The *Heat of Transport* is the energy rate that each component carries with it when diffuses at constant temperature per unit sectional area to this flow. It is one of the main transport properties of Irreversible Thermodynamics for it is the only contribution for heat transfer in an isothermal medium, that is in the absence of the conduction stream. So, for an isothermal system (details in Appendix A), it can be shown that

$$\underline{q} = \sum_{i=1}^{nc-1} \widehat{\underline{Q}}_i^{*,m} \underline{j}_i, \quad \text{where} \quad \nabla T = 0 \tag{4.29}$$

The constitutive equation for the heat of transport in binary mixtures proposed by Shukla and Firoozabadi (1998), based on the semi-empirical kinetic viscosity theories of Glasstone et al. (1941) and Dougherty and Drickamer (1955), was generalized by Ghorayeb and Firoozabadi (2000) for multicomponent mixtures:

$$Q_i^{*,M} = -\frac{\overline{U}_i}{\tau_i} + \left[\sum_{j=1}^{nc} x_j \frac{\overline{U}_j}{\tau_j}\right] \frac{\overline{V}_i}{\sum_{j=1}^{nc} x_j \overline{V}_j}, \quad \text{for} \quad i = 1, \ldots, nc \tag{4.30}$$

where $Q_i^{*,M}$ is the net heat of transport of component i in relation to the average molar velocity axis (hence the superscript M); $\overline{U}_i$ is the partial molar internal energy, $\overline{V}_i$ is the partial molar volume of component i (both can be obtained using an equation of state) and $\tau_i = \frac{\Delta U_i^{\text{vap}}}{\Delta U_i^{\text{vis}}}$. ΔU_i^{vap} and ΔU_i^{vis} are, respectively, the vaporization internal energy and the viscous flow energy of pure component i (Glasstone et al., 1941). Generally, τ_i is considered as an adjustable parameter of the model. For hydrocarbons, $\tau_i = 4.0$ is a good guess recommended by Firoozabadi et al. (2000). Complete derivation of Eq. (4.30) is presented in Appendix B. We should emphasize that the net heat of transport in Eq. (4.30) is expressed in molar basis ($Q_i^{*,M}$); therefore, it is necessary to convert it into mass basis ($Q_i^{*,m}$) before using it in Eq. (4.28). This can be done using Eq. (4.18), replacing $\underline{j}_i$ by $Q_i^{*,m}$ and $\underline{J}_k$ by $Q_k^{*,M}$.

Leahy-Dios and Firoozabadi (2007) propose also an important empirical correlation for the direct calculation of the molecular-diffusion coefficients matrix $\underline{\underline{D}}^m$ as a function of T, P, and $\underline{x}$. This correlation is presented in Appendix C.

With $\underline{\underline{D}}^m$ already calculated, matrix $\underline{\underline{L}}$ might then be obtained by Eq. (4.24):

$$\underline{\underline{L}} = (\underline{\underline{D}}\,\underline{\underline{M}})^{-1} \underline{\underline{D}}^m (\underline{\underline{W}}\,\underline{\underline{F}})^{-1} \tag{4.31}$$

and then the vector of pressure diffusion coefficients $\underline{D}^{P,m}$ (or $\underline{D}^{P,M}$) might be obtained directly by Eq. (4.26). With vector $\widehat{\underline{Q}}_i^{*,m}$ (or $\widehat{\underline{Q}}_i^{*,M}$) with τ_i adjusted, the computation of the thermal diffusion coefficient $\underline{D}^{T,m}$ (or $\underline{D}^{T,M}$) is straightforward by Eq. (4.25), finishing then the computation of the total diffusion flux.

Proceeding on our literary review, Firoozabadi et al. (2000) also compare the thermal diffusion coefficients simulated based on the heat of transport with those experimentally measured for ternary mixtures. They also simulate some real examples in fields with natural convection, that is, reservoirs with lateral temperature gradient, consequently with fluid distribution in 2D. The most fascinating application of this research line is the field of Yufutsu, in Japan (Ghorayeb et al., 2000), where the effects of thermal diffusion, clearly in the opposite direction of the gravitational field, inverted the positions of gas and liquid in equilibrium within the reservoir. This example, which obviously needed reestimation of τ_i parameters, will be discussed in detail in Chapter 5.

Later, Hoteit and Firoozabadi (2006) reformulate the equations of the individual mass balance system, using the so-called *Pressure Equation,* on which this variable shows in the transient term, and applying the Hybrid Finite Elements Method coupled with Galerkin's discontinuous one for problems of injection and production in fractured reservoirs at constant temperature. The use of this numerical method enhances considerably the precision of the calculations compared to Finite Difference Method for a fixed mesh on the grid. This discussion is beyond the scope of this book.

Artola and Rousseau (2007) propose a microscopic model for the Soret coefficient (S_T) of binary mixtures modeled with Lennard–Jones potential. Their goal is to obtain the influence of the so-called chemical effect on the dependence of S_T on composition. These authors also compare their molecular simulations with the results obtained by Shukla and Firoozabadi (1998), stating the good performance of the macroscopic semi-empirical model of Eq. (4.30).

Yan et al. (2008) and Abbasi et al. (2009, 2010, 2011) propose alternative forms of estimating the τ_i-values in Eq. (4.30). These proposals do not show to be universally applicable to all kinds of thermal diffusion problems. In other words, one has to estimate the parameters of Abbasi et al. (2009) from case to case.

Moortgat et al. (2013) study the injection of CO_2 in a carbonate-rock plug saturated with a Brazilian oil, observing that the diffusion phenomena do influence recovery and gas breakthrough. Let us finally mention the interesting and recent work of Bolton and Firoozabadi (2014), who extend the problem to reservoirs originally in two phases and also with filling and leaking by migrating streams. These authors also update the temperature along simulation by solving the energy equation, which influences not only the final compositional grading, but also the position of the GOC along the domain, for example.

Parallelly to this research line leaded by Prof. Abbas Firoozabadi, Pedersen and Lindeloff (2003) proposed another method for the calculation of the Heat of Transport, based on the concepts of Irreversible Thermodynamics presented by Haase (1969). To these authors, the compositional grading in 1D nonisothermal reservoirs reduces to an isofugacity equation modified not only by the gravitational field like Eq. (3.27), but also by an additional temperature-dependent term, which adjustable parameters are the enthalpies of the pure components at the reference state of ideal gas. Starting from them, the residual-enthalpy expression of the equation of state provides the partial molar enthalpies necessary to the last correction term:

$$\hat{f}_{i,z} = \hat{f}_{i,z_{\text{ref}}} \exp\left[\frac{-M_i g(z - z_{\text{ref}})}{RT}\right] \exp\left[\frac{1}{RT}\int_{T_{z_{ref}}}^{T_z} M_i\left(\frac{\overline{H}_i}{M_i} - \frac{\overline{H}}{M}\right)\frac{dT}{T}\right], \text{ for } i = 1,\ldots,nc \tag{4.32}$$

Like Eq. (3.27), Eq. (4.32) consists of a nonlinear system of nc equations, which unknowns are the pressure and the $(nc - 1)$-independent mole fractions at level z, once one knows T, P, and $\underline{x}$ at z_{ref}, as well as the temperature gradient along the 1D column. Pedersen and Lindeloff (2003) suggest also to calculate the enthalpies of Eq. (4.32) at an average condition of temperature and pressure between z_{ref} and z, allowing them to get out of the integral and facilitating the resolution of the nonlinear system:

$$\hat{f}_{i,z} = \hat{f}_{i,z_{\text{ref}}} \exp\left[\frac{-M_i g(z - z_{\text{ref}})}{RT}\right] \exp\left[\frac{M_i\left(\frac{\overline{H}_i}{M_i} - \frac{\overline{H}}{M}\right)}{RT} \ln\left(\frac{T_z}{T_{z_{\text{ref}}}}\right)\right], \quad \text{for } i = 1,\ldots,nc \tag{4.33}$$

Eq. (4.33) is also completely derived in Appendix D and has the advantage of, replacing Eq. (3.27), provide an isofugacity expression analog to the isothermal case. Drawbacks are the set of adjustable parameters, the reference enthalpies, which should not vary from case to case; and the impossibility of generalizing the model for 2D and 3D systems.

At least for binary mixtures in 1D systems, the results obtained using the methodology of Pedersen and Lindeloff (2003) and updated by Pedersen and Hjermstad (2006) tend likely to enhance the compositional grading in relation to the isothermal case (isolated action of gravitational field). The research line of Pedersen and collaborators defends that, for reservoirs containing heavier fractions, like aromatics and asphaltenes, the effect of thermal diffusion at high temperatures makes these components, differently from intermediate n alkanes, tend to migrate to the hotter portion (the bottom), always intensifying segregation. For volatile oils and lighter n-paraffinic gas condensates, without the occurrence of such heavies, thermal diffusion would always tend to attenuate gradation by the experimental observation that the heavier n-alkanes, when submitted to a considerable geothermal gradient, would tend to migrate to the colder region (the top).

Blanco et al. (2008) also observe that, if in equimolar mixtures of n-alkanes the heavies tend to concentrate at the colder side, this behavior might be inverted when such components are in binary mixtures with aromatics (benzene, for example). In this case, the heavier n-alkane tends to migrate to the hotter side, while branched alkanes would keep the usual behavior, concentrating in the colder side. The prediction of the distribution of components of complex mixtures along depth under the influence of both gravity and thermal diffusion is still not completely dominated in the literature. Hashmi et al. (2016) have just added more important experimental data about the Soret effect in binary mixtures of polycyclic aromatic hydrocarbons.

Along this text, we will make some comparisons about the two approaches cited here, that is the research lines of both Firoozabadi (Ghorayeb and Firoozabadi, 2000a,b) and Pedersen (Pedersen and Lindeloff, 2003; Pedersen and Hjermstad, 2006), comparing their results for some basic examples in Chapter 5, and also trying to set a pattern for the examples studied in Chapter 6. Before that, one formulates some classical problems to settle the theory and guide computer implementation. Simplifying the equations used by Ghorayeb and Firoozabadi (2000a,b) for 1D systems and proceeding also to the discretization of the flux equations in 2D systems, we reach to the resolution method for these discretized equations by the so-called IMPEC (*Implicit Pressure Explicit Composition*) approach.

4.3 FORMULATION OF CLASSICAL PROBLEMS

Consider, for example, a 1D reservoir as those discussed in Chapter 3, but now submitted to a nonzero geothermal gradient, that is $(dT/dz) \neq 0$. In the absence of source terms, the mass-balance equation for component i (Eq. (4.10)) in steady state reduces to

$$\frac{d}{dz}\left[\phi(\underline{j}_i + w_i \rho \underline{v})\right] = 0, \quad \text{for} \quad i = 1, \ldots, nc \tag{4.34}$$

From the no-flow boundary condition, one can write for the limits of the domain:

$$\underline{j}_i = 0, \qquad \text{for} \quad i = 1, \ldots, nc - 1 \tag{4.35a}$$

$$w_i \rho \underline{v} = 0, \quad \text{for} \quad i = 1, \ldots, nc - 1 \tag{4.35b}$$

As the spatial derivatives of both diffusive and convective fluxes are zero by Eq. (4.34), they will also be zero for the rest of the domain.

And by the definition of Darcy's Law (Eq. (4.14)), if $\underline{v} = 0$, one has

$$\nabla P = \frac{dP}{dz} = -\rho \underline{g} \tag{4.36}$$

Therefore, Eqs. (4.35a) and (4.36) make up a system of nc nonlinear equations, which unknowns are the $nc - 1$ independent mass fractions plus the pressure.

On analyzing, therefore, Eq. (4.23) for this particular case, one will observe that matrices $\underline{\underline{D}}$, $\underline{\underline{M}}$, and $\underline{\underline{L}}$ cancel out and the flux expression, equated to zero, reduces to

$$\underline{\underline{D}}\,\underline{\underline{F}}\nabla\underline{x}+\frac{w_{nc}}{RT^2}\underline{\widehat{Q}}^{*,m}\nabla T+\underline{V}\nabla P=0 \tag{4.37}$$

Then, since we know pressure, temperature, and composition at the reference level, as well as the temperature gradient (generally consider constant), Eq. (4.37) might be integrated along depth to provide the $nc-1$ independent mass fractions:

$$\sum_{j=1}^{nc-1}\sum_{k=1}^{nc-1}W_{ik}F_{kj}\frac{dw_j}{dz}+\frac{w_{nc}}{RT^2}\underline{\widehat{Q}}_i^{*,m}\frac{dT}{dz}-\rho gV_i=0,\quad \text{for}\quad i=1,\ldots,nc-1 \tag{4.38}$$

where the hydrostatic relation $dP/dz=-\rho g$ has already been incorporated.

Approximating the derivatives of Eq. (4.38) by finite differences, one has

$$\mathscr{F}_i=\sum_{j=1}^{nc-1}\sum_{k=1}^{nc-1}W_{ik}F_{kj}(w_j^m-w_j^{m-1})+\frac{w_{nc}}{RT^2}\underline{\hat{Q}}_i^{*,m}(T_j^m-T_j^{m-1})-\rho gV_i(z_j^m-z_j^{m-1})$$
$$=0,\quad \text{for}\quad i=1,\ldots,nc-1 \tag{4.39}$$

Once we know the pressure at level $m-1$, the m-level pressure can be updated by

$$P^m=P^{m-1}-\rho^m g(z^m-z^{m-1}) \tag{4.40}$$

where ρ^m is calculated based on the initial estimate of composition at level m. This, in turn, might be updated, for example, by Newton–Raphson's method, applied to Eq. (4.39):

$$\underline{w}^{nr+1}=\underline{w}^{nr}-\left[\underline{\underline{Jac}}(\underline{w}^{nr})\right]^{-1}\underline{\mathscr{F}}(\underline{w}^{nr}) \tag{4.41}$$

where nr is the index of the Newtonian iteration and $\underline{\underline{Jac}}(\underline{w}^{nr})$ is the Jacobian matrix evaluated at iteration nr:

$$(Jac)_{ij}=\frac{\partial\mathscr{F}_i}{\partial w_j},\quad \text{for}\quad i,j=1,\ldots,nc-1 \tag{4.42}$$

being $\underline{w}$ the vector of independent mass fractions:

$$\underline{w}=\begin{bmatrix}w_1\\w_2\\\ldots\\w_{nc-1}\end{bmatrix} \tag{4.43}$$

At each Newtonian iteration, the pressure must be updated by Eq. (4.40), in order to update also the terms of the Jacobian provided by the equation of state in matrices $\underline{\underline{W}}$ and $\underline{\underline{F}}$, and in the vector $\underline{V}$.

In Chapter 5, we present comparisons between the results obtained by the proposals of Ghorayeb and Firoozabadi (2000) and Pedersen and Lindeloff (2003) for some examples of hydrocarbon mixtures submitted to both gravitational field and thermal diffusion in reservoirs considered unidimensional.

As previously discussed, for such reservoirs, the steady state hypothesis is assumed a priori and, therefore, once the thermodynamic model is well fitted, any deviation referent to the compositional experimental data along the 1D column might be due to source terms, chemical reactions, insufficient geological time to achieve steady state and other issues not considered in the problem's modeling.

For 2D or 3D domains, subject to temperature gradients, the natural-convection phenomena, that is the appearance of a Darcy's velocity resulting from the slight pressure unbalance due to fluid density differences between hot and cold regions, is also modeled by Eq. (4.10) or (4.13), this time with the transient term. In the case of reservoirs, the simulation is done in geological time until the properties calculated in all the points of the domain do not vary with time anymore, which characterizes the achievement of steady state. For laboratory assays, like the distribution of fluids in thermogravitational columns (see Blanco et al., 2000, for example), the simulation is done, obviously, in the order of magnitude of time necessary for the experiment to reach the steady state. Table 4.1 shows a resume of the equations to be solved, depending on the consideration of thermal gradient and also on the dimension of the simulation domain.

Now, consider a binary two-phase mixture subject to natural convection in a reservoir represented by a 2D domain ($x-z$ plane), for example, the Cartesian 4×4 grid sketched in Fig. 4.2.

It is necessary to write nc mass-balance equations, for $nc - 1$ independent mass fractions and pressure. One can write both individual per-component equations or, equivalently, use one per-component and the global mass balance, which is the sum of both individual ones. The latter formulation is more convenient because, in this sum, the diffusive fluxes cancel out, reducing the complexity of the discrete terms. Given the conditions measured in the reference point of the domain, the reservoir is initialized according to a predefined criterium of boundary and initial conditions and the simulation is carried out along time, obtaining mass fractions and pressure in all the cells of the domain till these variables do not change with time anymore.

Let us then recall Eq.(4.10), which, without the source-term, reads:

$$\frac{\partial(\phi\rho_k)}{\partial t} + \nabla\cdot\left[\phi(\underline{j}_k + w_k\rho\underline{v})\right] = 0, \quad \text{for} \quad k = 1, \ldots, nc \qquad (4.44)$$

TABLE 4.1 Summary of the Systems of Equations to be Solved, Depending on the Thermal Gradient and on the Dimension of the Simulation Domain

Case	System of Equations to Solve
Isothermal[a] ($\nabla T=0$)	$\hat{f}_i^z(T,P^z,\underline{x}^z)=\hat{f}_i^{z_{ref}}\exp\left[\frac{-M_ig(z-z_{ref})}{RT}\right]$, for $i=1,\ldots,nc$
Non-isothermal[b] ($\nabla T\neq 0$), 1D	$\underline{j}_i=0$, for $i=1,2,\ldots,nc-1$ $\nabla P=-\rho g$ (Ghorayeb and Firoozabadi, 2000)
	$\hat{f}_{i,z}=\hat{f}_{i,z_{ref}}\exp\left[\frac{-M_ig(z-z_{ref})}{RT}\right]\exp\left[\frac{M_i\left(\frac{\overline{H}_i}{M_i}-\frac{\overline{H}}{M}\right)}{RT}\ln\left(\frac{T_z}{T_{z_{ref}}}\right)\right]$, for $i=1,\ldots,nc$ (Pedersen and Lindeloff, 2003)
Non-isothermal[c] ($\nabla T\neq 0$), 1D, 2D or 3D	$\frac{\partial(\phi\rho_i)}{\partial t}+\nabla\cdot\left[\phi\left(\underline{j}_i+w_i\rho\underline{v}\right)\right]=0,\quad$ for $i=1,\ldots,nc$

[a] *Thermodynamic equilibrium assumed* a priori.
[b] *Steady state assumed* a priori.
[c] *Simulation done in transient regime until one reaches steady state.*

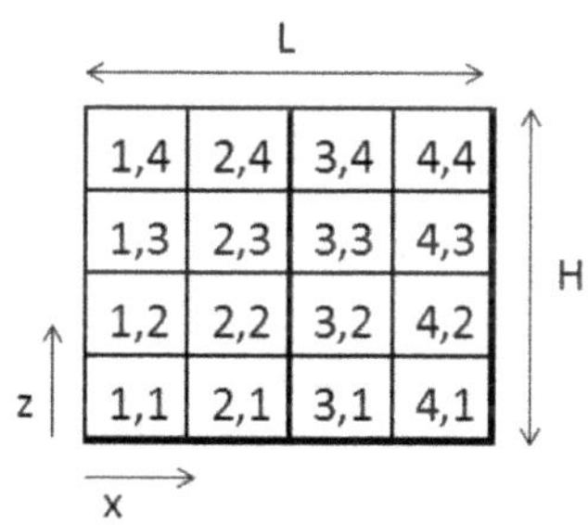

FIGURE 4.2 2D 4 × 4 mesh for the natural-convection problem.

Summing up over all components, one has:

$$\frac{\partial(\phi\rho)}{\partial t}+\nabla\cdot\left[\phi(\rho\underline{v})\right]=0 \tag{4.45}$$

The bulk velocity is given by Darcy's Law (Eq. (4.14)), simplified by the single-phase condition:

$$\underline{v}=-\frac{K}{\mu}(\nabla P+\rho\underline{g}) \tag{4.46}$$

And, finally, the mass-basis diffusive flux is given by Eq. (4.23), with temperature gradient specified a priori in both directions.

Beside a Dirichlet (first type) boundary condition, characterized by setting pressure and composition at the reference cell, one applies also a Neumann condition (second type, no-flow) at the walls, that is:

$$\underline{j}_k \cdot \underline{n} = 0, \quad \text{for} \quad k = 1, \ldots, nc - 1 \tag{4.47}$$

$$\underline{v} \cdot \underline{n} = 0 \tag{4.48}$$

which means:

$$\sum_{l=1}^{nc-1} D_{kl}^{m} \frac{\partial w_l}{\partial x} + D_{k}^{T,m} \frac{\partial T}{\partial x} + D_{k}^{P,m} \frac{\partial P}{\partial x} = 0, \quad \text{for} \quad k = 1, \ldots, nc - 1 \tag{4.49}$$

$$v_x = 0 \ (\text{or } \partial P/\partial x = 0), \quad \text{at } z = 0 \text{ and } z = H \tag{4.50}$$

and also:

$$\sum_{l=1}^{nc-1} D_{kl}^{m} \frac{\partial w_l}{\partial z} + D_{k}^{T,m} \frac{\partial T}{\partial z} + D_{k}^{P,m} \frac{\partial P}{\partial z} = 0, \quad \text{for} \quad k = 1, \ldots, nc - 1 \tag{4.51}$$

$$v_z = 0 \ (\text{or } \partial P/\partial z = -\rho g), \quad \text{at} \quad x = 0 \text{ and } x = L \tag{4.52}$$

As to initial conditions, given temperature, pressure, and composition at the reference cell, the geothermal gradient known in both directions will set up the temperature field throughout the domain. Once this is done, the 1D algorithm (with $\underline{j}_i = 0, \quad \text{for } i = 1, \ldots, nc$) is performed along the entire vertical column of the reference cell, obtaining compositions at each level, respecting the hydrostatic load. This condition might be reproduced for the rest of the domain (other vertical columns), which we call "1D initialization." Another option would be to repeat the bottom pressure of the reference column for all other neighbors in the bottom layer. Compositions of each cell of the bottom layer would be determined by the same condition (with $\underline{j}_i = 0$, for $i = 1, \ldots, nc$), once we know the temperatures. Then, from each cell of the bottom layer, the 1D algorithm would be applied toward the top, computing pressures and compositions in each cell. This procedure is called "convection-free initialization" and it is the most recommended in the literature. Details of the different types of initialization can be found in Bolton and Firoozabadi (2014).

With the problem of initial-value and boundary well posed, let us proceed to the formulation of its numerical resolution:

Discretizing Eq. (4.34) in finite differences, we have for the time derivative:

$$\frac{\partial(\phi\rho_k)}{\partial t} \sim \frac{(\phi\rho_k)_{i,j}^{n+1} - (\phi\rho_k)_{i,j}^{n}}{t^{n+1} - t^{n}}, \quad \text{for} \quad k = 1, \ldots, nc \tag{4.53}$$

where n is the index of the time step and i,j are the indexes of the directions x, z of the discrete grid, always applied on the center of the cells.

For the divergent of the convective flux, one has:

$$\nabla\cdot(\phi w_k\rho\underline{v})=\frac{\partial(\phi w_k\rho\underline{v})}{\partial x}+\frac{\partial(\phi w_k\rho\underline{v})}{\partial z}\sim\frac{(\phi w_k\rho\underline{v})^{n,n+1}_{i+1/2,j}-(\phi w_k\rho\underline{v})^{n,n+1}_{i-1/2,j}}{x_{i+1/2}-x_{i-1/2}}$$
$$+\frac{(\phi w_k\rho\underline{v})^{n,n+1}_{i,j+1/2}-(\phi w_k\rho\underline{v})^{n,n+1}_{i,j-1/2}}{z_{j+1/2}-z_{j-1/2}},\quad \text{for}\quad k=1,\ldots,nc \tag{4.54}$$

where indexes $i+\frac{1}{2}$, $i-\frac{1}{2}$, $j+\frac{1}{2}$, $j-\frac{1}{2}$ refer to the interfaces adjacent to the cell of center i,j. As to the index of the time step, the primary composition variables (in this case, w_k) will be computed in the previous step n, while pressure is updated in the current $n+1$. This choice of variable updating during the solution of the numerical algorithm characterizes the IMPEC (Implicit Pressure, Explicit Composition) method, which will be detailed in the following. Applying the Darcy's Law in velocity, one has:

$$\nabla\cdot(\phi w_k\rho\underline{v})\sim\frac{\left[\phi w_k\rho\frac{-K}{\mu}\nabla(P+\rho gz)\right]^{n,n+1}_{i+1/2,j}-\left[\phi w_k\rho\frac{-K}{\mu}\nabla(P+\rho gz)\right]^{n,n+1}_{i-1/2,j}}{x_{i+1/2}-x_{i-1/2}}$$
$$+\frac{\left[\phi w_k\rho\frac{-K}{\mu}\nabla(P+\rho gz)\right]^{n,n+1}_{i,j+1/2}-\left[\phi w_k\rho\frac{-K}{\mu}\nabla(P+\rho gz)\right]^{n,n+1}_{i,j-1/2}}{z_{j+1/2}-z_{j-1/2}},$$
$$\text{for}\quad k=1,\ldots,nc \tag{4.55}$$

Again applying the finite difference to the piezometric pressure gradient and knowing that, for x-direction, once one is at the same depth, the hydrostatic term ρgz cancels out, one can write:

$$\nabla(P+\rho gz)_{i+1/2}=\frac{P_{i+1}-P_i}{x_{i+1}-x_i} \tag{4.56}$$

$$\nabla(P+\rho gz)_{j+1/2}=\frac{(P+\rho gz)_{j+1}-(P+\rho gz)_j}{z_{j+1}-z_j} \tag{4.57}$$

Considering a uniform mesh and replacing Eqs. (4.56) and (4.57) in Eq. (4.55), one has:

$$\nabla\cdot(\phi w_k\rho\underline{v})\sim\frac{\left[\phi w_k\rho\frac{-K}{\mu}\right]_{i+1/2,j}^{n,n+1}(P_{i+1}-P_i)^{n+1}-\left[\phi w_k\rho\frac{-K}{\mu}\right]_{i-1/2,j}^{n,n+1}(P_i-P_{i-1})^{n+1}}{(\Delta x)^2}$$

$$+\frac{\left[\phi w_k\rho\frac{-K}{\mu}\right]_{i,j+1/2}^{n,n+1}\left[(P+\rho gz)_{j+1}-(P+\rho gz)_j\right]^{n+1}-\left[\phi w_k\rho\frac{-K}{\mu}\right]_{i,j-1/2}^{n,n+1}\left[(P+\rho gz)_j-(P+\rho gz)_{j-1}\right]^{n+1}}{(\Delta z)^2},$$

for $k=1,\ldots,nc$

(4.58)

The remaining interface indexes $i+\frac{1}{2}$, $i-\frac{1}{2}$, $j+\frac{1}{2}$, $j-\frac{1}{2}$ will still be computed based on the so-called *upwinding techniques*, that is, all properties of a certain interface will be obtained, for example, as being the average of the two cells separated by it or adopting the value of the adjacent cell that shows the greater pressure. Other numerical techniques can be found in Ertekin et al. (2001).

The divergent of the diffusive flux might be approximated by:

$$\nabla\cdot\underline{j}_k=\frac{\partial j_{k,i}}{\partial x}+\frac{\partial j_{k,j}}{\partial z}\cong\frac{j_{k_{i+\frac{1}{2},j}}-j_{k_{i-\frac{1}{2},j}}}{\Delta x}+\frac{j_{k_{i,j+\frac{1}{2}}}-j_{k_{i,j-\frac{1}{2}}}}{\Delta z},\quad \text{for}\quad k=1,\ldots,nc \tag{4.59}$$

where:

$$j_{k_{i+1/2,j}}=-\rho\left[\sum_{l=1}^{nc-1}(D_{kl}^m)_{i+\frac{1}{2},j}\frac{w_{l_{i+1,j}}-w_{l_{i,j}}}{\Delta x}+(D_k^{T,m})_{i+\frac{1}{2},j}\frac{T_{i+1,j}-T_{i,j}}{\Delta x}+(D_k^{P,m})_{i+\frac{1}{2},j}\frac{P_{i+1,j}-P_{i,j}}{\Delta x}\right],\quad \text{for}\quad k=1,\ldots,nc \tag{4.60a}$$

$$\nabla\cdot\underline{j}_k=\frac{\partial j_{k,i}}{\partial x}+\frac{\partial j_{k,j}}{\partial z}\sim\frac{j_{k_{i+\frac{1}{2},j}}-j_{k_{i-\frac{1}{2},j}}}{\Delta x}+\frac{j_{k_{i,j+\frac{1}{2}}}-j_{k_{i,j-\frac{1}{2}}}}{\Delta z},\quad \text{for}\quad k=1,\ldots,nc \tag{4.60b}$$

$$j_{k_{i,j+1/2}}=-\rho\left[\sum_{l=1}^{nc-1}(D_{kl}^m)_{i,j+1/2}\frac{w_{l_{i,j+1}}-w_{l_{i,j}}}{\Delta x}+(D_k^{T,m})_{i,j+1/2}\frac{T_{i,j+1}-T_{i,j}}{\Delta x}+(D_k^{P,m})_{i,j+1/2}\frac{P_{i,j+1}-P_{i,j}}{\Delta x}\right],\quad \text{for}\quad k=1,\ldots,nc \tag{4.60c}$$

$$j_{k_{i,j-1/2}} = -\rho\left[\sum_{l=1}^{nc-1}(D_{kl}^{m})_{i,j-1/2}\frac{w_{l_{i,j}} - w_{l_{i,j-1}}}{\Delta x} + (D_k^{T,m})_{i,j-1/2}\frac{T_{i,j} - T_{i,j-1}}{\Delta x}\right.$$
$$\left. + (D_k^{P,m})_{i,j-1/2}\frac{P_{i,j} - P_{i,j-1}}{\Delta x}\right], \quad \text{for} \quad k = 1,\ldots,nc \tag{4.60d}$$

For a binary mixture, putting the pressure in evidence, we get:

$$F_{1,i,j} = \frac{(\phi\rho_1)_{i,j}^{n+1} - (\phi\rho_1)_{i,j}^{n}}{t^{n+1} - t^{n}} + \frac{\left[-\phi w_1\rho\frac{K}{\mu} - \rho D_1^P\right]_{i+1/2,j}^{w_1^n, P_{i,j}^{n+1}, P_{i+1,j}^{n+1}}}{(\Delta x)^2} P_{i+1,j}^{n+1}$$

$$+ \frac{\left[-\phi w_1\rho\frac{K}{\mu} - \rho D_1^P\right]_{i,j+1/2}^{w_1^n, P_{i,j}^{n+1}, P_{i,j+1}^{n+1}}}{(\Delta z)^2} P_{i,j+1}^{n+1}$$

$$+\left(\frac{\left[-\phi w_1\rho\frac{K}{\mu} - \rho D_1^P\right]_{i+1/2,j}^{w_1^n, P_{i,j}^{n+1}, P_{i+1,j}^{n+1}}}{(\Delta x)^2} + \frac{\left[-\phi w_1\rho\frac{K}{\mu} - \rho D_1^P\right]_{i-1/2,j}^{w_1^n, P_{i,j}^{n+1}, P_{i-1,j}^{n+1}}}{(\Delta x)^2}\right.$$

$$\left. + \frac{\left[-\phi w_1\rho\frac{K}{\mu} - \rho D_1^P\right]_{i+1/2,j}^{w_1^n, P_{i,j}^{n+1}, P_{i,j+1}^{n+1},}}{(\Delta z)^2} + \frac{\left[-\phi w_1\rho\frac{K}{\mu} - \rho D_1^P\right]_{i,j-1/2}^{w_1^n, P_{i,j}^{n+1}, P_{i,j-1}^{n+1}}}{(\Delta z)^2}\right) P_{i,j}^{n+1}$$

$$+ \frac{\left[-\phi w_1\rho\frac{K}{\mu} - \rho D_1^P\right]_{i-1/2,j}^{w_1^n, P_{i,j}^{n+1}, P_{i-1,j}^{n+1}}}{(\Delta x)^2} P_{i-1,j}^{n+1} + \frac{\left[-\phi w_1\rho\frac{K}{\mu} - \rho D_1^P\right]_{i,j-1/2}^{w_1^n, P_{i,j}^{n+1}, P_{i,j-1}^{n+1}}}{(\Delta z)^2} P_{i-j-1}^{n+1}$$

$$+\xi\left(w_{1_{i+1,j}}^n, w_{1_{i,j}}^n, w_{1_{i-1,j}}^n, w_{1_{i,j+1}}^n, w_{1_{i,j}}^n, w_{1_{i,j-1}}^n, P_{i+1,j}^n, P_{i,j}^n, P_{i-1,j}^n, P_{i,j+1}^n, P_{i,j}^n, P_{i,j-1}^n\right) = 0 \tag{4.61}$$

where $F_{1,i,j}$ is the mass-balance function for component 1 in cell i,j and ξ is a long function of all the primary variables adjacent to cell i,j evaluated at the previous time step and therefore known. The coefficients of the pressures in evidence also depend on them, no matter the upwinding technique (see the superscripts of the interface terms). That is why the algebraic system resulting from this discretization is nonlinear and can be solved, for example, by the Newton–Raphson method. Before proceeding to the details of this solution, let us apply the same thought to the global mass-balance function, which is simpler due to the absence of the diffusive flux.

$$
\begin{aligned}
F_{G_{i,j}} = {} & \frac{(\phi\rho)_{i,j}^{n+1} - (\phi\rho)_{i,j}^{n}}{\Delta t} + \frac{\left(-\phi\rho\frac{K}{\mu}\right)_{i+1/2,j}^{w_1^n, P_{i,j}^{n+1}, P_{i+1,j}^{n+1}}}{(\Delta x)^2} P_{i+1,j}^{n+1} \\
& + \left[\frac{\left(\phi\rho\frac{K}{\mu}\right)_{i+1/2,j}^{w_1^n, P_{i,j}^{n+1}, P_{i+1,j}^{n+1}} + \left(\phi\rho\frac{K}{\mu}\right)_{i-1/2,j}^{w_1^n, P_{i,j}^{n+1}, P_{i-1,j}^{n+1}}}{(\Delta x)^2}\right. \\
& \left. + \frac{\left(\phi\rho\frac{K}{\mu}\right)_{i,j+1/2}^{w_1^n, P_{i,j+1}^{n+1}, P_{i,j}^{n+1}} + \left(\phi\rho\frac{K}{\mu}\right)_{i,j-1/2}^{w_1^n, P_{i,j}^{n+1}, P_{i,j-1}^{n+1}}}{(\Delta z)^2}\right] P_{i,j}^{n+1} \\
& + \frac{\left(-\phi\rho\frac{K}{\mu}\right)_{i-1/2,j}^{w_1^n, P_{i,j}^{n+1}, P_{i-1,j}^{n+1}}}{(\Delta x)^2} P_{i-1,j}^{n+1} + \frac{\left(-\phi\rho\frac{K}{\mu}\right)_{i,j+1/2}^{w_1^n, P_{i,j}^{n+1}, P_{i,j+1}^{n+1}}}{(\Delta z)^2} P_{i,j+1}^{n+1} \\
& + \frac{\left(-\phi\rho\frac{K}{\mu}\right)_{i,j-1/2}^{w_1^n, P_{i,j}^{n+1}, P_{i,j-1}^{n+1}}}{(\Delta z)^2} P_{i,j-1}^{n+1} \\
& + \xi_G\left(w_{1_{i,j}}^n, w_{1_{i,j-1}}^n, w_{1_{i,j+1}}^n, P_{i,j}^n, P_{i,j+1}^n, P_{i,j-1}^n\right)
\end{aligned}
\tag{4.62}
$$

where:

$$
\begin{aligned}
\xi_G\left(w_{1_{i,j}}^n, w_{1_{i,j-1}}^n, w_{1_{i,j+1}}^n, P_{i,j}^n, P_{i,j+1}^n, P_{i,j-1}^n\right) = {} & \frac{\left(-\phi\rho\frac{K}{\mu}\right)_{i,j+1/2}^{w_1^n, P_{i,j}^n, P_{i,j+1}^n}}{(\Delta z)^2} (\rho g z)_{i,j+1}^n \\
& + \left[\frac{\left(-\phi\rho\frac{K}{\mu}\right)_{i,j+1/2}^{w_1^n, P_{i,j}^n, P_{i,j+1}^n} + \left(-\phi\rho\frac{K}{\mu}\right)_{i,j-1/2}^{w_1^n, P_{i,j}^n, P_{i,j-1}^n}}{(\Delta z)^2}\right] (\rho g z)_{i,j}^n \\
& + \frac{\left(-\phi\rho\frac{K}{\mu}\right)_{i,j-1/2}^{w_1^n, P_{i,j}^n, P_{i,j-1}^n}}{(\Delta z)^2} (\rho g z)_{i,j-1}^n
\end{aligned}
\tag{4.63}
$$

where $F_{G_{i,j}}$ is the global mass-balance function in cell i,j.

For the discrete mesh of Fig. 4.2, one has, for each one of the 16 cells of the domain, two nonlinear equations, $F_{1,i,j}$ and $F_{G_{i,j}}$, for two unknowns, $w_{1_{i,j}}$ and $P_{i,j}$, making up a system of 32 nonlinear equations, which solution by Newton–Raphson method presents a tridiagonal sparse Jacobian matrix, as the one shown in Fig. 4.3. The nonzero positions represent the influence

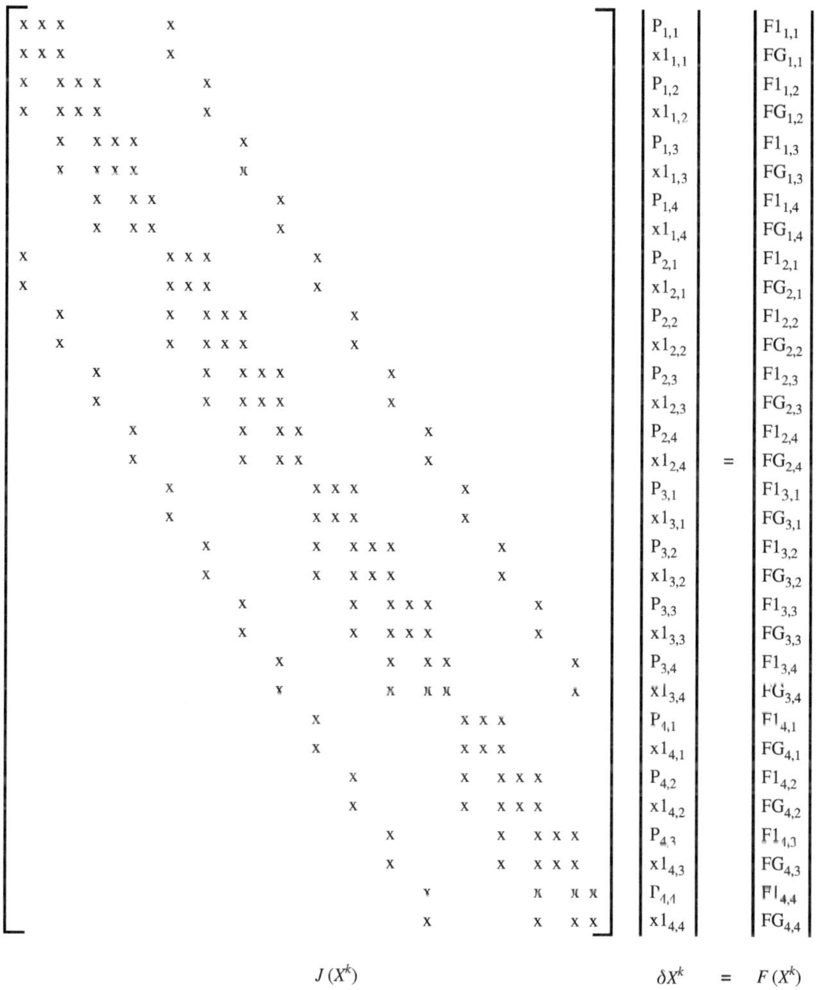

FIGURE 4.3 Structure of the sparse Jacobian matrix for Newton–Raphson method at kth iteration: $\delta \underline{X}^k = \underline{X}^{k+1} - \underline{X}^k$.

of the adjacent cells (or the *i,j* cell itself) in its primary variables. For example, in the first and second rows (individual and global material balance for cell *1,1*), only the pressures of cells 1,1, 2,1 and 1,2, beside the composition of cell 1,1 itself in the temporal term, contribute to these equations and therefore only these positions will have nonzero partial derivatives when putting the Jacobian together. Such derivatives, due to the complexity of the original equations, might be obtained by numerical differentiation, which increment size on function evaluation is left to be chosen by the user. Note that, at each time step, the diffusion coefficients are obtained by Eqs. (4.24)–(4.28) with the compositions of the previous time step, so the flux can be calculated in an explicit way and incorporated to the ξ-term in Eq. (4.61).

The most remarkable characteristic of the IMPEC formulation is the possibility of, making some elementary operations, decoupling the primary variables, that is, solving the system only for pressures and further obtaining the mass fractions by direct substitution in the remaining equations. In the present case, one can eliminate the position referent to the mass fraction of component 1 in one of the equations of cell *i,j* using the other equation. For example, through a single and simple Gaussian elimination, one can nullify the second position of row 1 (which from now on will have nonzero coefficients only for pressures), using row 2. The same thought holds for rows 3 and 4 and so on. At the end of these operations, we will have the modified Jacobian of Fig. 4.4. When detaching the modified rows, we will have the reduced system shown in Fig. 4.5, which is to be solved only for pressures. To solve this new system, one has to transform the matrix of Fig. 4.5 in an upper triangle matrix. To do so, one has to nullify both lower diagonals. Note that row 1 is the only one that can nullify the position in rows 2 and 5. When this happens, a new nonzero element will be inserted in the second position of row 5. Then, new row 2 will have the mission of nullifying the second position of lines 3, 5, and 6. Fig. 4.6 shows the final form of the upper triangle matrix. The zeros and the black diamonds (♦) are in positions repeatedly altered by the elementary operations. The pressure of cell 4,4 is the first to be obtained by back substitution.

$$
\left[\begin{array}{cccccccccccccccccccccccccccccccc}
&0&&&&&&&*&&&&&&&&&&&&&&&&&&&&&&&\\
x&x&x&&&&&&x&&&&&&&&&&&&&&&&&&&&&&&\\
&&&0&*&&&&&&*&&&&&&&&&&&&&&&&&&&&&\\
x&&x&x&x&&&&&&x&&&&&&&&&&&&&&&&&&&&&\\
&&*&&*&0&*&&&&&&*&&&&&&&&&&&&&&&&&&&\\
&&x&&x&x&x&&&&&&x&&&&&&&&&&&&&&&&&&&\\
&&&&*&&*&0&&&&&&&*&&&&&&&&&&&&&&&&&\\
&&&&x&&x&x&&&&&&&x&&&&&&&&&&&&&&&&&\\
&&&&&&&&&0&*&&&&&&*&&&&&&&&&&&&&&&\\
x&&&&&&&&x&x&x&&&&&&x&&&&&&&&&&&&&&&\\
&&*&&&&&&*&&*&0&*&&&&&&*&&&&&&&&&&&&&\\
&&x&&&&&&x&&x&x&x&&&&&&x&&&&&&&&&&&&&\\
&&&&*&&&&&&*&&*&0&*&&&&&&*&&&&&&&&&&&\\
&&&&x&&&&&&x&&x&x&x&&&&&&x&&&&&&&&&&&\\
&&&&&&*&&&&&&*&&*&0&&&&&&&*&&&&&&&&&\\
&&&&&&x&&&&&&x&&x&x&&&&&&&x&&&&&&&&&\\
&&&&&&&&*&&&&&&&&*&0&*&&&&&&*&&&&&&&\\
&&&&&&&&x&&&&&&&&x&x&x&&&&&&x&&&&&&&\\
&&&&&&&&&&*&&&&&&*&&*&0&*&&&&&&*&&&&&\\
&&&&&&&&&&x&&&&&&x&&x&x&x&&&&&&x&&&&&\\
&&&&&&&&&&&&*&&&&&&*&&*&0&*&&&&&&*&&&\\
&&&&&&&&&&&&x&&&&&&x&&x&x&x&&&&&&x&&&\\
&&&&&&&&&&&&&&*&&&&&&*&&*&0&&&&&&&*&\\
&&&&&&&&&&&&&&x&&&&&&x&&x&x&&&&&&&x&\\
&&&&&&&&&&&&&&&&*&&&&&&&&*&0&*&&&&&\\
&&&&&&&&&&&&&&&&x&&&&&&&&x&x&x&&&&&\\
&&&&&&&&&&&&&&&&&&*&&&&&&*&&*&0&*&&&\\
&&&&&&&&&&&&&&&&&&x&&&&&&x&&x&x&x&&&\\
&&&&&&&&&&&&&&&&&&&&*&&&&&&*&&*&0&*&\\
&&&&&&&&&&&&&&&&&&&&x&&&&&&x&&x&x&x&\\
&&&&&&&&&&&&&&&&&&&&&&*&&&&&&*&&*&0\\
&&&&&&&&&&&&&&&&&&&&&&x&&&&&&x&&x&x
\end{array}\right]
\left|\begin{array}{c}
P_{1,1}\\ x1_{1,1}\\ P_{1,2}\\ x1_{1,2}\\ P_{1,3}\\ x1_{1,3}\\ P_{1,4}\\ x1_{1,4}\\
P_{2,1}\\ x1_{2,1}\\ P_{2,2}\\ x1_{2,2}\\ P_{2,3}\\ x1_{2,3}\\ P_{2,4}\\ x1_{2,4}\\
P_{3,1}\\ x1_{3,1}\\ P_{3,2}\\ x1_{3,2}\\ P_{3,3}\\ x1_{3,3}\\ P_{3,4}\\ x1_{3,4}\\
P_{4,1}\\ x1_{4,1}\\ P_{4,2}\\ x1_{4,2}\\ P_{4,3}\\ x1_{4,3}\\ P_{4,4}\\ x1_{4,4}
\end{array}\right|
=
\left|\begin{array}{c}
F1^*_{1,1}\\ FG^*_{1,1}\\ F1^*_{1,2}\\ FG^*_{1,2}\\ F1^*_{1,3}\\ FG^*_{1,3}\\ F1^*_{1,4}\\ FG^*_{1,4}\\
F1^*_{2,1}\\ FG^*_{2,1}\\ F1^*_{2,2}\\ FG^*_{2,2}\\ F1^*_{2,3}\\ FG^*_{2,3}\\ F1^*_{2,4}\\ FG^*_{2,4}\\
F1^*_{3,1}\\ FG^*_{3,1}\\ F1^*_{3,2}\\ FG^*_{3,2}\\ F1^*_{3,3}\\ FG^*_{3,3}\\ F1^*_{3,4}\\ FG^*_{3,4}\\
F1^*_{4,1}\\ FG^*_{4,1}\\ F1^*_{4,2}\\ FG^*_{4,2}\\ F1^*_{4,3}\\ FG^*_{4,3}\\ F1^*_{4,4}\\ FG^*_{4,4}
\end{array}\right|
$$

$$J^*(X^k) \qquad \delta X^k \quad = \quad F^*(X^k)$$

FIGURE 4.4 Gaussian elimination of the positions relative to the mass fractions in the odd lines.

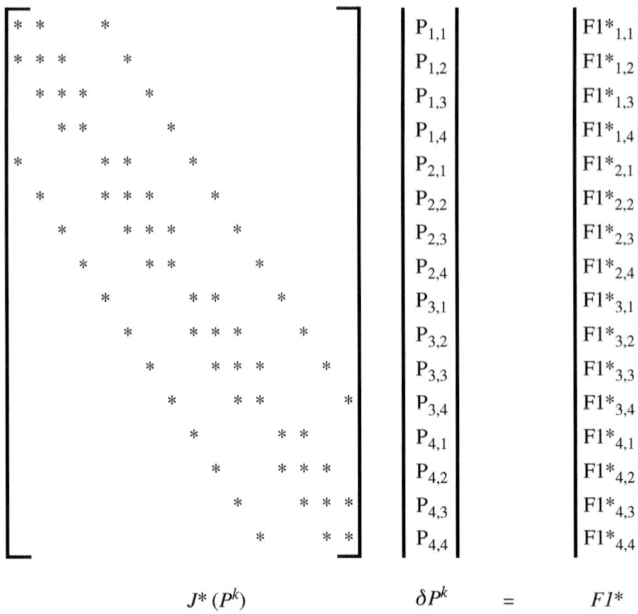

FIGURE 4.5 Newton–Raphson method for IMPEC formulation, with the Jacobian reduced to the pressure coefficients.

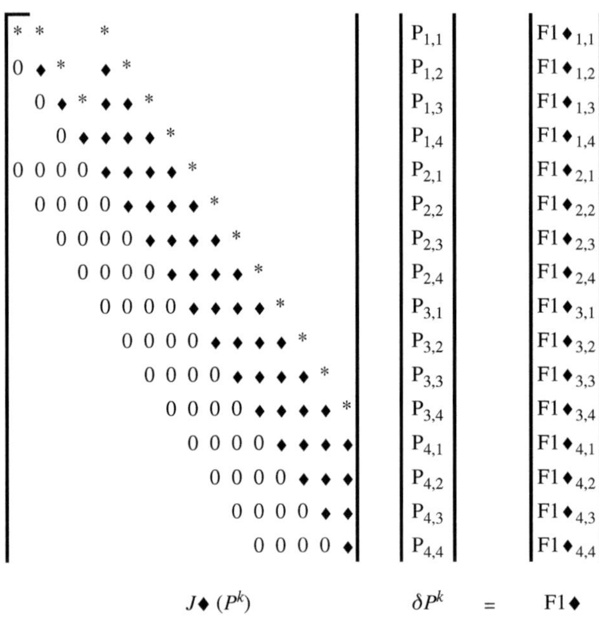

FIGURE 4.6 Final form of the upper triangle Jacobian matrix in Newton–Raphson method applied to IMPEC formulation. The zeroes and the black diamonds (♦) are in positions repeatedly altered by the elementary operations.

4.4 EXERCISES

The objective of this section is to get the reader familiar with the nomenclature found in some classic references of simplified case studies, like binary mixtures with or without the action of the gravitational field.

1. Starting from the general expression for the diffusive mass flux derived in Appendix A:

$$\underline{j}_i = -\rho D_{ic} a_{ic} \left[\begin{array}{l} \frac{k_{Ti}}{T} \nabla T + \frac{w_i M}{L_{ii}} \sum_{k=1}^{nc-1} L_{ik} \sum_{j=1}^{nc-1} \sum_{l=1}^{nc-1} \left(\frac{w_j + w_{nc} \delta_{jk}}{M_j} \right) \left(\frac{\partial \ln f_j}{\partial w_l} \right) \nabla w_l \\ + \frac{M w_i}{RTL_{ii}} \sum_{k=1}^{nc-1} L_{ik} \left(w_{nc} \tilde{v}_k + \sum_{j=1}^{nc-1} w_j \tilde{v}_j - \frac{1}{\rho} \right) \nabla P \end{array} \right],$$

$$\text{for} \quad i = 1, \ldots, nc-1 \tag{A.99}$$

and using the formula to convert it to the molar average velocity axis:

$$\underline{J}_i = \sum_{k-1}^{nc-1} \frac{\underline{j}_k}{M_k} \left[\delta_{ik} - \frac{x_i}{x_k} \left(x_k - \frac{w_k}{w_{nc}} x_{nc} \right) \right], \quad \text{for} \quad i = 1, \ldots, nc-1 \tag{A.125}$$

Derive the equation presented by Bird et al. (1960) and reproduced by Firoozabadi (1999) for the diffusive flux of component 1 in a binary mixture:

$$\underline{J}_1 = -\frac{c^2}{\rho} M_1 M_2 D_{12}^* \left[k_{T,1} \frac{\nabla T}{T} + \frac{\partial \ln f_1}{\partial \ln x_1} \nabla x_1 + \frac{M_1 x_1}{RT} \left(\frac{\overline{v}_1}{M_1} - \frac{1}{\rho} \right) \nabla P \right]$$

where c is the molar concentration and $D_{12}^* = D_{12} \dfrac{M}{M_1 M_2}$.

2. For the same binary system, prove that the equation above reduces to the one derived in Problem 3.1, considering the system in equilibrium, that is, $\nabla T = 0$ and $\underline{J}_i = 0, \quad for\ i = 1, \ldots, nc$.
3. For the same binary mixture at steady state ($\underline{J}_i = 0, \quad for\ i = 1, \ldots, nc$) and in the absence of the gravitational field ($\nabla P = 0$), derive the expression presented by Shukla and Firoozabadi (1998) for the so-called thermal diffusion factors: theoretical (α) and experimental (α_{exp}):

$$\alpha_{\text{exp}} = \frac{k_{T,1}}{x_1 x_2} \frac{\partial \ln x_1}{\partial \ln f_1} = \alpha \frac{\partial \ln x_1}{\partial \ln f_1}, \quad \text{where: } \alpha - \frac{k_{T,1}}{x_1 x_2}$$

$$\frac{\nabla x_1}{x_1 x_2} = \alpha_{\text{exp}} \frac{\nabla T}{T} \rightarrow \alpha_{\text{exp}} = \frac{\ln\left[(x_1/x_2)^H (x_2/x_1)^C \right]}{\ln\left(\frac{T^H}{T^C} \right)}$$

where H and C mean, respectively, the hot and cold sides of an experimental separation apparatus without the effect of the gravitational field. Discuss also the approximations involved in the integration of the equation above, as well as the signal conventions established by Shukla and Firoozabadi (1998).

4. For the same binary system at steady state, starting now from Eq. (A.168) (see Appendix A) for the mass-based diffusion flux:

$$\underline{j}^t = -\rho\left[\underline{\underline{D}}\,\underline{\underline{M}}\,\underline{\underline{L}}\,\underline{\underline{W}}\,\underline{\underline{F}}(\nabla\underline{w}^t)^t + \frac{w_{nc}}{RT^2}\underline{\underline{D}}\,\underline{\underline{M}}\,\underline{\underline{L}}\,\underline{\hat{Q}}^{*,m}(\nabla T)^t + \underline{\underline{D}}\,\underline{\underline{M}}\,\underline{\underline{L}}\,\underline{V}(\nabla P)^t\right] \tag{A.168}$$

and, using the same transformation to the average molar velocity axis:

$$\underline{J}_i = \sum_{k=1}^{nc-1}\frac{\underline{j}_k}{M_k}\left[\delta_{ik} - \frac{x_i}{x_k}\left(x_k - \frac{w_k}{w_{nc}}x_{nc}\right)\right], \quad \text{para} \quad i = 1,\ldots,nc-1 \tag{A.125}$$

obtain the expression that relates the thermal diffusion factor to the Net Heat of Transport related to the average molar velocity axis $Q_i^{*,M}$:

$$\alpha_{\exp} = \frac{Q_2^{*,M} - Q_1^{*,M}}{x_1\dfrac{\partial\mu_1}{\partial x_1}}$$

where: $Q_i^{*,M} = \frac{M_1M_2}{M}Q_i^{*,m}$ and $\alpha = \alpha_{\exp}\frac{\partial \ln f_1}{\partial \ln x_1} = \frac{Q_2^{*,M} - Q_1^{*,M}}{RT}$.

5. The Soret coefficient—see, for example, Platten (2006)—is given by the ratio between the thermal diffusion coefficient (D^T) and the molecular diffusion coefficient (D^m) in a binary mixture kept at constant pressure at steady state, when the mass-basis diffusive flux is written in the following way:

$$\underline{j}_1 = -\rho D^m\frac{\partial w_1}{\partial x} - \rho D^T w_1 w_2\frac{\partial T}{\partial x}, \quad \text{at steady state:} \frac{\partial w_1}{\partial x} = -S_T w_1 w_2\frac{\partial T}{\partial x},$$

where: $S_T = \frac{D^T}{D^m}$

Write the Soret coefficient as a function of the variables defined in Eqs. (A.111) and (A.112) (see Appendix A). Rewrite $\alpha_{\exp}$ defined in Problem 4.4.4 in mass basis and set up its relation with S_T. Also establish the relation between S_T and the mass-based Net Heats of Transport $Q_i^{*,m}$.

REFERENCES

Abbasi, A., Saghir, M.Z., Kawaji, M., 2009. A new proposed approach to estimate the thermodiffusion coefficients for linear-chain hydrocarbon binary mixtures. J. Chem. Phys. 131, 014502.

Abbasi, A., Saghir, M.Z., Kawaji, M., 2010. Theoretical and experimental comparison of the Soret effect for binary mixtures of toluene and n-hexane, and benzene and n-heptane. J. Nonequilib. Thermodyn. 35, 1–14.

Abbasi, A., Saghir, M.Z., Kawaji, M., 2011. Study of thermodiffusion of carbon dioxide in binary mixtures of n-butane & carbon dioxide and n-dodecane & carbon dioxide in porous media. Int. J. Therm. Sci. 50, 124–132.

Artola, P.A., Rousseau, B., 2007. Microscopic interpretation of a pure chemical contribution of the Soret effect. Phys. Rev. Lett. 98, 125901.

Bird, R.B., Stewart, W.E., Lightfoot, E.N., 1960. Transport Phenomena. John Wiley and Sons, Inc, New York.

Bolton, E.W., Firoozabadi, A., 2014. Numerical modeling of temperature and species distributions in hydrocarbon reservoirs. J. Geophys. Res.: Solid Earth 119, 18–31.

Coats, K.H., 1969. Elements of Reservoir Simulation. University of Texas.

De Groot, S.R., Mazur, P., 1962. Nonequilibrium Thermodynamics. North-Holland Publishing Co., Amsterdam.

Dougherty, E.L., Drickamer, H.G., 1955. A theory of thermal diffusion in liquids. J. Chem. Phys. 23 (2).

Ertekin, T., Abou-Kassem, J.H., King, G.R., 2001. Basic applied reservoir simulation. SPE Textbook Ser. 7.

Firoozabadi, A., 1999. Thermodynamics of Hydrocarbon Reservoirs. McGraw-Hill.

Firoozabadi, A., Ghorayeb, K., Shukla, K., 2000. Theoretical model of thermal diffusion factors in multicomponent mixtures. AIChE J. 46 (May 5), 892–900.

Fitts, D.D., 1962. Nonequilibrium Thermodynamics. McGraw-Hill Series in Advanced Chemistry.

Ghorayeb, K., Firoozabadi, A., 2000. Modeling multicomponent diffusion and convection in porous media. SPE J. 5 (June (2)), 158–171.

Ghorayeb, K., Anraku, T. and Firoozabadi, A., Interpretation of the fluid distribution and GOR behavior in the Yufutsu fractured gas-condensate field, SPE 59437, SPE Asia Pacific Conference, Yokohama, Japan, April/2000.

Glasstone, S., Laidler, K.J., Eyring, H., 1941. The Theory of Rate Processes. McGraw-Hill Book Co, New York.

Haase, R., 1969. Thermodynamics of Irreversible Processes. Addison-Wesley.

Hashmi, Sara M., Senthilnathan, S., Firoozabadi, A., 2016. Thermodiffusion of polycyclic aromatic hydrocarbons in binary mixtures. J. Chem. Phys. 145, 184503.

Hoteit, H., Firoozabadi, A., 2006. Compositional modeling by the combined discontinuous Galerkin and mixed methods. SPE J. 11 (1), 19–34.

Leahy-Dios, A., Firoozabadi, A., 2007. Unified model for nonideal multicomponent molecular diffusion coefficients. AIChE J. 53 (11), 2932–2939.

Moortgat, J., Firoozabadi, A., Li, Z., Espósito, R.O., 2013. CO_2 injection in vertical and horizontal cores: measurements and numerical simulation, SPE 135563. SPE J. 1–14.

Pedersen, K.S., Hjermstad, H.P., Modeling of large hydrocarbon compositional gradient, SPE 101275, Abu Dhabi International Petroleum Exhibition and Conference, November/2006.

Pedersen, K.S., Lindeloff, N., Simulations of compositional gradients in hydrocarbon reservoirs under the influence of a temperature gradient, SPE 84364, SPE Annual Technical Conference and Exhibition, Denver, Colorado, October/2003.

Platten, J.K., 2006. The Soret effect: a review of recent experimental results. J. Appl. Mech. 73, 5–15.

Shukla, K., Firoozabadi, A., 1998. A new model of thermal diffusion coefficients in binary hydrocarbon mixtures. Ind. Eng. Chem. Res. 37, 3331–3342.

Yan, Y., Blanco, P., Saghir, M.Z., Bou-Ali, M.M., 2008. An improved theoretical model for thermal diffusion coefficient in liquid hydrocarbon mixtures: comparison between experimental and numerical results. J. Chem. Phys. 129, 194507.

Chapter 5

Classic Examples From Literature

This chapter starts presenting examples of separation by thermal diffusion of simple mixtures studied in the literature. Some of them are reproduced by simulating their respective experiments with the theories proposed in Chapter 4, Irreversible Thermodynamics Applied to Reservoir Engineering. Some examples about the influence of thermal diffusion on the compositional grading of real petroleum reservoirs are also presented. We believe that our interpretation of these examples may serve as a starting point for additional studies, as well as for specific examples, which will be presented in Chapter 6, Case Studies.

5.1 HEAT OF TRANSPORT FOR HYDROCARBON BINARY MIXTURES

Shukla and Firoozabadi (1998) proposed the theoretical model of the heat of transport (detailed in Appendix B) for binary mixtures of hydrocarbons in simplified systems kept at constant pressure. Solving the problems proposed in Chapter 4, it is possible to understand the procedure used by Shukla and Firoozabadi (1998). They compared the results of their simulations for the thermal diffusion factor (α_{exp}) as a function of the heat of transport with the values experimentally measured by other investigators, who obtained the steady-state resulting compositions in both hot and cold sides of an experimental apparatus without the influence of the gravitational field. Translating into equations,

$$\alpha_{\text{exp}} = \frac{\ln\left[(x_1/x_2)^H (x_2/x_1)^C\right]}{\ln \dfrac{T^H}{T^C}} = \frac{Q_2^{*,M} - Q_1^{*,M}}{x_1 \left(\dfrac{\partial \mu_1}{\partial x_1}\right)_{T,P}} \tag{5.1}$$

where α_{exp} is the *thermal diffusion factor* (or *Soret coefficient*), experimentally obtained from temperature and compositions measurements at cold and hot sides of the equipment. The chemical potential variation with composition

Compositional Grading in Oil and Gas Reservoirs. DOI: http://dx.doi.org/10.1016/B978-0-12-812452-9.00005-8

TABLE 5.1 Ideal Gas (IG) Entalpies used by Pedersen and Lindeloff (2003)

Component	$h_i^{IG}/(M_iR)$ (K/g)
N_2	1.0
CO_2	17.0
C_1	0.0
C_2	3.9
C_3	15.8
C_4	7.1
C_5	37.3
C_6	48.4
C_{7+}	50.0

$\left(\frac{\partial \mu_1}{\partial x_1}\right)_{T,P}$ might be given by a well-fitted equation of state (EoS). Shukla and Firoozabadi (1998) used, for the C_1-C_3 system, the Peng–Robinson EoS, with the value of 0.01 for the binary interaction parameter ($k_{C_1-C_3}$). As for parameter τ_i from the net heat-of-transport model, the authors used the recommended value of 4.0. For the C_7-C_{12} system, it varied between 3.5 and 4.0.

The adjustable parameters of the model of Pedersen and coworkers (Pedersen and Lindeloff, 2003; Pedersen and Hjermstad, 2006) described in Appendix D are ideal gas (IG) enthalpies which, added to the residual term of the well-fitted EoS, provide the real enthalpies to be used in the heat of transport equation, or, equivalently, in the expression of the modified fugacity (Eq. (D.28)). Table 5.1 shows the IG enthalpies used by Pedersen and Lindeloff (2003).

Pedersen and Hjermstad (2006) used the following correlation as a function of the hydrocarbon molar mass (methane is the reference component and has, by hypothesis, its IG enthalpy set to zero).

$$\frac{h_i^{IG}}{RT_0} = 0.2806\, M_i - 4.5011, \quad \text{for } i = 1, \ldots, nc \tag{5.2}$$

where $T_0 = 298.15$ K is the reference temperature for the IG enthalpies.

Figs. 5.1–5.4 present the comparison between the results obtained with the models of Shukla and Firoozabadi (1998), Pedersen and Lindeloff (2003),

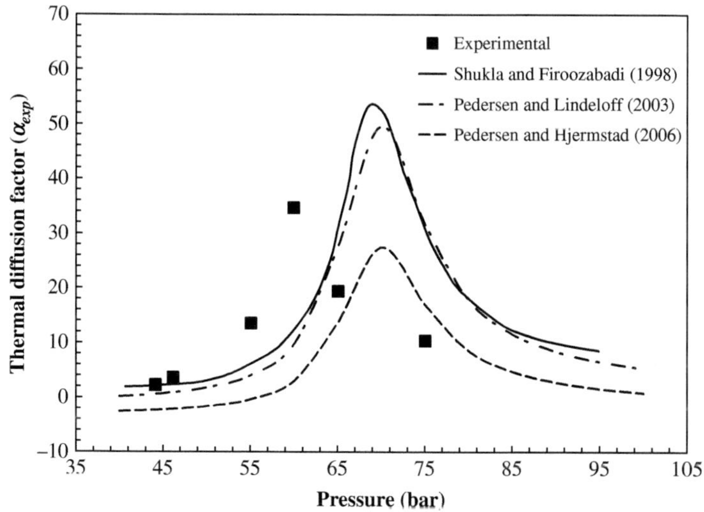

FIGURE 5.1 Thermal diffusion factor α_{exp} of the binary mixture C_1-C_3 ($x_{C_1} = 0.34$) at 346 K as a function of pressure.

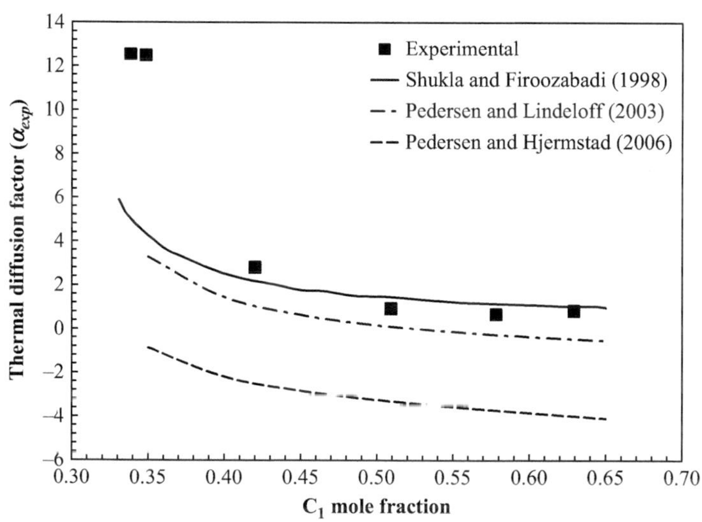

FIGURE 5.2 Thermal diffusion factor α_{exp} of the binary mixture C_1-C_3 ($T = 346$ K and $P = 55$ bar) as a function of C_1 mole fraction.

and Pedersen and Hjermstad (2006). By using the models of the latter two, these results can be reproduced by two ways: (1) replacing, in Eq. (5.1), the expressions for the enthalpies that appear in the heats of transport presented in Appendix D; (2) using directly Eq. (D.28) with a height difference small

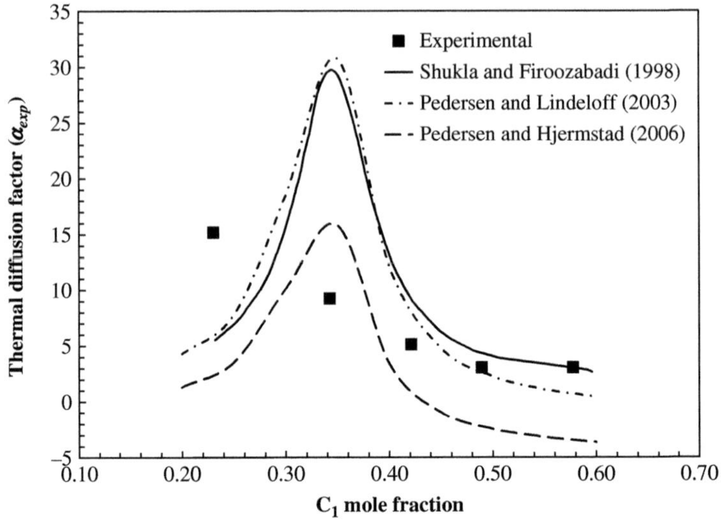

FIGURE 5.3 Thermal diffusion factor α_{exp} of the binary mixture C_1–C_3 ($T = 346$K and $P = 75$ bar) as a function of C_1 mole fraction.

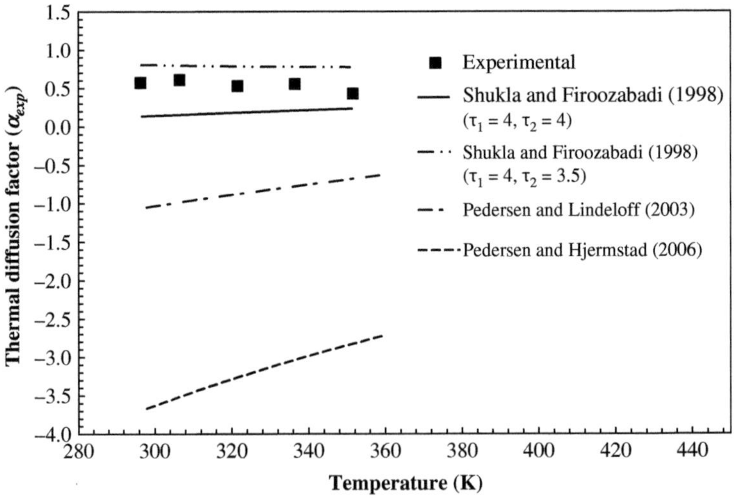

FIGURE 5.4 Thermal diffusion factor α_{exp} of the binary mixture C_7–C_{12} ($x_{C_7} = 0.5$) as a function of temperature at $P = 1.01325$ bar.

enough to eliminate the effect of gravity. As to the temperature gradient, note that α_{exp} is independent of it and assumed to be constant, as the compositional gradient ∇x_1 is directly proportional to $\nabla T/T$, where T is the reference temperature of the experiment, intermediate (but close enough) to T^H and T^C.

For the binary system C_1-C_3, the mole fraction $x_{C_1} = 0.34$ leads the mixture to the critical region at 346 K. Not considering some phenomena taking place under these conditions (discussed further) makes Shukla and Firoozabadi's heat-of-transport model fail to predict experimental data, at least with default parameters, improving its performance for compositions away from the critical point, like the case of C_7-C_{12} mixture. The same behavior can be observed at least qualitatively for the models of Pedersen and Lindeloff (2003) and Pedersen and Hjermstad (2006), suggesting that the original parameters adopted by these authors must be reestimated from case to case, which is not convenient regarding IG enthalpies. Maybe, there it is the main limitation of the heat-of-transport model proposed by Pedersen and collaborators. One of the main limitations of Shukla and Firoozabadi's model is the unacceptance of negative partial molar volumes, occurrence probability of which is higher at the critical region (note that $\bar{v}_{C_3} < 0$ in this example) and also could lead to nonfeasible values for the ψ_i fraction, defined in Appendix B. The authors opted for using the module of the partial molar volume when computing ψ_i. Without reestimating the parameter τ_i, this option certainly impairs the representation of the separation by thermal diffusion in the critical region. We here point out an interesting research line in this area for new investigators: bringing up a more robust heat-of-transport model for near-critical multicomponent systems. In the next examples along this text, you will find some parameter reestimation proposals that try to improve compositional grading predictions by the existing models already discussed.

5.2 RESERVOIR WITH PHASE INVERSION

One of the most fascinating examples of the literature in the last years about thermal diffusion in petroleum reservoirs, the field of Yufutsu, located in Hokkaido, Japan, shows extremely low permeability, producing by fractures and then eliminating the phenomenon of natural convection in compositional grading studies. Its atypical thermodynamic behavior was studied by Ghorayeb et al. (2000, 2002, 2003), who published three main papers about the subject in 2000, 2002, and 2003.

Figs. 5.5 and 5.6 show respectively the geographic location and a mapping of the reservoir with some nonidentified perforated wells.

In Yufutsu, the high temperature, around 140°C, with an ordinary thermal gradient (−0.025°C/m upward) applied on a light near-critical reservoir fluid, made thermal diffusion act strong and conversely to the gravitational field, the lights migrating to the hot bottom region and the heavies to the cold top region. So, the critical transition took place inverted, with the liquid phase floating on a gas phase. As one moves from bottom to top, density increases and gas-oil ratio (GOR) decreases. These two experimental observations were proved not only by accurate pressure-gradient measurements in various wells, but also by a significant reduction in production GOR in time

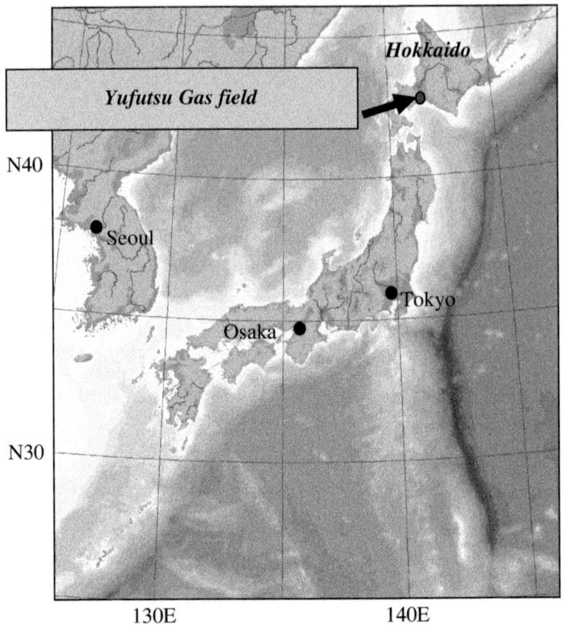

FIGURE 5.5 Geographic location of Yufutsu field, in Hokkaido, Japan. *Source: The image is a copyright of Japan Petroleum Exploration Co., Ltd. Used with permission.*

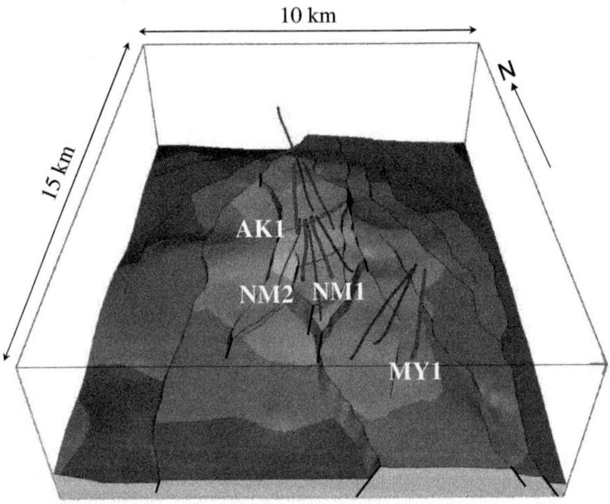

FIGURE 5.6 Mapping of Yufutsu reservoir with some perforated wells in perspective. *Source: The image is a copyright of Japan Petroleum Exploration Co., Ltd. Used with permission.*

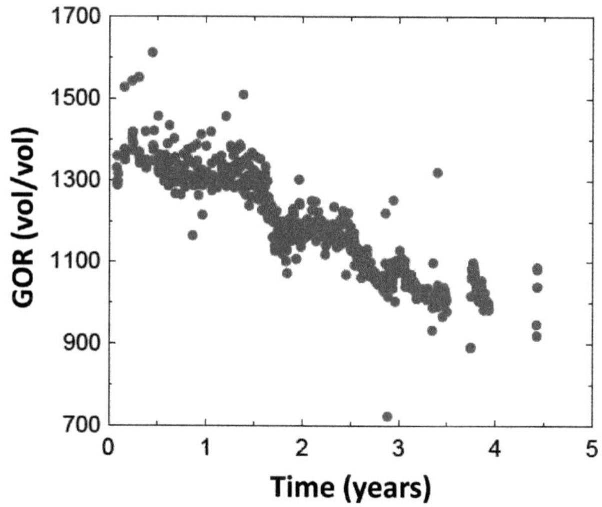

FIGURE 5.7 History of GOR in Yufutsu field after 5 years of production. *Reproduced from Ghorayeb, K., Firoozabadi, A., Anraku, T., Modeling of the Unusual GOR Performance in a Fractured Gas Condensate Reservoir, SPE 75258-MS, Improved Oil Recovery Symposium, Tulsa-Oklahoma, April/2002.*

due to flow of the denser fluid from higher portions toward the producer wells. Fig. 5.7, reproduced from Ghorayeb et al. (2002), shows GOR history in the former 5 years of production.

Following the thought of the research line of Pedersen and Lindeloff (2003), Pedersen and Hjermstad (2006), this is a typical example in which thermal diffusion attenuated compositional grading due to high temperature and absence of heavier aromatic components. The migration of the lighter components to the bottom of the structure and that of the heavies to the top reached a point of inverting the distribution by the isothermal gravitational field. This is maybe the one and only experimental observation of the kind reported in the literature. The peculiar characteristics of the fluid distribution in Yufutsu, with methane increasing to the bottom and C_{7+} to the top, were reasonably modeled by Ghorayeb et al. (2000, 2003), using the theoretical development shown in Chapter 4, Irreversible Thermodynamics Applied to Reservoir Engineering.

The authors, however, had to reestimate parameters τ_i and also the properties of fraction C_{30+} along reservoir depth, which impairs the reproduction of their results, also introducing the drawback of having EoS parameters change with depth. Our interpretation is that parameters τ_i are not influenced by gravity, but by composition and temperature, which in turn vary considerably with depth in this example. Knowing also that the compositional grading of this magnitude is strongly influenced by the quality of the EoS fitting, we tested a new splitting of C_{20+} fraction, based on the quadrature points of Shibata et al. (1987), already introduced in Chapter 2, Phase Equilibrium Thermodynamics.

We also propose a correlation for τ_i parameters as a function of reduced temperature and molar mass of component i. The parameters of this correlation are estimated according to the experimental data of the compositional grading. We believe that this new methodology makes the representation of fluid distribution in Yufutsu easier compared to the one presented by Ghorayeb et al. (2000). We have also updated the code of the program CVS (Compositional Variation Software), provided by Reservoir Engineering Research Institute (RERI), directed by Professor Abbas Firoozabadi, so that the τ_i parameters can be updated as a function of temperature and composition of each level, starting from the reference depth.

From now on, we evaluate the performance of the modeling proposed here, starting by the reproduction of PVT analysis of correlation wells "AK-1" and "MY-1", the main targets in the study of Ghorayeb et al. (2000). Fig. 5.8 is a simplification of Fig. 5.6 shown by Ghorayeb et al. (2003), emphasizing the

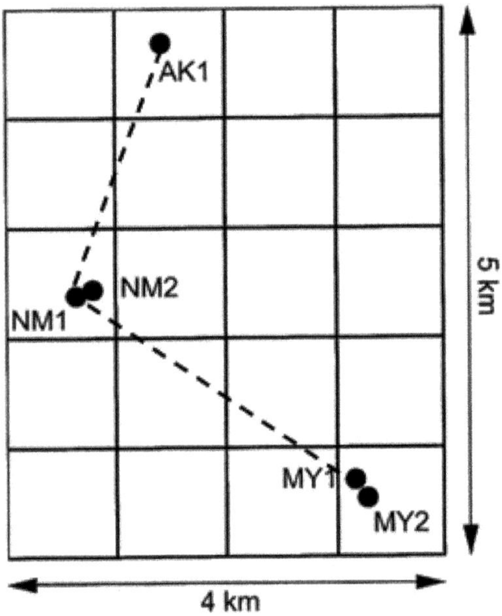

FIGURE 5.8 Sketch of Yufutsu's map, locating some of the main wells. *Adapted from Ghorayeb, K., Firoozabadi, A., Anraku, T., Interpretation of the unusual fluid distribution in the Yufutsu gas condensate field, SPE 84953, SPE Journal, pp. 114–123, June/2003.*

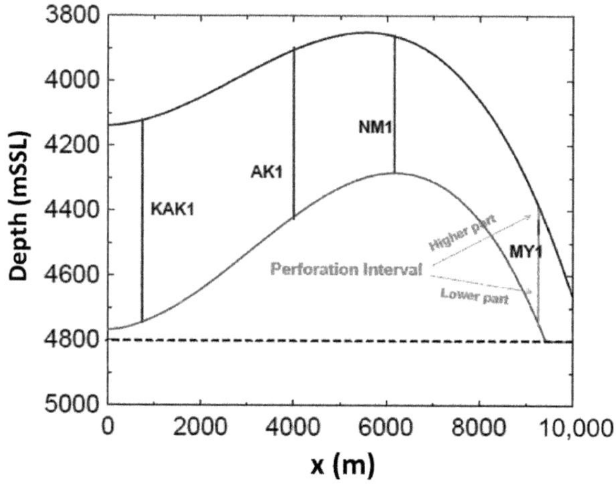

FIGURE 5.9 Section of Yufutsu reservoir represented by the dashed line of Fig. 5.8, showing the relative positions of the main correlation wells. We have used, in this work, the data of wells AK-1 and MY-1. Black and gray lines represent the upper and lower limits of the reservoir, respectively. *Reproduced from Ghorayeb, K., Firoozabadi, A., Anraku, T., Modeling of the Unusual GOR Performance in a Fractured Gas Condensate Reservoir, SPE 75258-MS, Improved Oil Recovery Symposium, Tulsa-Oklahoma, April/2002.*

location of the main wells. Fig. 5.9 shows the section represented by the dashed line in Fig. 5.8, with the relative distances between these wells.

The parameters of Peng–Robinson EoS were fitted on the basis of the compositions of the samples at the reference levels indicated by Ghorayeb et al. (2000): 4032 m for well AK-1 (sample 4) and 4614.5 m for well MY-1 (sample 2).

Tables 5.2 and 5.3 show the final parameters of the regression, which also altered the heavy pseudocomponents mole fractions generated by the generalized Gaussian quadrature of Shibata et al. (1987). Overall composition was normalized on methane.

Fig. 5.10 shows liquid volumes of constant volume depletion in sample 4 of well AK-1 at 136.6°C. Figs. 5.11 and 5.12 show liquid volumes in the constant composition expansions of sample 2 of well MY-1, carried out at 315°F (157.2°C) and 275°F (135°C), respectively.

We propose here a correlation for τ_i-parameters as a function of reduced (or pseudoreduced) temperature (T_r) and molar mass of the component i (M_i), based on the compositional grading data observed in these two wells. The following functional relation is proposed:

$$\tau_i = M_i^{\beta} \exp(\alpha), \quad \text{for } i = 1, \ldots, nc \tag{5.3}$$

TABLE 5.2 Pseudocomponent Properties of Yufutsu Fluid and Molar Compositions at the Reference Levels Adjusted for the Peng–Robinson EoS

Components	P_c (bar)	T_c (K)	ω	M (g/mol)	Volume Shift	Composition (mol %)	
						AK-1 (4032 m)	MY-1 (4614.5 m)
N_2–C_1	45.86	189.75	0.00840	16.19	−0.1802	79.33	77.4
C_2	48.84	305.40	0.09800	30.07	−0.1218	7.58	8.15
C_3–nC_4	40.49	387.38	0.16350	49.10	−0.0932	4.89	6.09
iC_5–C_7	33.88	519.30	0.21849	101.12	−0.0564	3.05	2.54
C_8–C_{12}	26.34	588.41	0.44977	141.50	−0.0305	2.32	3.42
C_{13}–C_{19}	17.62	709.59	0.70845	231.05	0.0170	1.62	1.73
$QC_{22.4}$	13.49	800.41	0.97764	326.22	0.0519	0.905	0.22
$QC_{36.3}$	6.96	990.61	1.03521	605.05	0.0649	0.305	0.45

Note: The latter two pseudocomponents refer to the splitting of C_{20+} fraction by generalized Gaussian quadrature (Shibata et al., 1987) and show, respectively, the hypothetic single carbon numbers of 22.4 and 36.3.

TABLE 5.3 Binary Interaction Coefficients ($k_{C_1-C_n}$) for Component N_2-C_1. All Others are Zero

Component	$k_{C_1-C_n}$
N_2-C_1	0.0000
C_2	0.0357
C_3-nC_4	0.0376
iC_5-C_7	0.0411
C_8-C_{12}	0.0452
$C_{13}-C_{19}$	0.0541
$QC_{22.4}$	0.0636
$QC_{36.3}$	0.1278

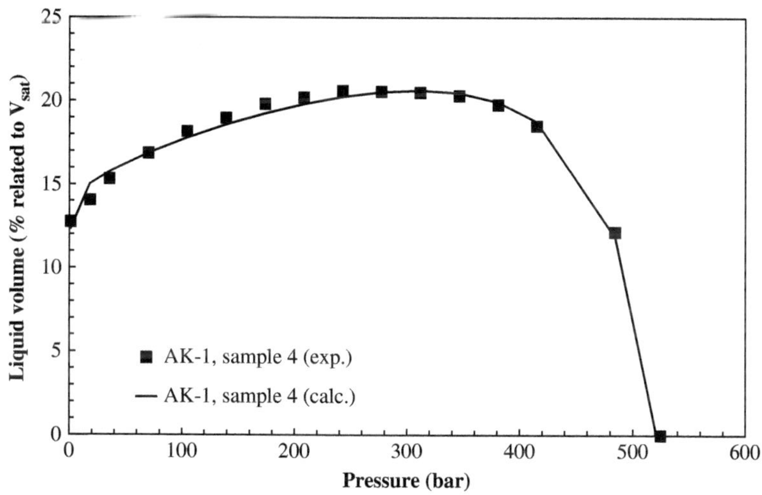

FIGURE 5.10 Liquid volumes in the constant volume depletion for the reference-level sample of well AK 1 (sample 4, 4032 m).

where α and β are functions of $T_r = T/T_c$ (where T_c is the critical temperature of the mixture), given by

$$\alpha = \alpha_0 + \alpha_1 T_r \tag{5.4}$$

$$\beta = \beta_0 + \frac{\beta_1}{T_r} \tag{5.5}$$

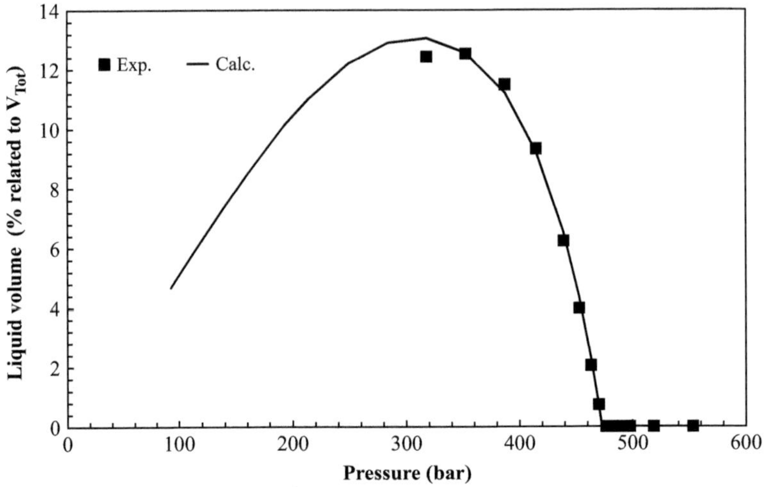

FIGURE 5.11 Liquid volumes in the constant composition expansion of well MY-1, sample 2 (4614.5 m). $T = 315°F$ (157.2°C).

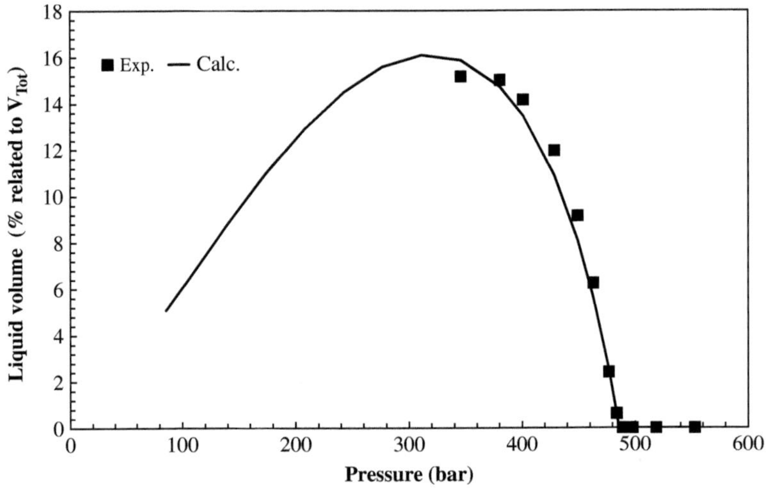

FIGURE 5.12 Liquid volumes in the constant composition of well MY-1, sample 2 (4614.5 m). $T = 275°F$ (135°C).

The parameters α_0, α_1, β_0, and β_1 are constants; estimated values of which for this example are as follows:

$$\begin{aligned} \alpha_0 &= -0.302752 \\ \alpha_1 &= 2.083215 \\ \beta_0 &= -1.120000 \\ \beta_1 &= 1.035688 \end{aligned} \tag{5.6}$$

It is important to emphasize that, in this work, we used the pseudoreduced temperature based on two main reasons: (1) Applying a robust algorithm for the calculation of the critical point at each discrete depth level would increase computational cost; (2) the possible nonexistence of a real critical point for some compositions could impair the convergence of the code in the parameter estimation process. Pseudoreduced temperature might be calculated, for example, by Kay's rule (average of the individual critical temperatures, weighted by the respective mole fractions, that is, $\left(T_c = \sum_{i=1}^{nc} x_i T_{c,i}\right)$, regardless the existence of the critical point or the robustness of any algorithm used for its computation.

Values of 4.0 and 0.9 were also set as upper and lower limits for τ_i, respectively. In other words, $0.9 \leq \tau_i \leq 4.0$, for $i = 1, \ldots, nc$. These limits were taken for the sake of coherence with the values presented by Ghorayeb et al. (2000) in their direct estimation. Let us see the results of this correlation: In the case of well AK-1, there are also three samples plus the reference-level one (4032 m): 4375.5 m (sample 1), 4217.5 m (sample 2), and 4108.5 m (sample 3). Figs. 5.13–5.15 show the density of the reservoir fluid and the mole fractions of methane and C_{20+} along depth, respectively. For comparison with experimental data, we also show the results of the following simulations: isothermal; nonisothermal with $\tau_i = 4$, for $i = 1, \ldots, nc$; nonisothermal with heat of transport calculated by the enthalpies of Pedersen and Lindeloff (2003) and Pedersen and Hjermstad (2006); and finally, nonisothermal with τ_i values of Ghorayeb et al. (2000). The performance of our correlation is comparable to the latter, both being much superior to all others. The isothermal simulation, as well as those using

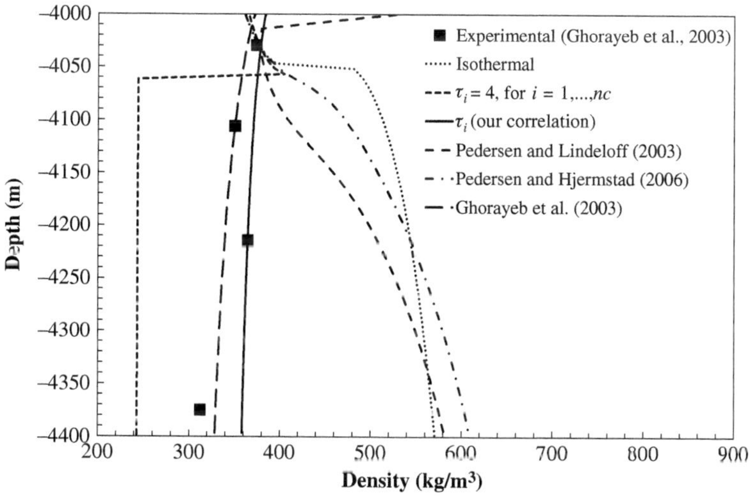

FIGURE 5.13 Density of the reservoir fluid along depth for well AK-1. Comparison between different calculation methods.

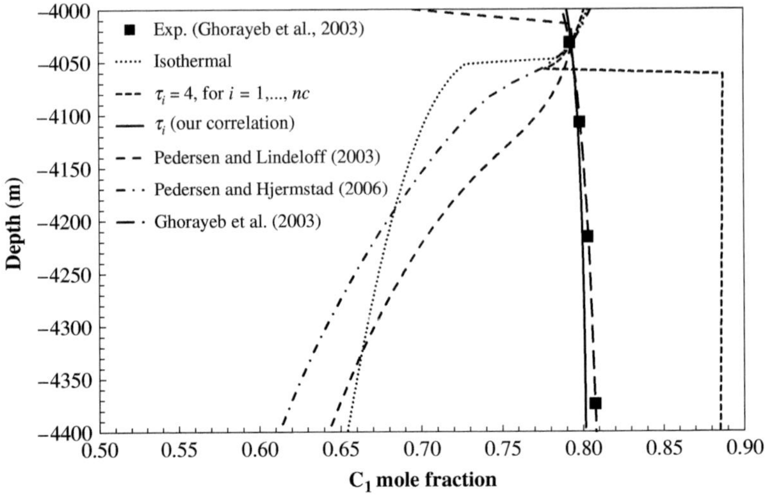

FIGURE 5.14 Methane's mole fraction along depth for well AK-1. Comparison between different calculation methods.

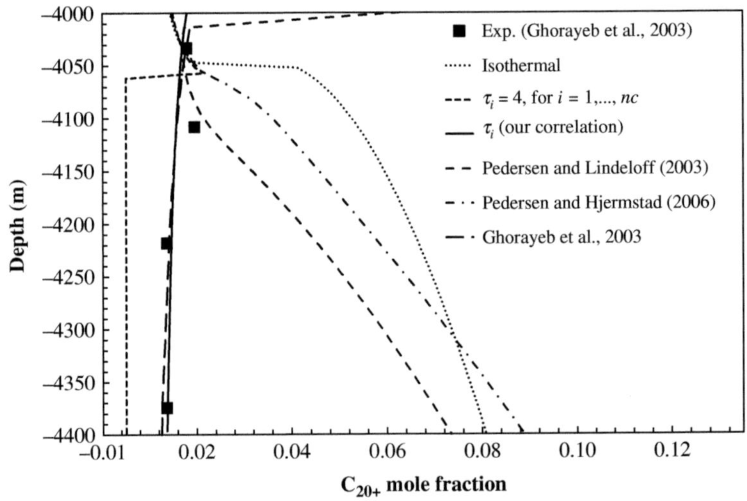

FIGURE 5.15 C_{20+} mole fraction along depth for well AK-1. Comparison between different calculation methods.

Pedersen's heat-of-transport model, predicts a compositional grading more exaggerated in the direction of the gravitational field, showing also spurious gas–oil contacts (GOCs). Values of $\tau_i = 4$ manage to invert the positions of gas and liquid, though miscalculating the gas density, showing also a spurious GOC below the reference level.

Figs. 5.16–5.18 show the simulated liquid volumes in the constant volume depletion for the compositions obtained in the other three sampling depths of well AK-1 by the calculation of the compositional grading. The agreement with the experimental data of the respective PVT analysis is very reasonable

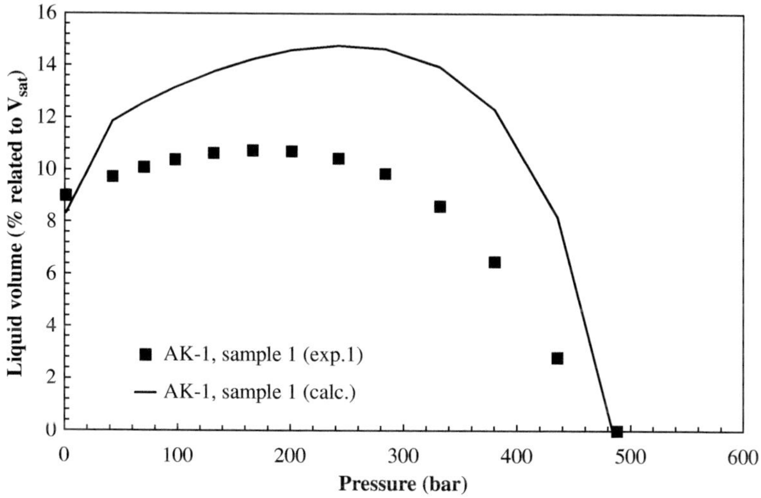

FIGURE 5.16 Liquid volumes in the constant volume depletion at 142.7°C, well AK-1, sample 1 (4375.5 m). Simulated curve obtained with the composition calculated for this depth from the reference level.

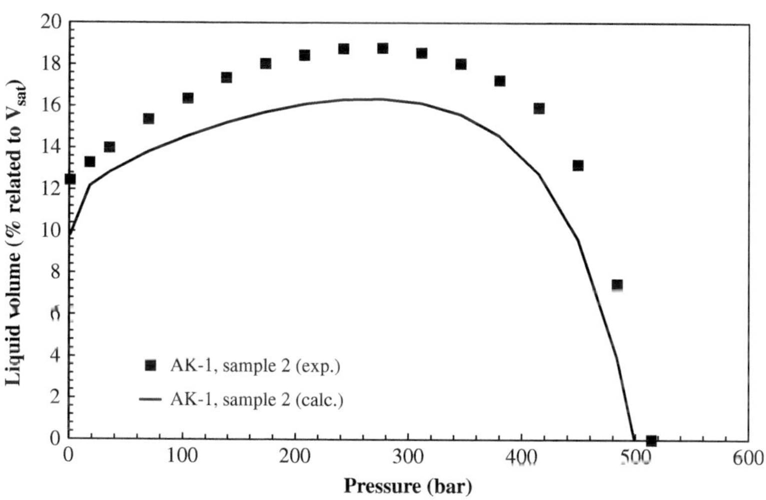

FIGURE 5.17 Liquid volumes in the constant volume depletion at 138.7°C, well AK-1, sample 2 (4217.5 m). Simulated curve obtained with the composition calculated for this depth from the reference level.

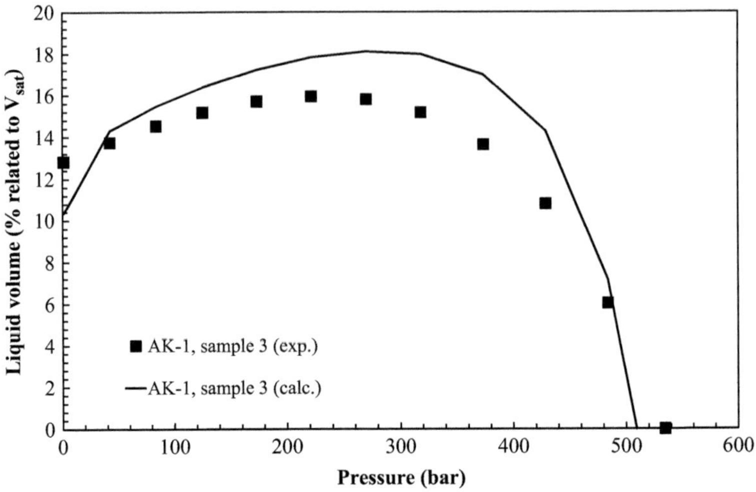

FIGURE 5.18 Liquid volumes in the constant volume depletion at 139.4°C, well AK-1, sample 3 (4108.5 m). Simulated curve obtained with the composition calculated for this depth from the reference level.

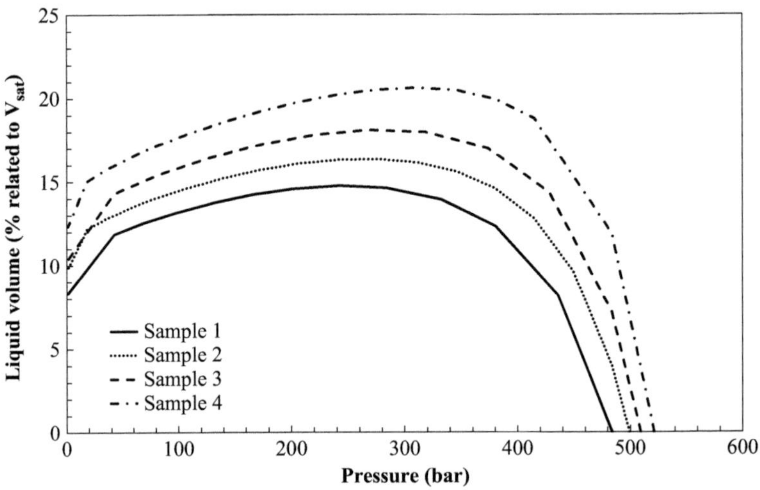

FIGURE 5.19 Simulated liquid volumes curves in the CVDs of well AK-1, superposed to evidence the tendency of liquid dropout to increase towards the top of the structure. The experimental data of samples 2 and 3 do not follow this tendency, a fact observed (but not explained) by Ghorayeb et al. (2000).

and can still be improved, for example, with mole fraction fine tuning in an outer loop. Fig. 5.19 repeats the four simulated curves, confirming the tendency of the liquid volume to increase toward the top of the formation. However, according to an observation made by Ghorayeb et al. (2000), there is an inversion on this tendency (not explained by the authors) in the experimental data of samples 2 and 3. By the time of the releasing of this book, this pendency remains under discussion with Japex's technical staff.

In the case of well MY-1, the follow-up of the evolution of the static pressure data on the fluid remaining within the column of production (CoP), stabilized after its formation test, allowed us to compute density along the entire length of the tubing. This measured density values were the main information used in the estimation of our parameters τ_i. More surprising than the samples taken at the formation itself, this fluid, once stabilized within the column (then submitted to the same thermal gradient of the reservoir), revealed a gravity segregation that presented not only the inversion on the critical transition previously observed in well AK-1, but also a well-defined GOC at around 2000 m above the reference level. The occurrence of a leaner gas (far from its critical point) at the top of the CoP was confirmed. Fig. 5.20 shows such a gravity segregation represented through a color sketch along the 5000-m-depth CoP. From bottom to top, note the critical transition showing the rich gas condensate by the reference level and the volatile oil a little bit above, the latter showing a well-defined interface (GOC around 2500 m) with the leaner top gas. Fig. 5.21 shows the comparison between the several models tested for the density profiles along MY-1's CoP. Note that the correlation proposed here and the τ_i values suggested by Ghorayeb et al. (2000) are, once more, the ones that present the best performance, predicting with excellent agreement the inverted critical transition at the bottom, the location of the ordinary GOC, and the properties of the top gas. There are no data of components' distribution along the tubing, but the profiles obtained by our correlation for methane and C_{20+} in Figs. 5.22 and 5.23 are in reasonable agreement with those coherently predicted by Ghorayeb et al. (2000).

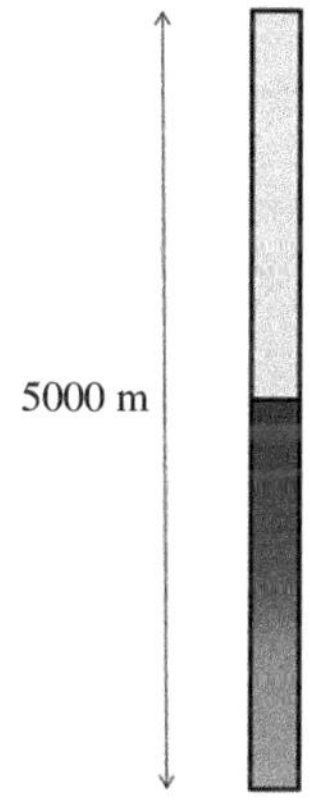

FIGURE 5.20 Sketch of the gravity segregation of the stabilized fluid in the CoP of well MY-1. Critical transition inverted, with rich gas condensate at the bottom and the volatile oil just above, showing an ordinary well-defined GOC around 2500 m.

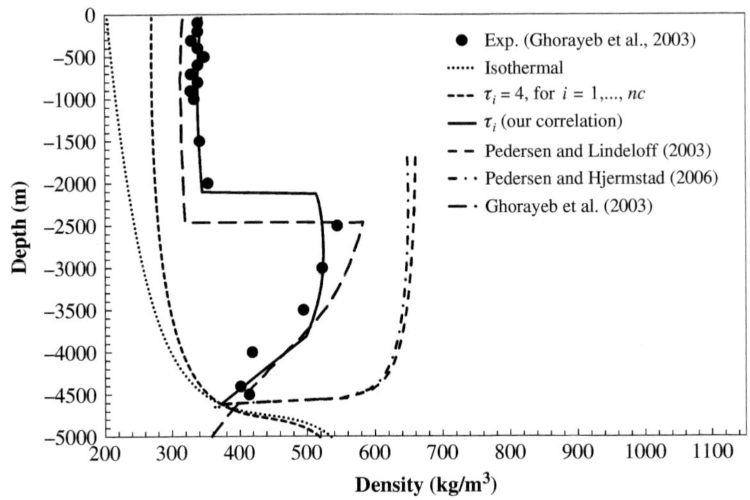

FIGURE 5.21 Density profile along the column of production (CoP) of well MY-1. Comparison between some calculation methods.

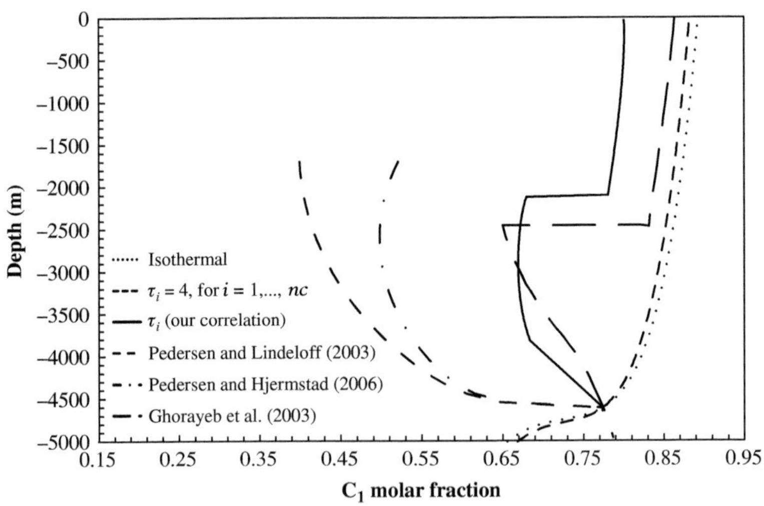

FIGURE 5.22 Methane distribution along the column of production (CoP) of well MY-1. Comparison between some calculation methods.

Fig. 5.24 shows quantitatively the phase envelopes of the three fluids segregated in the CoP of well MY-1, constructed from compositions chosen at the three regions after the calculation of the gravity segregation with both EoS and τ_i correlation tuned. The rich gas condensate at the bottom

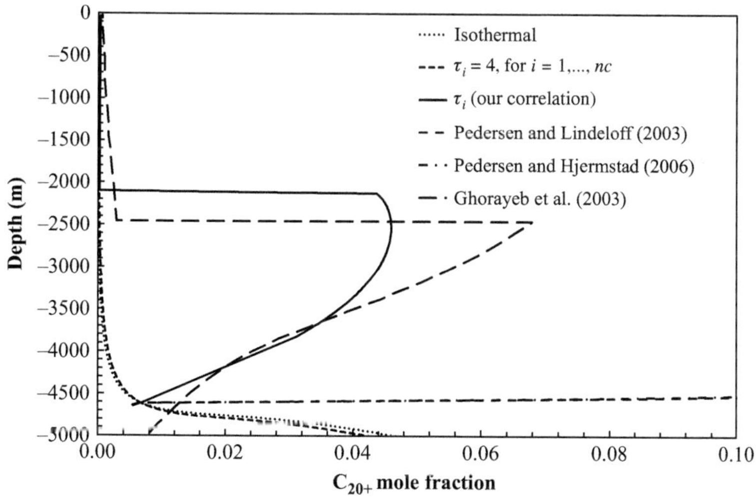

FIGURE 5.23 C_{20+} distribution along the column of production (CoP) of well MY-1. Comparison between some calculation methods.

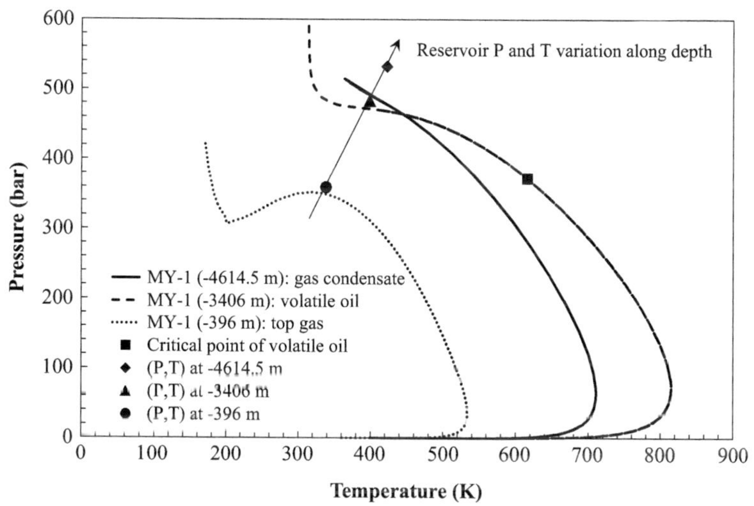

FIGURE 5.24 Phase envelopes of the fluids within the column of production (CoP) of well MY-1.

(GOR = 1378 m^3 std/m^3 std), following the inverted critical transition to the volatile oil just above the reference level (GOR = 375 m^3 std/m^3 std), and, at the top, a retrograde gas, leaner and farther from the critical point (GOR = 2848 m^3 std/m^3 std).

In addition to the conclusion that thermal diffusion acted contrarily to the gravitational field, attenuating compositional grading up to invert the fluids' positions in the critical transition at reservoir conditions, we note that, at this temperature level, the parameters of the heat-of-transport models must necessarily be reestimated. Yufutsu's reservoir fluid is very light and paraffinic, which evidences the inversion tendency. Most likely, that is why the models of Pedersen and Lindeloff (2003) and Pedersen and Hjermstad (2006) did not show satisfactory performance in this example, as their parameters (IG enthalpies) were estimated in a context of heavier oils whose segregation was, by hypothesis, enhanced by thermal diffusion in the same direction as the gravitational field. In the next example, two oils from North Sea studied by Pedersen and collaborators are discussed, for which one expects the models of these authors to perform better than those fitted for Yufutsu.

5.3 COMPOSITIONAL GRADING ENHANCED BY THERMAL DIFFUSION

After Pedersen and Lindeloff (2003) have presented the theoretical development of a heat-of-transport model, which is fully derived in Appendix D, Pedersen and Hjermstad (2006) have studied the gravity segregation in a high-temperature reservoir in the North Sea. The reservoir—not identified by the authors—had its compositional grading enhanced by thermal diffusion. In order to reproduce both fluid and properties distributions along depth, the authors proposed the correlation described in Eq. (5.2). Table 5.4 shows considerable composition variation of this fluid within a 45-m-depth interval.

Table 5.5 shows the main fluid properties published by the authors. Complete PVT analyses were not available. Pedersen and Hjermstad (2006) have used SRK-EoS with volume translation but did not make any EoS fitting based on these data. Their modeling was based on the splitting procedure of fraction C_{10+} detailed by Pedersen et al. (2014). The latter suggest their own correlations for T_c, P_c, and ω once the pseudocomponents of the splitting are lumped. The thermodynamic modeling of Pedersen and Hjermstad (2006) is presented in Table 5.6.

Note, in the thermodynamic model by Pedersen and Hjermstad (2006), an incoherence from the characterizing procedure of Pedersen et al. (2014): Both acentric factor and volume shift (VS) decrease from a certain molar mass value, when one expected a monotonic behavior. These authors also recommend null binary interaction coefficients between hydrocarbons, without considering the different interactions between methane (the lighter and predominant component) and the other components, as recommended by Arbabi and Firoozabadi (1995). A counter-argument to these assumptions would be the fact that the volume shift does not have a physical meaning, being only a mathematic correction introduced by Peneloux et al. (1982), not necessarily monotonic. As for the binaries, these would be only

TABLE 5.4 Compositional Variation (in mol %) of the North Sea Reservoir as a Function of Depth

Depth (m)	3638.2	3644.3	3651.1	3661.6	3676.0	3682.8
Fluid	Gas	Gas	Oil	Oil	Oil	Oil
N_2	0.431	0.295	0.358	0.331	0.337	0.395
CO_2	2.752	2.834	2.332	2.455	2.363	2.060
C_1	68.861	68.546	56.142	55.261	54.253	53.871
C_2	8.427	8.341	8.094	8.025	7.961	7.589
C_3	5.198	5.212	5.535	5.481	5.494	5.575
iC_4	0.847	0.892	1.001	0.995	1.000	1.009
nC_4	1.885	2.100	2.439	2.433	2.454	2.514
iC_5	0.587	0.675	0.879	0.877	0.889	0.900
nC_5	0.752	0.866	1.184	1.182	1.202	1.396
C_6	0.921	0.981	1.504	1.504	1.539	1.557
C_7	1.482	1.519	2.474	2.520	2.579	2.630
C_8	1.595	1.610	2.583	2.667	2.777	2.823
C_9	1.031	1.048	1.695	1.779	1.869	1.897
C_{10+}	5.231	5.080	13.780	14.491	15.282	15.783
M of C_{10+} (g/mol)	211.3	216.8	281.6	284.3	291.8	297.2
ρ of C_{10+} (g/cm^3)	0.8425	0.8440	0.8800	0.8825	0.8847	0.8868

Adapted from Pedersen, K.S., Hjermstad, H.P., Modeling of large hydrocarbon compositional gradient, SPE 101275, Abu Dhabi International Petroleum Exhibition and Conference, November/2006.

TABLE 5.5 Properties of the North Sea Reservoir Fluid by Pedersen and Hjermstad (2006)

Depth (m)	3638.2	3644.3	3651.1	3661.6	3676.0	3682.8
Fluid	Gas	Gas	Oil	Oil	Oil	Oil
Pressure (bar)	377.8	377.9	378.2	378.8	379.6	380.2
P_{sat} (bar)	375.5	372.8	364.2	364.5	360.1	353.8
GOR (m^3 std/m^3 std)	1086.0	1105.0	323.0	311.9	285.2	268.3
Density (g/cm^3)	0.367	0.376	0.574	0.581	0.595	0.601
Temperature (°C)	137.5	137.7	137.8	138.1	138.5	138.7

TABLE 5.6 Thermodynamic Modeling (SRK-EoS) of Pedersen and Hjermstad (2006) for the North Sea Reservoir Fluid

Component	Composition (in mol %)	M (g/mol)	T_c (K)	P_c (bar)	ω	VS (cm^3/mol)	H^{IG} (J/mol)
N_2	0.395	28.0	126.15	33.94	0.0400	−4.23	8320
CO_2	2.060	44.0	304.25	73.76	0.2250	−1.64	19,450
C_1	53.871	16.0	190.55	46.00	0.0080	−5.20	0
C_2	7.589	30.1	305.45	48.84	0.0980	−5.79	9781
C_3	5.575	44.1	369.85	42.46	0.1520	−6.35	19,520
iC_4	1.009	58.1	408.15	36.48	0.1760	−7.18	29,259
nC_4	2.514	58.1	425.25	38.00	0.1930	−6.49	29,259
iC_5	0.900	72.2	460.45	33.84	0.2270	−6.20	39,067
nC_5	1.396	72.2	469.65	33.74	0.2510	−5.12	39,067
C_6	1.557	86.2	507.45	29.69	0.2960	1.39	48,806
C_7	2.630	96.0	536.45	29.45	0.3370	6.70	55,623
C_8	2.823	107.0	558.05	27.42	0.3740	10.80	63,275
C_9	1.897	121.0	582.05	25.06	0.4210	14.52	73,014
C_{10}–C_{13}	4.406	152.9	631.45	21.32	0.5280	24.55	95,205
C_{14}–C_{16}	2.479	205.1	697.25	17.85	0.6860	26.56	131,517

(Continued)

TABLE 5.6 (Continued)

Component	Composition (in mol %)	M (g/mol)	T_c (K)	P_c (bar)	ω	VS (cm^3/mol)	H^{IG} (J/mol)
C_{17}–C_{19}	1.941	249.6	746.85	16.16	0.8120	25.75	162,472
C_{20}–C_{22}	1.520	289.5	788.35	15.17	0.9170	22.30	190,228
C_{23}–C_{26}	1.526	336.9	834.75	14.33	1.0290	15.13	223,201
C_{27}–C_{31}	1.325	399.7	892.75	13.55	1.1510	1.78	266,887
C_{32}–C_{38}	1.145	481.5	964.25	12.89	1.2540	−19.81	323,790
C_{39}–C_{48}	0.830	595.7	1058.85	12.31	1.2710	−55.16	403,231
C_{49}–C_{80}	0.610	811.2	1234.05	11.73	0.8750	−124.90	553,140

The Overall Composition is the One at the Reference Level (Chosen by the Authors at 3682.8 m). The *Volume Shift* (VS) is Not Divided by Covolume *b*. So it Has Dimensions of Molar Volume. The Ideal Gas Enthalpies Follow Eq. (5.2), Except Methane, the Reference Component ($H^{IG} = 0$). All Binaries are Zero.

adjustable parameters of a modification proposed to the classical mixing rule of van der Waals, recovered when one sets all k_{ij} equal to zero. Nevertheless, the acentric factor is to be always monotonic, according to Pitzer's definition.

Anyway, apart from the representativeness of thermodynamic modeling, it is true that Pedersen and Hjermstad (2006) did good on reproducing the data shown in Table 5.5. As will be seen later in this text, their IG enthalpies provided the fine tuning necessary for the action of thermal diffusion to reproduce the GOC and profiles shown in Table 5.4 with good agreement.

We compare the results of Pedersen and Hjermstad (2006) with those obtained applying the expression of Firoozabadi et al. (2000) for the heat of transport, using the Peng–Robinson EoS fitted after the splitting of the C_{10+} fraction by the methodology of Shibata et al. (1987). As for τ_i parameters, we will use both the default values of 4.0 and those obtained by the correlation adjusted for Yufutsu. These sets of parameters will be compared with a new adjustment of τ_i carried out specifically for this example.

Tables 5.7 and 5.8 present the thermodynamic modeling that we propose in this book for the example of Pedersen and Hjermstad (2006). It is based

TABLE 5.7 Parameters of Peng–Robinson EoS Adjusted for the Example of Pedersen and Hjermstad (2006)

Component	P_c (bar)	T_c (K)	ω	M (g/mol)	VS (cm^3/mol)
CO_2	73.76	304.20	0.2250	44.01	−0.0718
N_2	33.94	126.20	0.0400	28.01	−0.1284
C_1	46.00	190.60	0.0080	16.04	−0.2108
C_2	48.84	305.40	0.0980	30.07	−0.1581
C_3	42.46	369.80	0.1520	44.10	−0.1262
iC_4	36.48	408.10	0.1760	58.12	−0.1028
nC_4	38.00	425.20	0.1930	58.12	−0.1028
iC_5	33.84	460.40	0.2270	72.15	−0.0841
nC_5	33.74	469.60	0.2510	72.15	−0.0842
C_6	32.89	507.50	0.2750	86.00	−0.0688
C_7	31.38	543.20	0.3083	96.00	−0.0595
C_8	29.51	570.50	0.3513	107.00	−0.0502
C_9	27.30	598.50	0.3908	121.00	−0.0395
$QC_{13.67}$	20.25	694.89	0.5306	205.29	−0.0017
$QC_{31.69}$	11.65	898.26	1.1310	458.22	0.0960
$QC_{62.88}$	5.58	1085.09	1.7359	896.00	0.1063

TABLE 5.8 Nonzero Binary Interaction Coefficients for Peng–Robinson EoS for the Example of Pedersen and Hjermstad (2006)

Component	CO_2	C_1
CO_2	0.0000	0.1500
N_2	−0.0200	0.0000
C_1	0.1500	0.0000
C_2	0.1455	0.0000
C_3	0.1427	0.0023
iC_4	0.1399	0.0046
nC_4	0.1399	0.0046
iC_5	0.1371	0.0069
nC_5	0.1371	0.0069
C_6	0.1343	0.0091
C_7	0.1323	0.0108
C_8	0.1301	0.0126
C_9	0.1273	0.0148
$QC_{13.67}$	0.1101	0.0289
$QC_{31.69}$	0.0595	0.0702
$QC_{62.88}$	0.0000	0.1300

on the generalized quadrature technique of Shibata et al. (1987) for the splitting of C_{10+} fraction at the reference level (3682.8 m) with three pseudocomponents. From the initial estimates of T_c, P_c, and ω by the correlations of Cavett (1962), we reestimated these parameters based on the properties informed by the authors (density, saturation pressure, and GOR) not only at the reference level, but also in the gas zone. It was assumed that the heaviest pseudocomponent ($QC_{62.88}$) does not distribute up to the gas, so we avoid mass balance problems using only two components for C_{10+} fraction in the gas zone.

The proposed model gives similar results as compared to those of Pedersen and Hjermstad (2006), though without the issues about parameters' trends. We emphasize that, even in the isothermal simulation, we predict a GOC close enough to the actual location, which does not happen in the original model proposed by the authors. Based on density and composition data

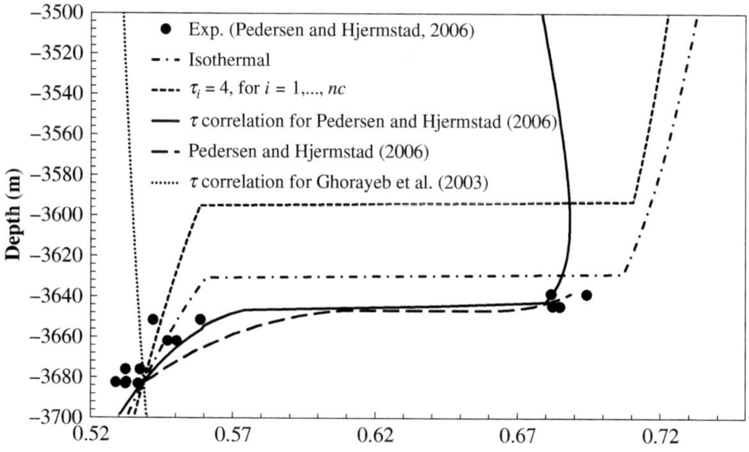

FIGURE 5.25 Methane distribution in the North Sea reservoir studied by Pedersen and Hjermstad (2006). Comparison between different methods.

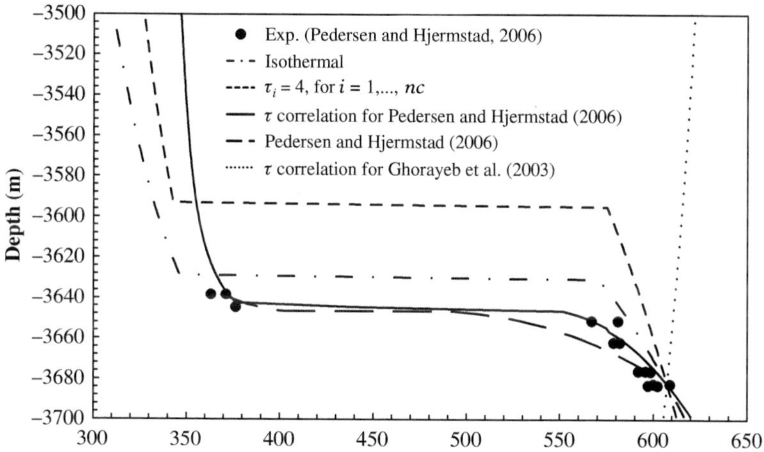

FIGURE 5.26 Density of the North Sea reservoir fluid studied by Pedersen and Hjermstad (2006). Comparison between different methods.

along depth, we could also reestimate the parameters of the τ_i correlation proposed for Yufutsu (Eqs. (5.3)–(5.6)) for this example in particular, matching then the actual GOC location.

Figs. 5.25 and 5.26 show the results of methane mole fraction and density along depth for both isothermal and nonisothermal grading. For the latter, besides the IG enthalpies of Pedersen and Hjermstad (2006), H^{IG} given by Eq. (5.2), we also applied the generalized heat-of-transport model of Firoozabadi et al. (2000) for multicomponent mixtures (Appendix B), using different sets for the τ_i values:

- $\tau_i = 4$, for $i = 1, \ldots, nc$;
- τ_i reestimated for Yufutsu, Eqs. (5.3)–(5.6)
- τ_i reestimated for this example, parameters of Eq. (5.6) adjusted on the basis of the composition and density experimental data:

$$\begin{aligned} \alpha_0 &= -12.114876 \\ \alpha_1 &= 2.973253 \\ \beta_0 &= 7.350923 \\ \beta_1 &= 3.331057 \end{aligned} \tag{5.7}$$

- Furthermore, we show the results of the IG enthalpies of Pedersen and Hjermstad (2006), H^{IG}, given by Eq. (5.2).

It is important to emphasize the improvement of the agreement with experimental data if one compares the model proposed here with the one of Pedersen and Hjermstad (2006). On the other hand, we should remark that the absence of complete PVT analysis along the reservoir depth for the regression of the model parameters increases the uncertainties related to our interpretation about whether thermal diffusion effectively enhanced or attenuated compositional grading. Although it is obvious that the parameters estimated for Yufutsu do not apply to this example, as the default values of $\tau_i = 4.0$ (for $i = 1, \ldots, nc$) have lifted the GOC location about 30 m compared to the isothermal simulation, suggesting that gravity segregation was attenuated. The locally optimized parameters of the τ_i correlation pointed it down correctly 15 m below the isothermal prediction, suggesting that it was enhanced. On the other hand, the methane and density distribution profiles above the GOC given by our tuned model are considerably less severe than those obtained with the isothermal simulation and with default values of τ_i ($\tau_i = 4.0$, for $i = 1, \ldots, nc$), suggesting attenuation of the compositional grading. In other words, as this is a case in which thermal diffusion was to enhance compositional grading, one would expect higher methane contents, and consequently lower densities, in comparison to the isothermal case, as one moves to the top of the structure. The absence of any experimental property in this region of the gas cap farther from GOC is likely to be the main cause of this incoherence. Note that, fortunately, this behavior does not repeat in the oil zone, where the optimized τ_i-value curve has a slope greater than the isothermal curve, which in turn is more pronounced than the original-parameter ($\tau_i = 4$, for $i = 1, \ldots, nc$) one, which suggests once more the enhancement of the compositional grading by thermal diffusion.

Finally, although its experimental value was not informed by the authors, the API degree of the dead oil can be simulated by the fitted EoSs, if one carries out a flash to the standard conditions. Both ours and Pedersen and Hjermstad (2006) models give approximately 33°API for this North Sea oil, suggesting that this is a low enough value to contain aromatic and/or branched heavy components that allow the phenomenon of enhancement of composition grading by thermal diffusion to occur. This conclusion must be validated with more detailed studies on the theme, using examples with more information available.

5.4 RESERVOIR NOT MUCH INFLUENCED BY THERMAL DIFFUSION

This example was already approached in Chapter 3 (Section 3.6), considering the reservoir isothermal. Now, we just include the thermal diffusion effect, assuming a typical thermal gradient of -0.025K/m, and proceeding further to the 1D simulations with the same thermodynamic model, using, for the heat of transport, the model of Shukla and Firoozabadi (1998)—generalized by Firoozabadi et al. (2000)—with default values (4.0) for τ_i. We compare the results with the ones obtained by the correlation for τ_i optimized in Yufutsu and also with the IG-enthalpy model of Pedersen and Hjermstad (2006).

Figs. 5.27–5.29 show the comparison of all these nonisothermal models with the isothermal case, regarding the experimental data of methane mole fraction, C_{7+} mole fraction and C_{7+} molar mass along the depth related to the reference level of Creek and Schrader (1985). One notes clearly that Pedersen and Hjermstad's (2006) methodology, with their original enthalpies, overestimate the compositional grading observed in practice, even showing a spurious

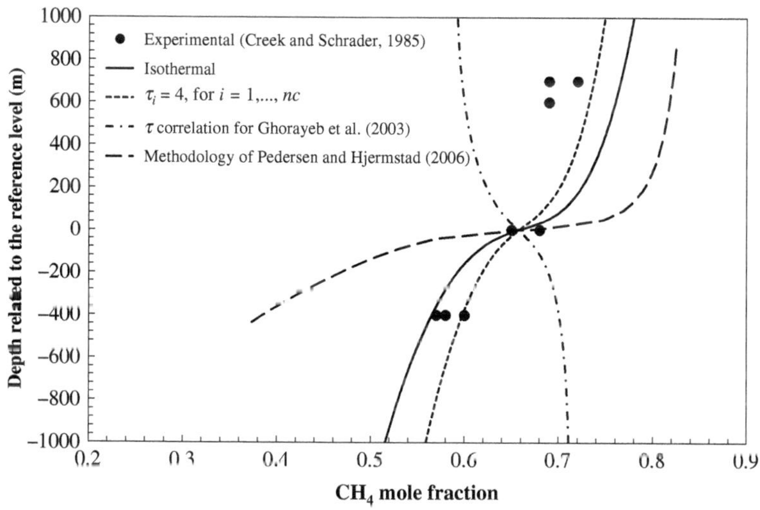

FIGURE 5.27 Influence of thermal diffusion in the variation of methane mole fraction as a function of the depth related to the reference level of East Painter reservoir (Creek and Schrader, 1985).

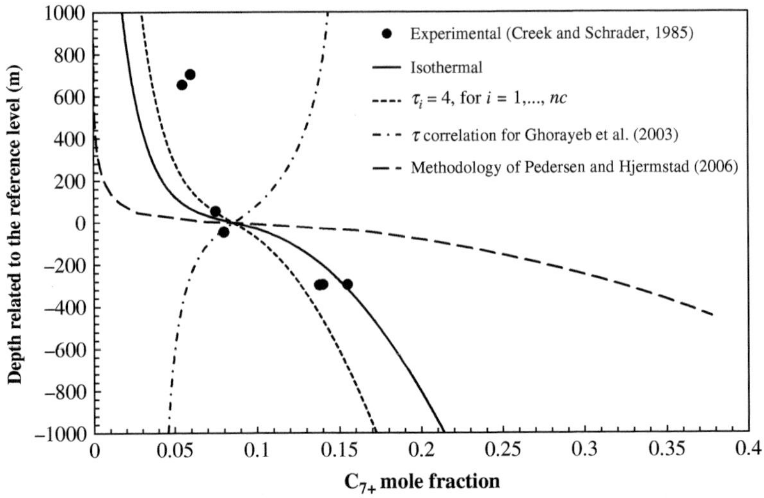

FIGURE 5.28 Influence of thermal diffusion in the variation of C_{7+} mole fraction as a function of the depth related to the reference level of East Painter reservoir (Creek and Schrader, 1985).

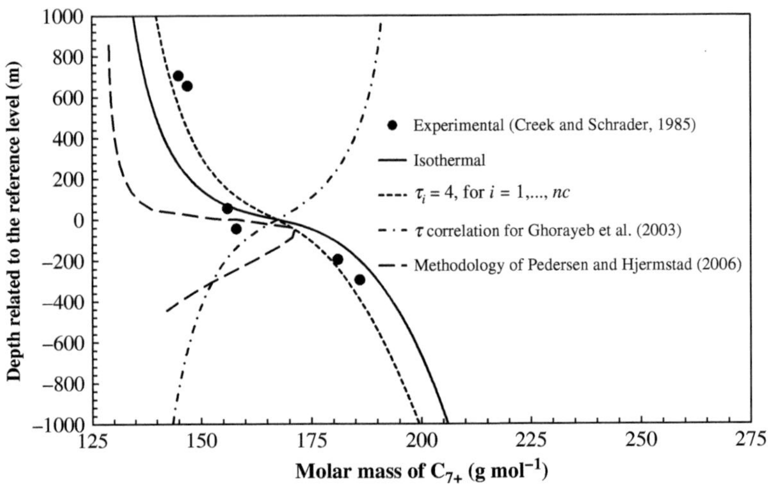

FIGURE 5.29 Influence of thermal diffusion in the variation of the molar mass of C_{7+} as a function of the depth related to the reference level of East Painter reservoir (Creek and Schrader, 1985).

inversion in the tendency of C_{7+} molar mass. A parameter reestimation in Eq. (5.2) seems to be necessary but inconvenient, as it is not reasonable that reference ideal gas enthalpies might vary from case to case.

The heat-of-transport model of Firoozabadi et al. (2000), with $\tau_i = 4.0$, for $i = 1, \ldots, nc$, provides results very similar to those of the isothermal case, which, in turn, is already reasonably close to the experimental data, suggesting that thermal diffusion's influence is not so important in this example. The discrepancies still observed between simulated and experimental data are likely due to the lack of PVT reports of the samples collected in the original paper of Creek and Schrader (1985). Measurement error bars are also not shown, but composition and properties' uncertainties may have been considerable at that time for recombined samples. Our PR-EoS fitting quality might have been limited because of the sparse data of saturation pressure and density published by Creek and Schrader (1985), the only ones used in our regression.

Our interpretation, following the point of view of Pedersen's group is that, in East Painter reservoir (Creek and Schrader, 1985), both temperature (around 90°C) and heavies' mole fractions (like asphaltenes, aromatics, and long-chain paraffins) are not high enough to enhance compositional grading. Being a light oil, for which the API of C_{7+} fraction spams within 38 and 57—see Figure 2 of the original paper of Creek and Schrader (1985)—the effect of thermal diffusion tends to attenuate segregation softly, reestimation of τ_i default parameters of Firoozabadi et al. (2000) being unnecessary. Therefore, this is an intermediate case between Yufutsu (Ghorayeb et al., 2000)—in which the high temperature (around 140°C) together with an ordinary thermal gradient (around -0.025 K/m upward) was enough to lead the lights to the hot bottom, inverting gas and liquid positions along the critical transition, requiring also reestimation of τ_i—and the North Sea reservoir fluid studied by Pedersen and Hjermstad (2006), where, for the same temperature levels of Yufutsu, compositional grading was likely enhanced by the presence of these components. That is why the parameters estimated by Pedersen and Hjermstad (2006) to reproduce the behavior of the North Sea reservoir may overestimate the segregation in any other case in which the absence of such heavier components would lead to a just ordinary distribution, very similar to the one observed in an isothermal gravitational field.

5.5 FINAL REMARKS

In this chapter, the applications of the main thermal diffusion models of the literature on practical examples were studied. Their theoretical limitations were discussed, and a temperature- and composition-dependent correlation was proposed for reestimating the main parameter of the heat-of-transport model of Firoozabadi et al. (2000), specifically for the case of Yufutsu. Regardless the peculiarities of this field, one expects the thermal diffusion influence to be as important as the reservoir conditions are close to the critical point of the fluid. We noticed that this correlation is not universal, and its parameters must be estimated case to case, or, at least, regionally in other examples. The composition dependency is in the pseudoreduced temperature, which saves computational effort due to the simplicity of the pseudocritical temperature compared to the calculation of the actual critical point of the mixture at each discrete level of the domain.

The same nonuniversal characteristic applies to the correlation of Pedersen and Hjermstad (2006) for the IG enthalpies. However, as these are intrinsic properties of the components, evaluated at a fixed reference temperature, one does not recommend its adjustment for each example approached.

One also verifies that, depending on the composition (chemical nature) of the reservoir fluid dealt with, the Soret effect may attenuate or enhance the grading observed in the isothermal calculation. Qualitatively, at least in the examples we analyzed, we observed that thermal diffusion will act more in the opposite direction of the gravitational field the lighter the fluid is. One emphasizes that the quality of the EoS fitting based on PVT data is crucial to minimize the characterization uncertainties when one is to determine the fluid distribution in the reservoir. In the next chapter, these questions are discussed in new examples, reestimating the parameters of the proposed correlation in a regional way.

REFERENCES

Arbabi, S., Firoozabadi, A., 1995. Near-critical phase behavior of reservoir fluids using equations of state. SPE-24491-PA Adv. Technol. 3 (1). Available from: https://doi.org/10.2118/24491-PA.

Cavett, R.H., Physical data for distillation calculations—vapor–liquid equilibrium, Proc. 27th Meeting, API, San Francisco, pp. 351–366, 1962.

Creek, J.L., Schrader, M.L., East Painter reservoir: an example of a compositional gradient from a gravitational field, SPE 14411, 60th Annual Technical Conference and Exhibition, Las Vegas-NV, September/1985.

Firoozabadi, A., Ghorayeb, K., Shukla, K., 2000. Theoretical model of thermal diffusion factors in multicomponent mixtures. AIChE J. 46 (5), 892–900.

Ghorayeb, K., Anraku, T., Firoozabadi, A., Interpretation of the fluid distribution and GOR behavior in the Yufutsu fractured gas-condensate field, SPE 59437, SPE Asia Pacific Conference, Yokohama, Japão, April/2000.

Ghorayeb, K., Firoozabadi, A., Anraku, T., Modeling of the unusual GOR performance in a fractured gas condensate reservoir, SPE 75258-MS, Improved Oil Recovery Symposium, Tulsa-Oklahoma, April/2002.

Ghorayeb, K., Firoozabadi, A., Anraku, T., 2003. Interpretation of the unusual fluid distribution in the Yufutsu gas condensate field, SPE-84953-PA. SPE J. 8 (2). Available from: https://doi.org/10.2118/84953-PA.

Pedersen, K.S., Christensen, P.L., Shaikh, J.A., 2014. Phase Behavior of Petroleum Reservoir Fluids, 2nd Edition, CRC Press, Boca Raton, FL.

Pedersen, K.S., Hjermstad, H.P., Modeling of large hydrocarbon compositional gradient, SPE 101275, Abu Dhabi International Petroleum Exhibition and Conference, November/2006.

Pedersen, K.S., Lindeloff, N., Simulations of compositional gradients in hydrocarbon reservoirs under the influence of a temperature gradient, SPE 84364, SPE Annual Technical Conference and Exhibition, Denver, Colorado, October/2003.

Peneloux, A., Rauzy, E., Freze, R., 1982. A consistent correction for Redlich–Kwong–Soave volumes. Fluid Phase Equilibria 8, 7–23.

Shibata, S.K., Sandler, S.I., Behrens, R.A., 1987. Phase equilibrium calculations for continuous and semicontinuous mixtures. Chem. Eng. Sci. 42 (8), 1977–1988.

Shukla, K., Firoozabadi, A., 1998. A new model of thermal diffusion coefficients in binary hydrocarbon mixtures. Ind. Eng. Chem. Res. 37, 3331–3342.

Chapter 6

Case Studies

In this chapter, a number of unpublished case studies are presented in reservoirs which show noticeable compositional grading. Based on the theories developed in Chapter 4, and applied in Chapter 5, one tries to reproduce the composition profiles observed in some new case-study fields. These fields were carefully selected because of their peculiarities opened up to interpretations, rather to inspire new strategies in fluid modeling and reservoir engineering than to establish definite answers for such problems.

One starts by a high-temperature accumulation (above 120°C), for which one expects thermal diffusion to be relevant. On reproducing the experimental data of gravity segregation for this field, we formed a basis to establish a new set of parameters for the empirical correlation that provides the values of the parameter τ_i in the generalized Heat-of-Transport model of Firoozabadi et al. (2000). We observe that the parameters adjusted, for example, for Yufutsu, do not apply to this case. However, once they are reestimated, these new parameters will be used for the other reservoirs selected, increasing prediction reliability in thermal-diffusion influenced system with similar characteristics.

The second example uses the correlation generated by the first one, but in a much colder 1D reservoir. We do not expect thermal diffusion to be so important in this case. Then, we intend not only to predict the reservoir-fluid distribution, but also to infer how it should be if the reservoir was subjected to a hypothetical higher temperature.

In the third example, the compositional grading of a low-temperature (below 60°C) reservoir fluid is considered, even though it has not achieved the steady state below a certain depth. One also tries to reproduce the compositional grading above that depth, noticeably a critical transition for which the isothermal calculations are compared with those influenced by thermal diffusion, using again, among other premises, the parameters adjusted in the first example.

In the fourth example, we deal with another reservoir modeled in 2D. In this field, the differences on the thickness of the cap rock along a certain direction caused a considerable lateral temperature gradient, with a 90–120°C range over more than 20 km of extension. Once overall full connection is confirmed despite heterogeneities, our goal is to compare the fluid distribution within the 2D domain to the one that would be observed in the absence of natural convection. We use only default parameters of thermal diffusion.

Compositional Grading in Oil and Gas Reservoirs. DOI: http://dx.doi.org/10.1016/B978-0-12-812452-9.00006-X

The fifth and last example deals with a reservoir containing a light oil subjected to a lateral temperature gradient. One tries to confirm a possible global connection between two different sandstones in a 2D domain. One intends to find out whether the fluids sampled quite far from a reference point would make up a continuous critical transition with a huge compositional variation in a single full-connected reservoir.

6.1 HIGH-TEMPERATURE RESERVOIR, MODELED AS ONE-DIMENSIONAL (1D)

In this example, we consider the fluid data obtained by sampling the so-called Well-1, drilled exactly on the border line between two fields of a certain area of concession, as shown in Fig. 6.1. The borders of each field's concession do not obviously match the geological limits between the accumulations. Limiting wells are important not only to determine the geological limits, but also to establish the properties of the fluids involved in the so-called unitization processes, where one negotiates legal premises about sharing goods, rights, and obligations among attendees with respect to reservoir-fluid production involving two or more contracts simultaneously. The legal aspects of this discussion are beyond the scope of this book. In this example, it would be interesting to determine the quantities and fluid properties of the reservoir to the South of Well-1 that invades the border of the neighbor concession to the North, regardless of their geological limits (not shown in Fig. 6.1). The same thought would apply to the West border, for example, but we will not consider other drilled wells in this case study. Well-1 was sampled at two distinct depths, showing, on the shallowest one, a gas condensate of about 1200 m^3 std/m^3 std GCR (*Gas–Condensate Ratio*) and 41°American

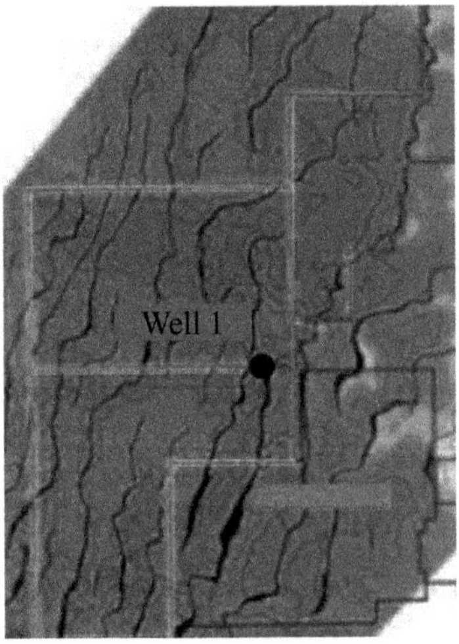

FIGURE 6.1 Well-1 sketched on the border line of different fields. The high-temperature reservoir fluid to the South of Well-1 is modeled by both phase behavior and compositional grading observed along this well.

Petroleum Institute (API)[1]. The deepest sample is a 200 m^3 std/m^3 std gas-oil ratio (GOR) and 30-API oil. One wants to know whether the fluids are connected or not and the location of the gas–oil contact (GOC), if it exists. Due to the huge thickness of the cap rock throughout the area extension, one believes that there are not lateral temperature gradients, which means that the compositional grading can be considered as 1D, that is, properties vary only with depth. One intends to obtain this distribution, which will hold for all the extension of the South field up to its geological limits with the adjacent reservoirs, assuming the effective connection of the whole porous matrix above the water–oil contact (WOC). Both GOC- and critical-transition possibilities are sketched in Fig. 6.2. Porosity and permeability data become irrelevant with the premise of 1D globally connected reservoir, since the convective flux is zero and steady-state hypothesis (assumed a priori) automatically eliminates the transient term of the flow equation.

On dealing with a high-temperature reservoir (about 130°C), even with an ordinary thermal gradient of −0.025°C/m, it is reasonable to suppose that thermal diffusion will have considerable influence on composition distribution along depth. So, we apply again the strategy of fitting the Peng–Robinson equation of state based on the Pressure, Volume and Temperature (PVT) analysis, with splitting of the C_{20+} fraction by the methodology of Shibata

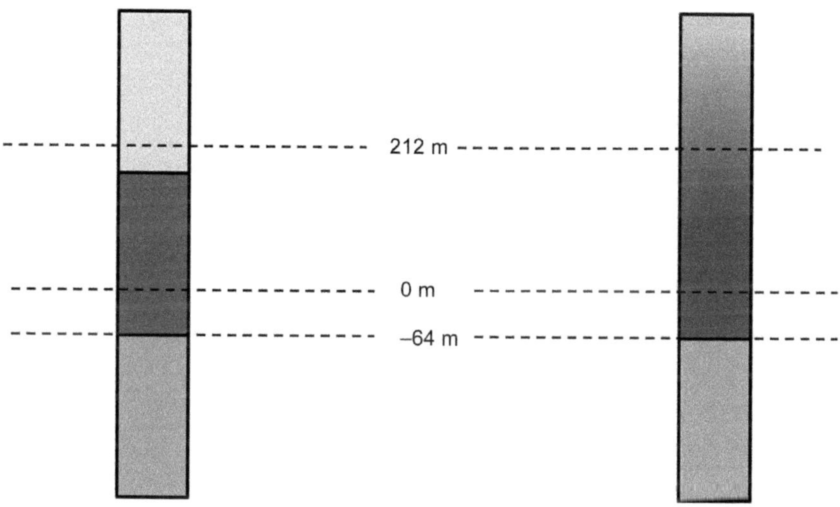

FIGURE 6.2 Possibilities of fluid distribution in the high-temperature reservoir to the South of Well-1: gas cap on the left (GOC depth still arbitrary) and critical transition to the right. Also showing both oil (reference point, 0 m) and gas (212 m above) sampling depths, as well as the WOC (64 m below).

1. The American Petroleum Institute gravity, or API gravity, is a measure of how heavy or light a petroleum liquid is compared to water.

et al. (1987). Then, we do the isothermal calculation of the compositional grading, assuming the oil sampling depth as the reference level. We will proceed further to consider the geothermal gradient, but using the default parameters ($\tau_i = 4$, for $i = 1, \ldots, nc$) in the Heat of Transport model. Finally, we reestimate the parameters of the correlations presented in Eqs. (5.3)–(5.6) to optimize τ_i based on the experimental data of methane and density distribution, according to the PVT report of the gas sample. We also compare these three simulations with that obtained using the adjusted values for the Yufutsu field in Chapter 5 (Eq. (5.6)), in order to verify the extension of the influence of thermal diffusion relative to the limiting case of the Japanese field.

The equations to be solved in both isothermal and nonisothermal cases are summarized in Table 4.1. As to the methodology of Pedersen and Lindeloff (2003), the nonisothermal results using the default ideal-gas enthalpies proposed by these authors and modified later by Pedersen and Hjermstad (2006) are shown in separate plots. This is to provide the reader a better visual comprehension of their performance when compared to the same isothermal calculations and experimental data of the example.

Table 6.1 shows the mole-fraction ranges of the pseudocomponents, with their respective molar masses after the splitting–lumping procedure of Shibata et al. (1987). Figs. 6.3–6.5 show, respectively, the comparison

TABLE 6.1 Compositions of the Reservoir Fluids for the Samples Collected in Well-1. The Gas Sample is 212 m Above the Oil zone (Reference)

Component	M (g/mol)	Composition Range (mol %)
CO_2	44	2–4
N_2–C_1	16	55–80
C_2	30	6–8
C_3–nC_5	53	7–9
C_6–C_{12}	124	5–10
C_{13}–C_{19}	217	2–5
	2 Quadrature Points	
$QC_{27.8}$	380	0–5
$QC_{60.1}$	833	0–2
	3 Quadrature Points	
$QC_{24.3}$	337	0–3
$QC_{43.7}$	609	0–3
$QC_{70.3}$	981	0–1

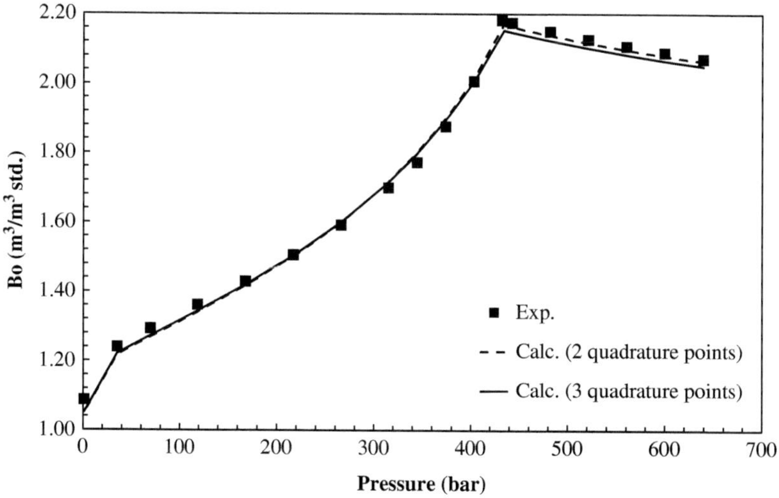

FIGURE 6.3 Comparison between experimental and calculated Bo of the differential liberation at 130°C for two and three quadrature points.

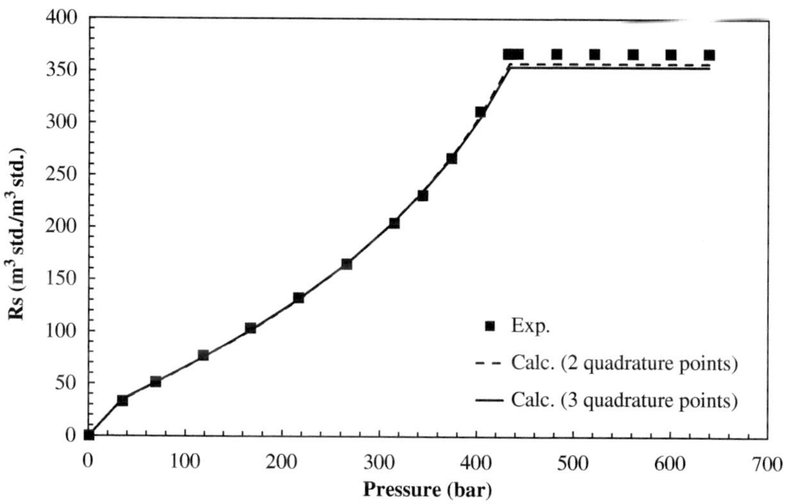

FIGURE 6.4 Comparison between experimental and calculated Rs of the differential liberation at 130°C for two and three quadrature points.

between calculated and experimental data of Bo, Rs and density on the differential liberation of the oil sample at 130°C for two and three quadrature points. Table 6.2 shows the comparison between experimental and calculated values of gas-oil ratio (GOR) and API for both flash and differential liberation. Due to reservoir's high temperature, the latter shows more gas evolving, around 20% more in relation to the flash, overwhelming the effect of light-

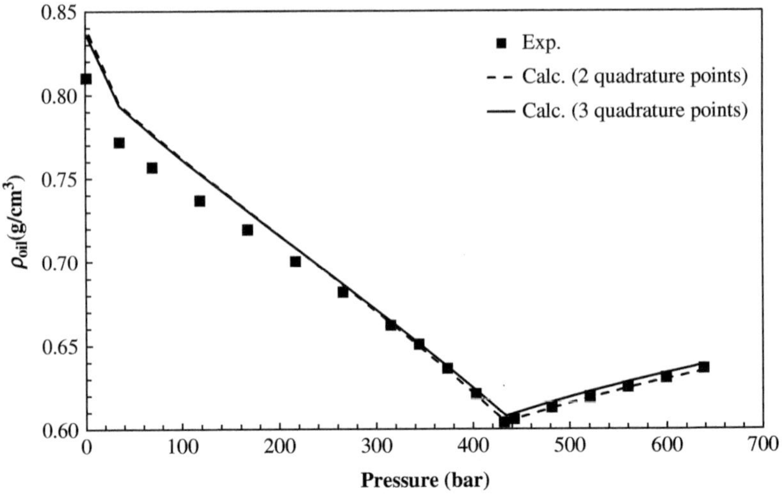

FIGURE 6.5 Comparison between experimental and calculated density of the differential liberation at 130°C for two and three quadrature points.

TABLE 6.2 GOR and API (Experimental and Calculated) in Both Flash and Differential Liberation (DL) for Two (2q) and Three (3q) Quadrature Points

Property	Experimental	Calculated (2q)	Calculated (3q)
Flash GOR (m^3 std/m^3 std)	302.0	305.6	302.0
Flash API	30.8	31.7	31.5
DL GOR (m^3 std/m^3 std)	367.1	357.3	354.0
DL API	28.9	28.9	29.5

fraction retaining due to numerous equilibrium stages. Despite the good agreement between calculated properties and experimental data for the oil in Table 6.1, an inconsistency due to the use of the minimum number of quadrature points is observed: they are not enough to cover up the variation of the molar mass from oil to gas. The molar mass of the lightest pseudocomponent of C_{20+} in the oil is already greater than the molar mass of the whole C_{20+} fraction in the gas, not attending the overall mass balance of the compositional grading.

The average value of 512 g/mol, when represented by three quadrature points, expands more the range of molar mass in the discretized exponential-decay distribution, generating a pseudocomponent with molar mass lower than 345 g/mol, which makes the characterization of the gas possible with the same pseudocomponents of the oil.

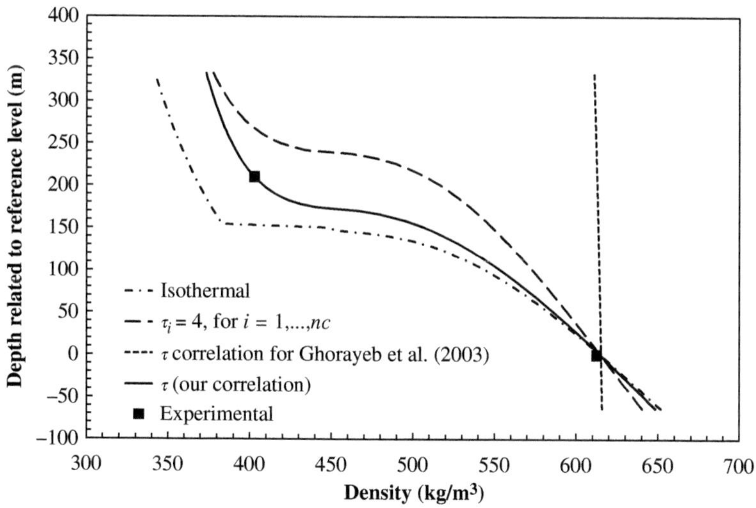

FIGURE 6.6 Density profile along depth.

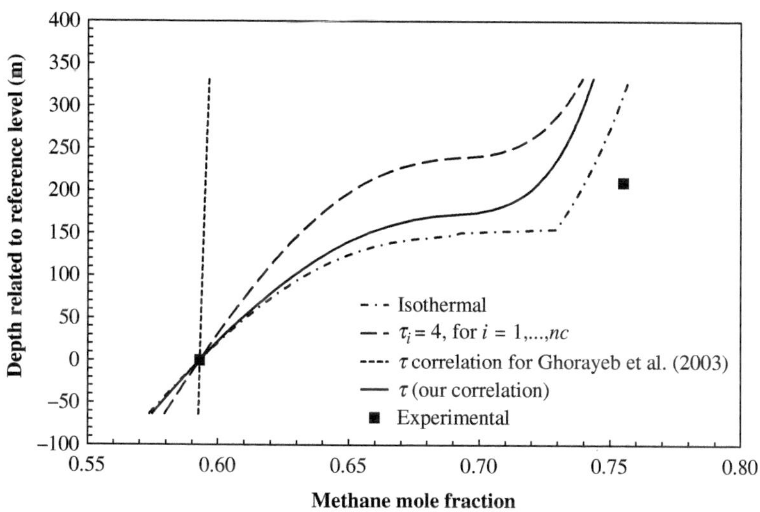

FIGURE 6.7 Methane's mole fraction along depth.

Fulfilling the mass balance assures that the constant-temperature calculation (Eq. (3.27)) will already provide good results, the closer to reality the smaller the influence of thermal diffusion. From this point, we discuss only the results obtained with three quadrature points on C_{20+} fraction splitting.

Figs. 6.6–6.8 show, respectively, the profiles of density, methane and C_{20+} along depth, using the heat-of-transport model of Firoozabadi et al. (2000) and its various reestimations, compared with the experimental values

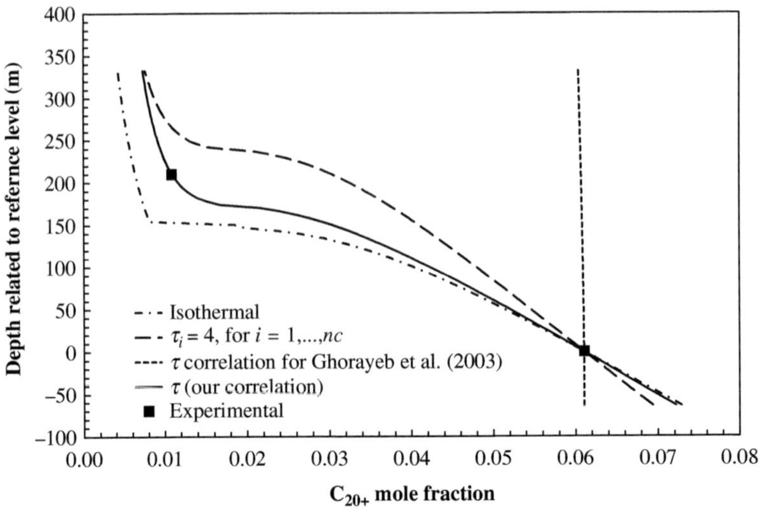

FIGURE 6.8 C_{20+} mole fraction along depth.

of the gas sample. Results are consistent, with the largest deviation observed for methane, the most abundant component.

The saturation-pressure profile along depth, shown in Fig. 6.9, does not touch the static-pressure curve, suggesting critical transition without GOC. Fig. 6.10 shows the critical temperature variation along depth, compared to the geothermal gradient. The curves intercept each other at about 180 m above the reference level, suggesting occurrence of gas condensate from this point (richer) up to the top (leaner). Figs. 6.11–6.14 present comparisons between the PVT properties calculated with the composition obtained at the gas sampling depth and the experimental data of this sample. The agreement is good enough to validate the modeling and its premises. Therefore, it confirms the critical transition without GOC for the fluids sampled in Well-1 (right-hand side of Fig. 6.2). One expects this behavior to spread throughout the extension of the reservoir in the South of Well-1.

The parameters of the empirical correlation for the parameter τ_i, presented in Eq. (5.6), were reestimated for the experimental data of this example, leading to the following values:

$$\begin{aligned} \alpha_0 &= -2.17078 \\ \alpha_1 &= 4.07908 \\ \beta_0 &= -2.71298 \\ \beta_1 &= 3.39140 \end{aligned} \tag{6.1}$$

The correlation adjusted and applied to this example showed good performance. As there are few experimental data in this reservoir, it is not straightforward to carry on a more detailed evaluation.

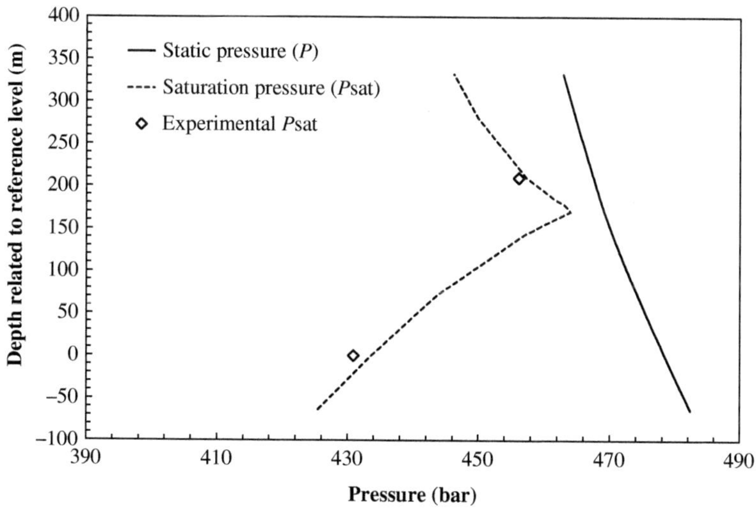

FIGURE 6.9 Saturation (P_{sat}) and static pressure (P) variation along depth, suggesting critical transition without GOC.

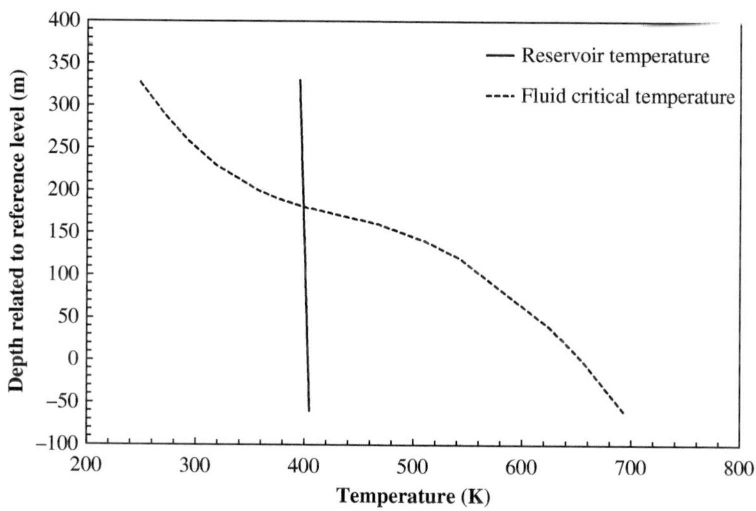

FIGURE 6.10 Variation of both reservoir and critical temperature of the fluid along depth. Transition to gas condensate observed approximately at 180 m above the reference level in Well-1.

From the calculations of compositional grading in Well-1, one concludes that, besides the confirmation of hydraulic connection and the critical transition, according to the heat-of-transport model of Firoozabadi et al. (2000), thermal diffusion attenuates compositional grading, acting conversely to the

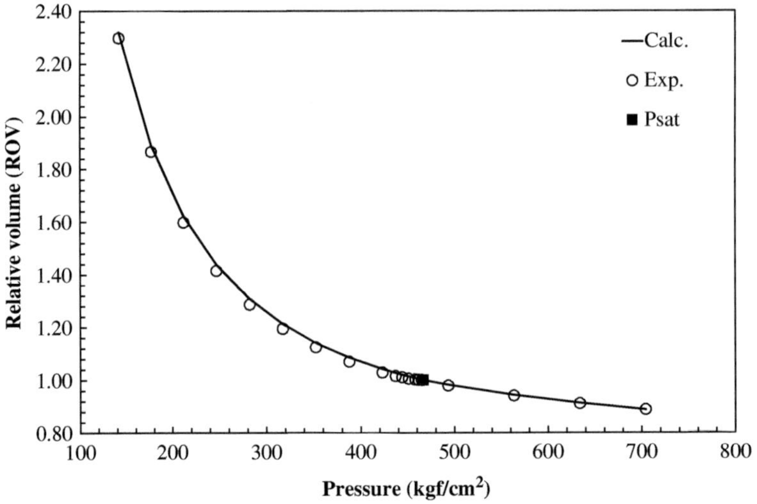

FIGURE 6.11 Relative volume in the constant composition expansion of the gas at 125°C. Theoretical values refer to the composition obtained at the sampling depth by the grading calculation.

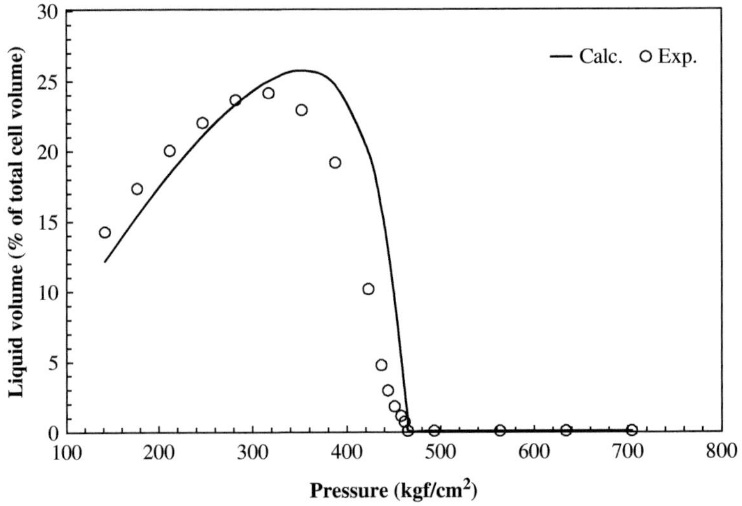

FIGURE 6.12 Retrograde liquid volume in the constant composition expansion of the gas at 125°C. Theoretical values refer to the composition obtained at the sampling depth by the grading calculation.

gravitational field. Such attenuation is slight in relation to the isothermal calculation, much weaker than the effect caused in Yufutsu, at very similar temperature, pressure, and geothermal gradient conditions. Maybe we are facing a limiting case for which, depending on the concentration of aromatics and other high molar mass hydrocarbons, thermal diffusion starts tending to act toward the gravitational field. As the occurrence of these components shows

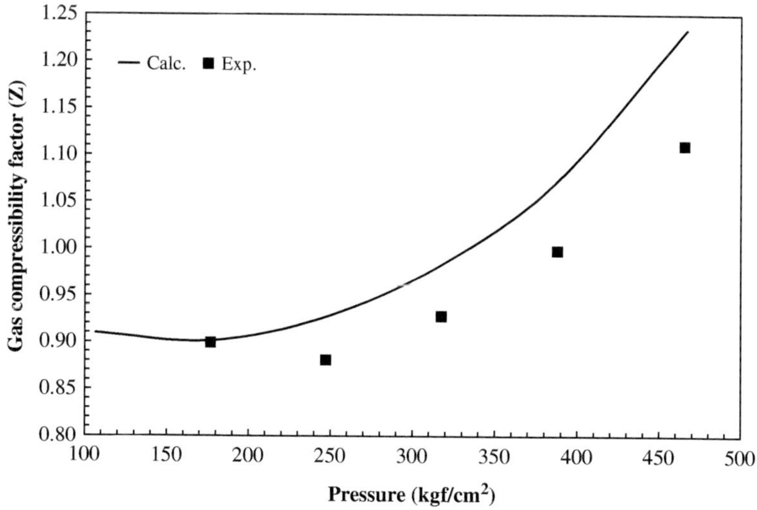

FIGURE 6.13 Compressibility factor of the gas in the constant volume depletion at 125°C. Theoretical values refer to the composition obtained at the sampling depth by the grading calculation.

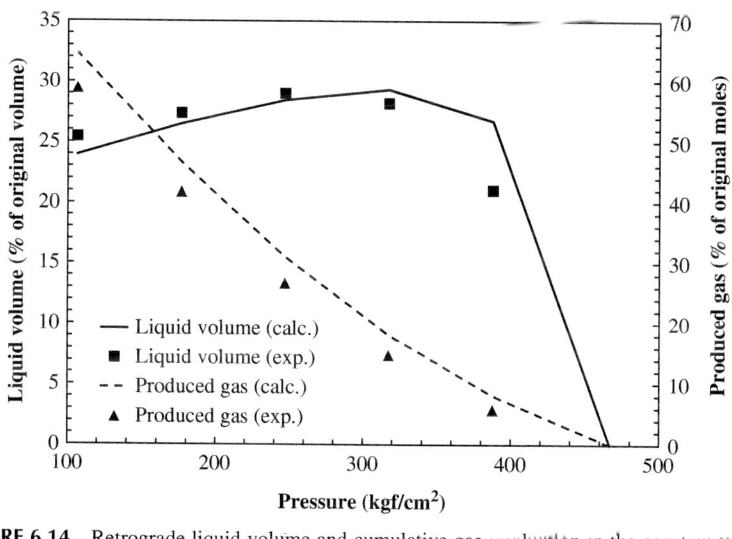

FIGURE 6.14 Retrograde liquid volume and cumulative gas production in the constant volume depletion at 125°C. Theoretical values refer to the composition obtained at the sampling depth by the grading calculation.

up mainly on the dead oil's API, we present, in Fig. 6.15, the variation of this property with depth. One notes the wide range that encompasses the 33-mark, established in example presented in Section 5.3 at the reference level of the gas-cap reservoir (far from the critical point) studied by Pedersen and Hjermstad (2006).

In fact, now considering the modeling of Pedersen and Lindeloff (2003) and their parameters' reestimation by Pedersen and Hjermstad (2006) in the calculations presented in Figs. 6.16–6.18 (density, and methane and C_{20+} mole fractions with depth, respectively), *we see a slightly enhanced grading*

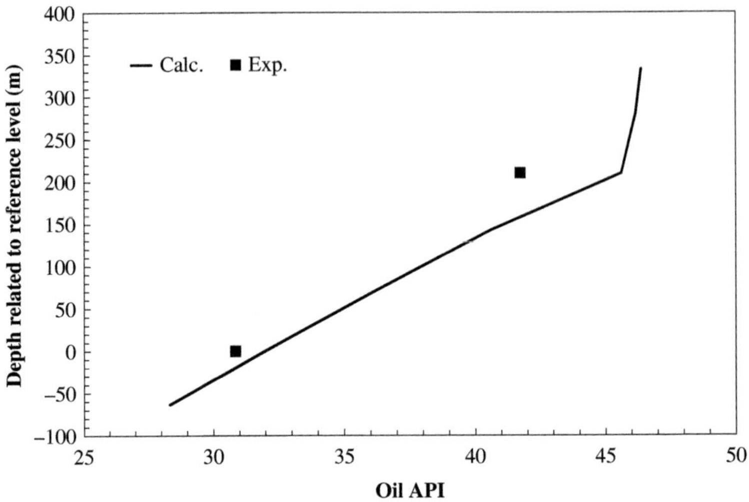

FIGURE 6.15 Dead oil (and condensate) API of flash liberation as a function of depth. Variation encompasses the mark of 33, suggested in example presented in Section 5.3 of Chapter 5 as the limit for inverting the tendency of grading attenuation by thermal diffusion.

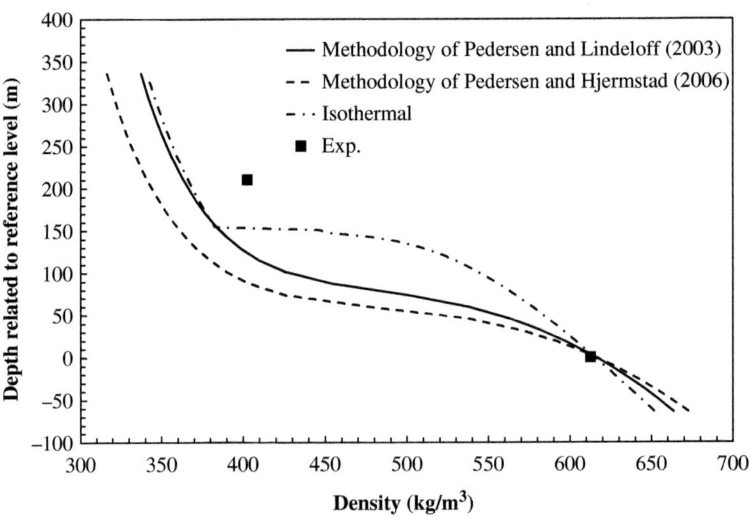

FIGURE 6.16 Density variation along depth according to the heat-of-transport modeling of Pedersen and collaborators.

compared to the isothermal calculation. Predictions of density and C_{20+} along depth are relatively exaggerated, but methane's is in better agreement than the profile obtained by the generalized Heat of Transport model of Firoozabadi et al. (2000). As to the saturation pressure profile in Fig. 6.19, it does not

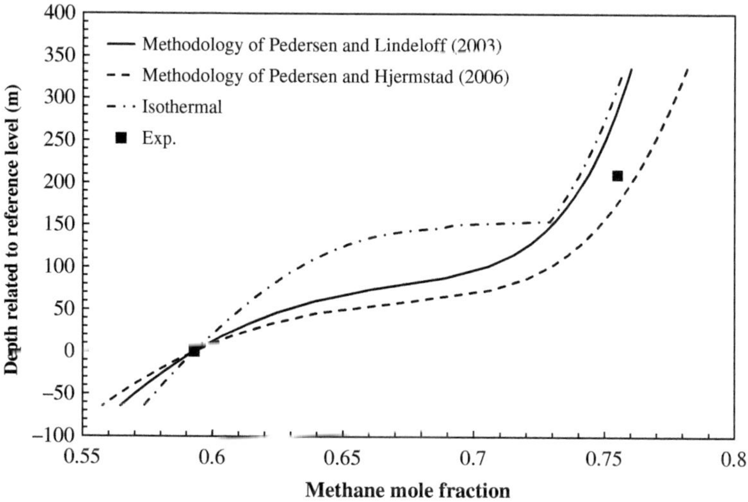

FIGURE 6.17 Methane's mole fraction along depth according to the heat-of-transport modeling of Pedersen and collaborators.

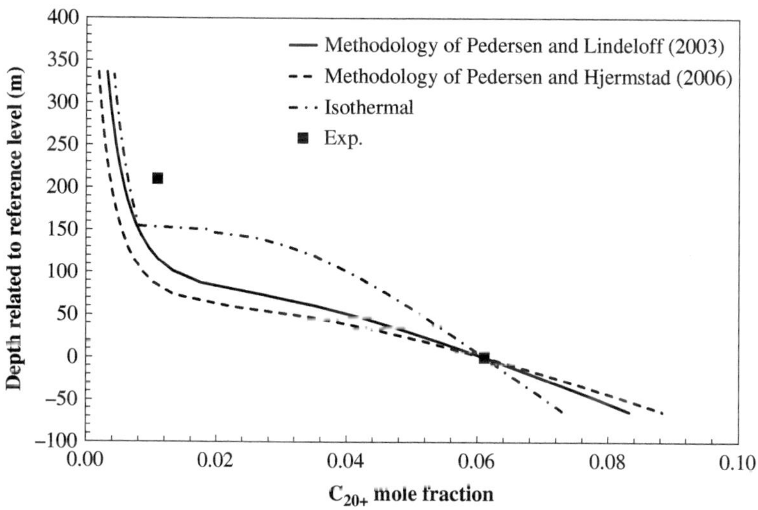

FIGURE 6.18 C_{20+} mole fraction along depth according to the heat-of-transport modeling of Pedersen and collaborators.

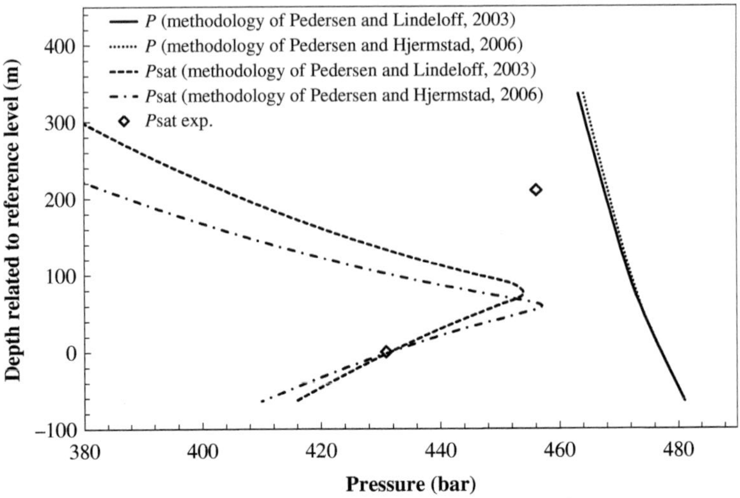

FIGURE 6.19 Saturation (P_{sat}) and static pressure (P) profiles along depth according to the heat-of-transport modeling of Pedersen and collaborators.

represent this property in the gas zone very well. The same happens with the other PVT properties, not shown. It is worth to emphasize that the necessity of reestimating the ideal gas enthalpies of Pedersen and Hjermstad (2006), due to the calculated fluid properties, does not change our interpretation that thermal diffusion contribution to the compositional grading may be at the border line of enhancing or compensating the gravitational-field action in this example.

6.2 TESTING THE CORRELATION: LOW-TEMPERATURE, 1D RESERVOIR

In this example, the correlation developed in Chapter 5 (Eqs. (5.3)−(5.5)) with the parameters presented in Eq. (6.1) is tested in the calculation of the 1D compositional grading in a colder reservoir (around 60°C) with a heavier fluid, below the 33-API mark suggested in previous examples. One expects that the relatively low temperature reduces the influence of thermal diffusion in relation to the isothermal case. We considered four testing wells, for which one collected samples along depth. Peng−Robinson equation of state, splitting the C_{20+} fraction in two pseudocomponents (quadrature points), was fitted based on the PVT analysis. Figs. 6.20−6.22 show, respectively, the experimental and calculated values of Bo, Rs, and density for the reference sample at 58.2°C. Agreement is very reasonable and was observed similarly for the other well's samples. Fig. 6.23 presents the qualitative distribution of the fluids in this reservoir, with the WOC's line and the interfaces suggested by the compositional data along depth. The zero reference level (545.8 bar and 58.2°C) was set close to the apex of the structure in

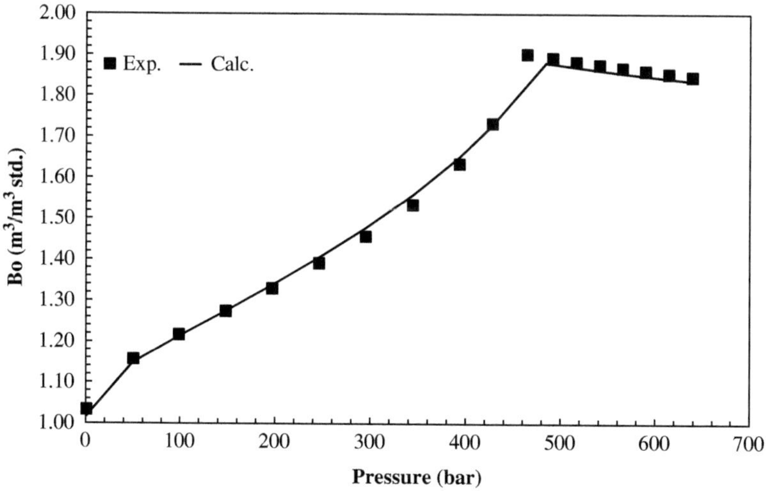

FIGURE 6.20 Experimental and calculated Bo for the reference sample in Well-1.

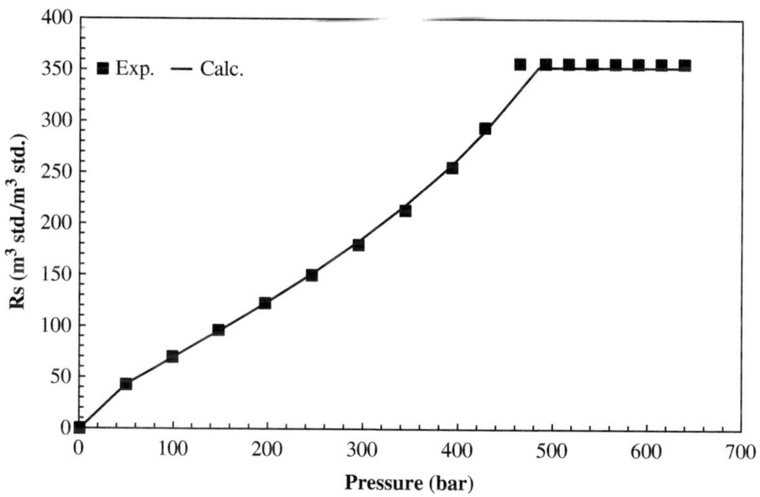

FIGURE 6.21 Experimental and calculated Rs for the reference sample in Well 1.

Well-1, where one observed a sudden increase in the methane distribution along depth, coherently followed by GOR. Same phenomenon was observed at the structural top of Well-4. Table 6.3 presents the molar masses of the pseudocomponents and the variation range of their composition with depth. C_{20+} fraction was modeled with two quadrature points, according to the methodology of Shibata et al. (1987). Results of the grading calculation for methane, C_{20+}, GOR, and API distributions (the latter two only for the actual temperature of the reservoir) are shown, respectively, in

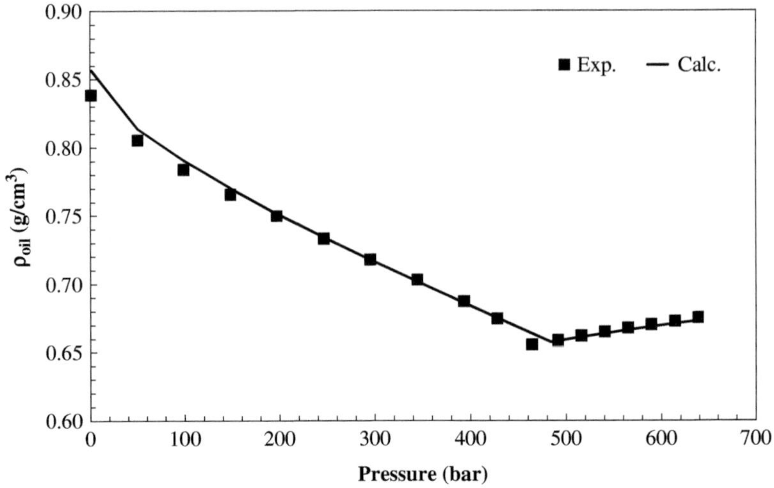

FIGURE 6.22 Experimental and calculated density for the reference sample in Well-1.

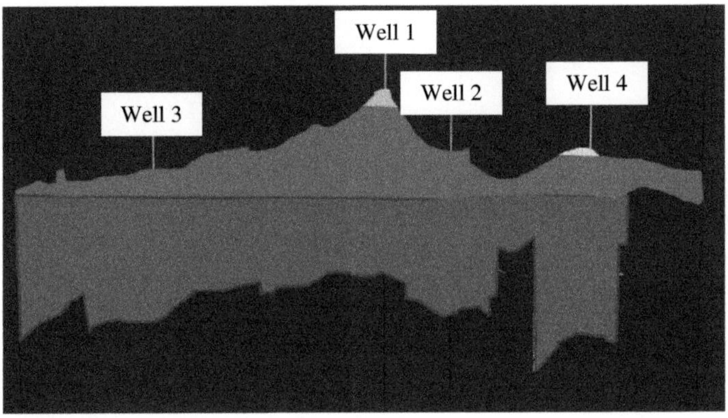

FIGURE 6.23 Well positions and qualitative fluid distribution in the studied section. We suggest a distinct methane-rich fluid on the formation tops. Right above the WOC, there is an incipient thin layer of an asphaltene-rich second liquid phase. Its modeling is beyond the scope of this example.

Figs. 6.24–6.27. Note that the abrupt variation on the slope of the distribution profiles in Wells 1 and 4 was *not* predicted by the isothermal simulation, nor using the geothermal gradient of the field (−0.025°C/m, τ_i given by our correlation with the parameters presented in Eq. (6.1)), which slightly attenuated the grading. Even when we set the exaggerated and hypothetical value of 140°C to the reference temperature, *we did not detect a possible gas cap, not even a critical transition that could reproduce the experimental compositional data.* Note, however, that this latter calculation intensified the grading

TABLE 6.3 Molar Masses and Composition Ranges Along Depth for the Fluid Adjusted With Two Quadrature Points

Component	*M* (g/mol)	Composition Range (mol %)
CO_2	44	0–1
N_2–C_1	16	60–70
C_2–nC_5	42	15–17
C_6–C_{12}	116	8–10
C_{13}–C_{19}	207	4–5
$QC_{27.0}$	381	0–5
$QC_{60.1}$	834	0–2

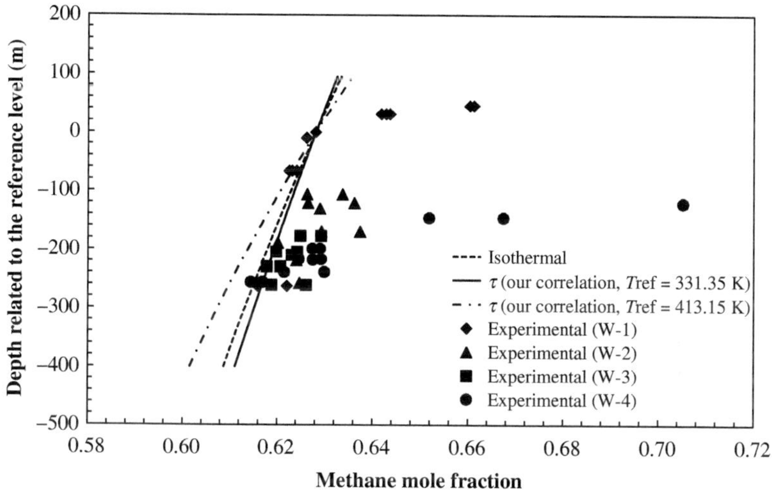

FIGURE 6.24 Methane distribution in the reservoir. Curves refer to isothermal and nonisothermal calculations at 58.2°C (331.35K), setting also a hypothetical 140°C (413.15K) reference temperature.

in relation to the isothermal case, as was expected for a lower-than 32-API oil throughout the domain. Similar results were obtained by setting the arbitrary reference level on the other testing wells. Therefore, we suggest the presence of a distinct and methane-reach fluid on structural tops, connected or not to the fluid below the reference level. In the case of no barriers, one suggests that there may have been different pulses within a geological time not long enough for steady state to be reached. We can also suggest that the influence of the interface between both reservoir fluids on the slope of

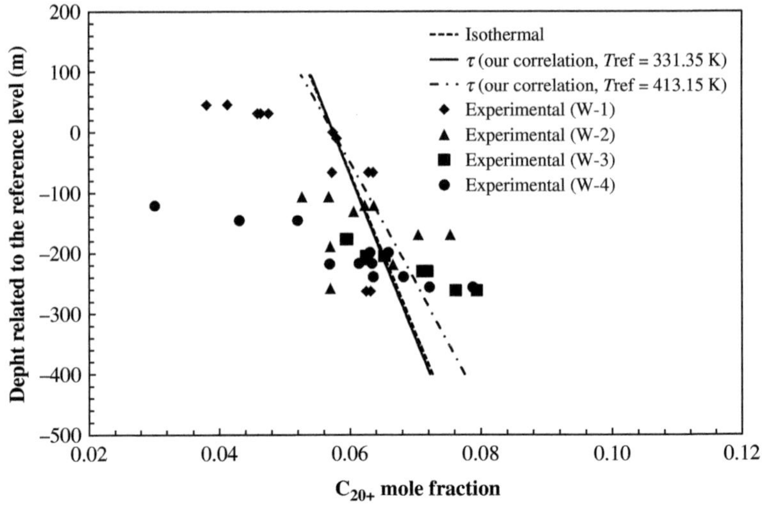

FIGURE 6.25 C_{20+} distribution in the reservoir. Curves refer to isothermal and nonisothermal calculations at 58.2°C (331.35K), setting also a hypothetical 140°C (413.15K) reference temperature.

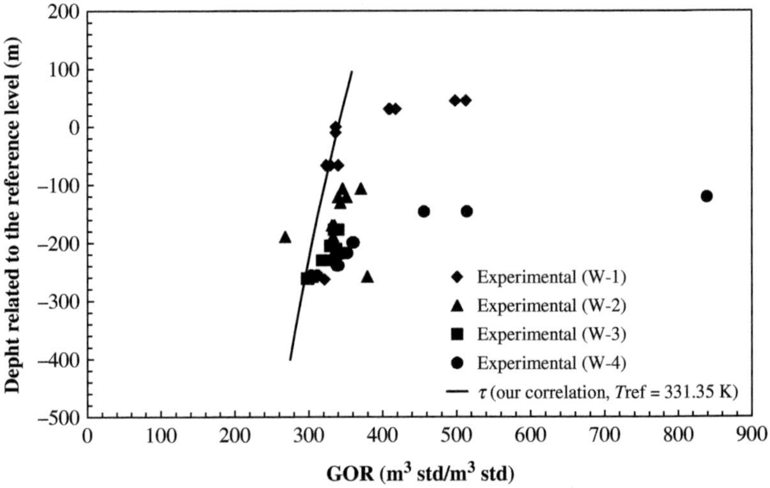

FIGURE 6.26 GOR variation along depth in the reservoir, only for the nonisothermal case setting the actual reservoir temperature (331.35K).

methane and C_{20+} profiles was more pronounced far from the apex (Wells 2, 3, and 4) due to the deeper locations. Finally, the abrupt decrease of API close to the WOC suggests also the occurrence of a thin layer made up by a possible second more viscous liquid phase, resulting likely from the asphaltene instability at the bottom of the oil zone. Due to its low thickness and

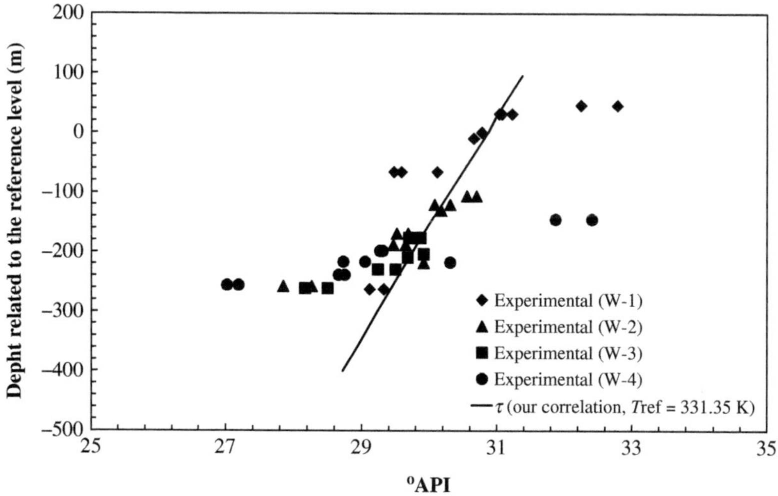

FIGURE 6.27 API variation along depth in the reservoir, only for the nonisothermal case setting the actual reservoir temperature (331.35K).

for being likely still on formation, its modeling is beyond the scope of this example. In Chapter 7, we study a three-phase system showing an asphaltene-rich second liquid phase well established at reservoir conditions, to be modeled by CPA EoS.

6.3 LOW-TEMPERATURE RESERVOIR WITH CO_2/WATER PALEOCONTACT

This example presents an interesting feature and that is why it is among the examples chosen for this book: the distribution profiles of the most important components (CO_2, CH_4, and C_{20+}) have their tendencies noticeably changed below a certain depth, named *paleocontact*. The paleocontact is defined as the depth at which the ancient CO_2-water interface was located. These were, by hypothesis, the only fluids existing in the reservoir before the oil filling. Back then, the top CO_2 may have likely carried out a so-called lights extraction to the high zone, heavies being restricted to the lower zone below the paleocontact. Due to the lower viscosity (and higher CO_2 content) of the top lighter fluid, one believes that the diffusive effects have had enough geological time for steady state to be achieved above the paleocontact, establishing (because of CO_2) a critical transition similar to those studied in our previous examples. By hypothesis, in the lower zone, the geological time has not been enough to establish steady state, due to either higher viscosity or dense CO_2 fingering. The latter would also undergo diffusive fluxes downwards, facing the resistance of both more viscous oil and lower quality rock in terms of

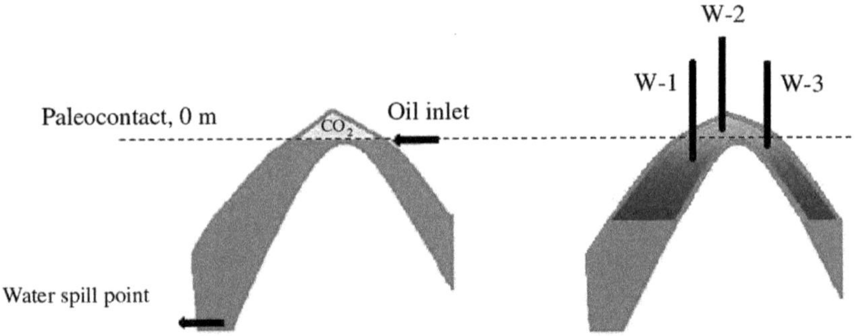

FIGURE 6.28 Qualitative fluid distribution in the reservoir before (left-hand side) and after (right-hand side) the oil filling to the trap, which displaced water through the spill point and generated a critical transition above the paleocontact due to lights extraction by CO_2. W-2 is hidden by projection and is not an apex well.

porosity and permeability. The pressure and temperature of this reservoir are around 550 bar and 60°C, at which the density of CO_2 (greater than 0.9 g/cm^3) is considerably higher than the oil, leading to irregular fronts toward the bottom (see, e.g., Moortgat et al., 2013). One takes three testing wells (W-1, W-2, and W-3) to set the reference levels for 1D calculations. Such levels will all have their depths referred to the 0-m paleocontact. Because one does not see any lateral temperature gradient due to the high thickness of the cap rock, the independent per-well calculations of compositional grading in 1D are enough to predict the profiles of all three wells above the paleocontact, using the same geothermal gradient (about −0.026°C/m) observed in the area. Fig. 6.28 shows, on the left-hand side, the CO_2-water interface before the oil filling to the trap. The right-hand side shows the current situation, including the well relative positions on projection. The WOC is deeper and the region below the paleocontact has not reached steady state yet.

There are reasons to believe that the oil has filled the trap by the east side around the paleocontact depth, that is why CO_2 concentrations decrease to the left and to the bottom. Therefore, it is necessary to carry out a simulation in geological time, using, for example, the original configuration of the section and introducing a source term responsible by the filling of CO_2-less oil and a sink term responsible for the water spill point to keep the reservoir pressure. Such a simulation would obviously be semi-quantitative, not only because of the limited 2D domain, but also due to the uncertainties of the geological time and the composition of the CO_2-free oil. The latter can be obtained through an average of the chromatography of all samples, after excluding CO_2 and renormalization of the hydrocarbon mole fractions. One expects to verify the nonvariation of composition and properties with time in the upper zone, much before the lower one, which would show CO_2 fingering toward the base, depending on the rock quality (porosity and permeability distribution along the domain).

TABLE 6.4 Molar Masses and Composition Variations Along Depth for the Fluid Adjusted with Two Quadrature Points

Components	*M* (g/mol)	Composition Ranges (mol %)		
		W-1	W-2	W-3
CO_2	44	10–21	16–18	10–16
N_2–C_1	16	48–52	48–52	47–51
C_2–C_5	43	14–16	15–17	15–18
C_6–C_{12}	139	6–8	7–9	7–9
C_{13}–C_{19}	238	2–4	3–5	3–5
QC_{26}	381	3–5	2–5	4–8
QC_{54}	693	0–3	1–3	1–4

Temperature and pressure at the reference level (paleocontact = 0 m):
W-1 (74 m): 54.8°C and 534.1 bar
W-2 (28 m): 58.3°C and 549.2 bar
W-3 (73 m): 54.8°C and 531.5 bar

Simulation of the trap filling by the oil in geological time is beyond the scope of this book. As seen in Chapter 7 and Chapter 8, solubility of CO_2 in water and the accurate calculation of the properties of the phases that will occur in local equilibrium require a thermodynamic model to describe polar components. The goal here is just to reproduce the fluid distribution above the reference level at steady state, having the paleocontact hypothesis as a premise. One observes then the performance and limitations of Peng–Robinson equation of state, which parameters were fitted to the PVT experimental data available along the entire oil zone in this reservoir.

The splitting of C_{20+} was carried out using two quadrature points, according to Shibata et al. (1987). Table 6.4 presents the composition variations of the testing wells with depth and the pseudocomponent molar masses after the EoS fitting. Figs. 6.29–6.31 show the comparisons between the experimental data and calculated values of Bo, Rs, and density in the differential liberation of the sample of Well-2. The EoS fitting has performed similarly for the other wells.

Figs. 6.32–6.34 show the distribution of CO_2 along the reservoir depth for the three testing wells, starting from their respective reference levels (zero is set at the paleocontact). We performed isothermal and nonisothermal calculations using the geothermal gradient of the field, the latter having applied both default ($\tau_i = 4$, for $i = 1,\ldots,nc$) and the correlation proposed in Chapter 5 (Eqs. (5.3)–(5.5)) with parameters given by Eq. (6.1). For CO_2, we have used $\tau_{CO_2} = 0.5$, optimized round value validated in Chapter 7 for a

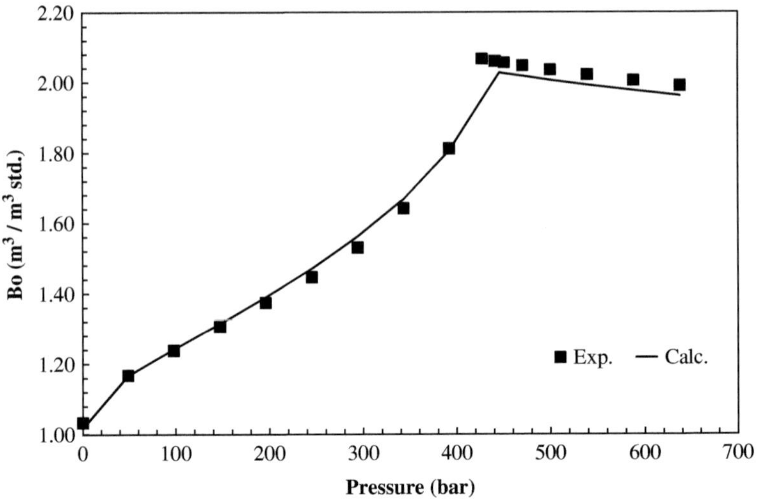

FIGURE 6.29 Comparison between calculated values and experimental data of Bo in the differential liberation of Well-2's sample at 58.3°C, after EoS tuning.

CO_2-rich reservoir. The testing wells, one at a time, have their experimental data evidenced by different symbols (■), all the rest (not shown in Fig. 6.28) represented by the same symbol (+). Figs. 6.35–6.37 and 6.38–6.40 show, respectively, the distribution of methane and C_{20+} following the same criteria.

As an analytical exercise, let us suppose that the 1D domain is investigated up to, say, 500 m above the paleocontact. Let us redo some calculations up to this level. Fig. 6.41 reproduces, for example, the methane distribution in Well-1 already shown in Fig. 6.35, but now extending the domain to evaluate the critical transition updip. We see that the shallowest sample of Well-1 (108 m above the paleocontact) already shows as a gas condensate, although the model still predicts a bubble point for its calculated composition (see comparison in Fig. 6.42). One can see also that the transition to gas condensate occurs above 170 m. Despite this distance, the predictions of properties like GOR and API still show very reasonable (always considering only the upper portion above the paleocontact), regardless the correct identification of the reservoir fluid. That is a common situation when one is dealing with critical transitions. Figs. 6.43–6.48 compare the results of calculations in the testing wells, W-1, W-2, and W-3, respectively, with the available GOR and API data of the field.

The results presented for this reservoir corroborate and help validating the hypothesis of the paleocontact existence, since one notices a considerable increase on dispersion, followed by a trend switch of some composition and property data below level zero (paleocontact). Above it, the nonisothermal 1D calculations with prefitted parameters are very reasonable, suggesting

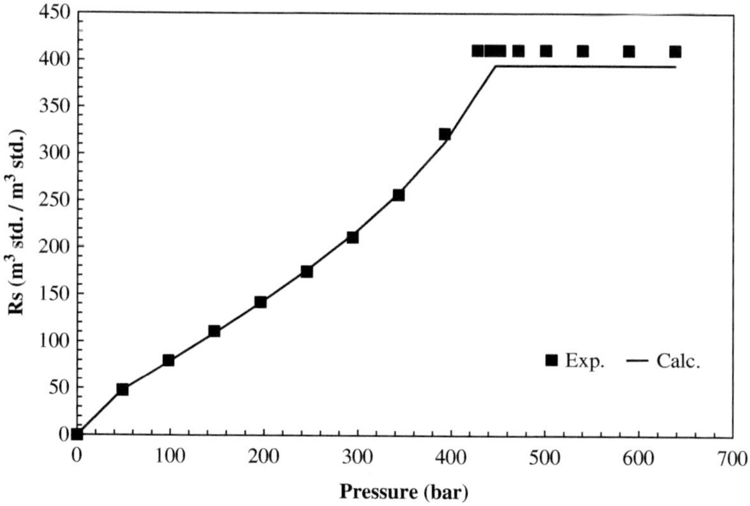

FIGURE 6.30 Comparison between calculated values and experimental data of Rs in the differential liberation of Well-2's sample at 58.3°C, after EoS tuning.

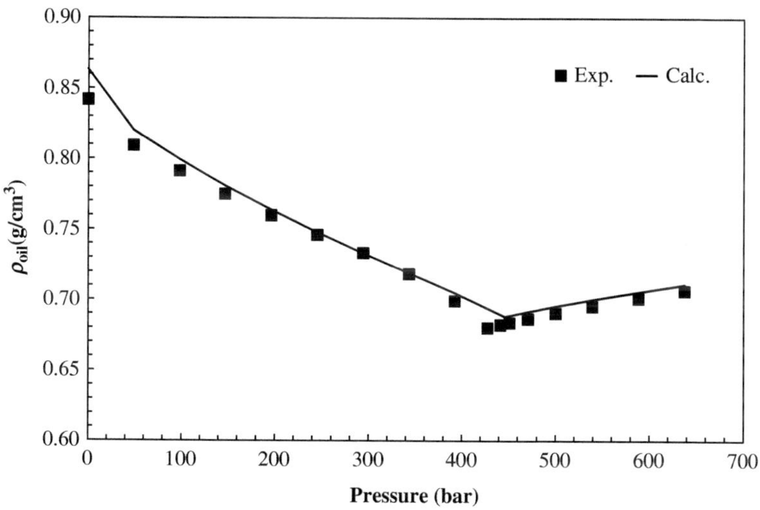

FIGURE 6.31 Comparison between calculated values and experimental data of density in the differential liberation of Well-2's sample at 58.3°C, after EoS tuning.

fully established steady state. It is possible that only part of the reservoir has achieved steady state and then only this portion can be predicted or modeled by time-invariant compositional grading algorithms. One reinforces that this diagnostic must be confirmed by performing a geological-time simulation

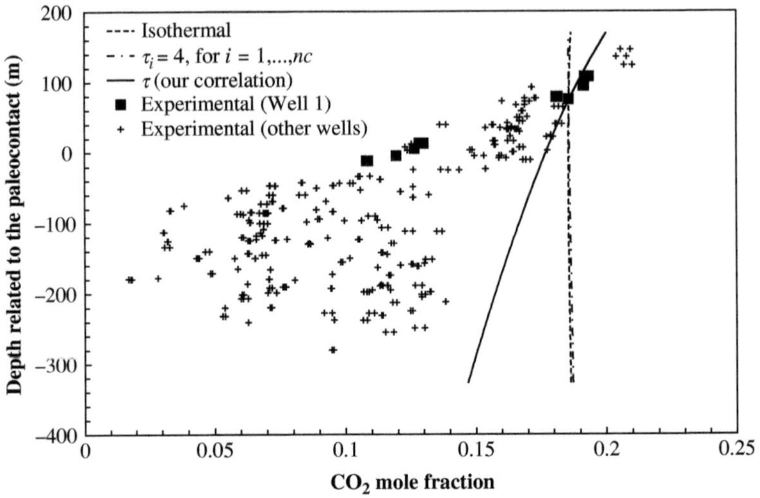

FIGURE 6.32 Distribution of CO_2 along depth for Well-1.

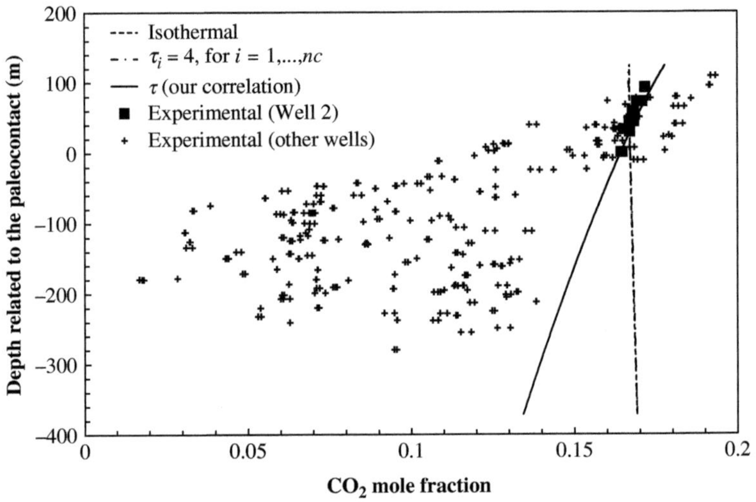

FIGURE 6.33 Distribution of CO_2 along depth for Well-2.

about trap filling with CO_2-free oil. Calculations of multiphase flash and thermodynamic modeling of polar components are needed. The composition of this oil and the order of magnitude of the elapsed time, together with the simplifications of the geological model, turn this task only semi-quantitative, just showing the time-invariant properties (steady-state achievement) on the upper portion, above level zero, much before the lower zone.

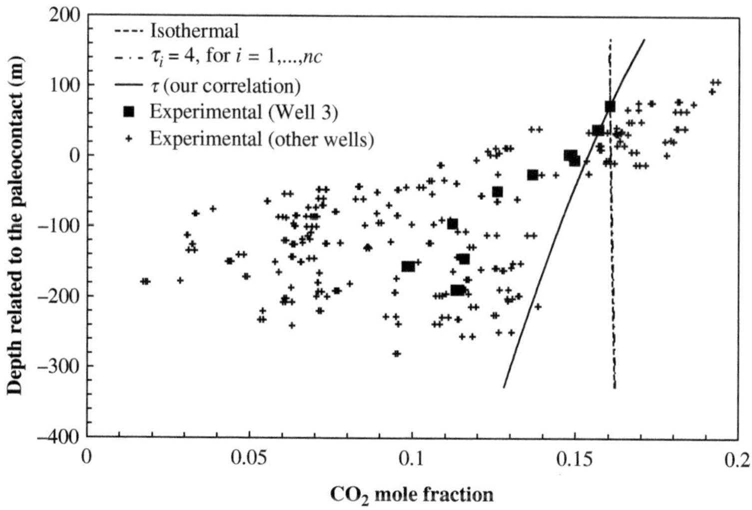

FIGURE 6.34 Distribution of CO_2 along depth for Well-3.

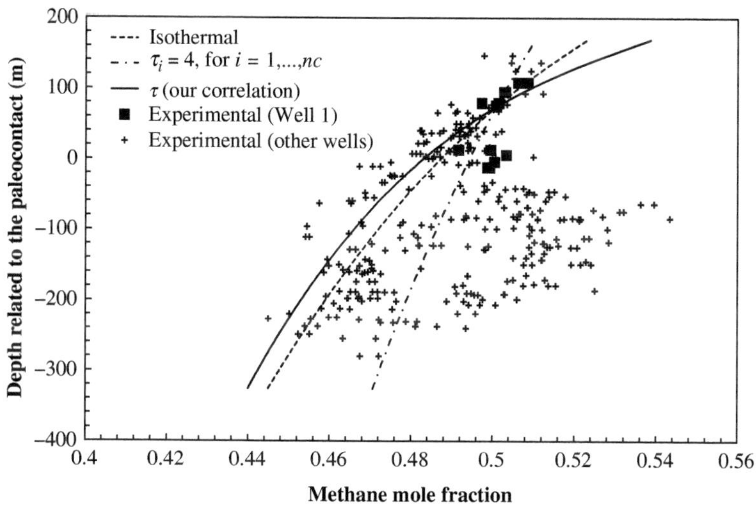

FIGURE 6.35 Distribution of CH_4 along depth for Well-1.

It is important to emphasize that, in this example, for which the maximum API is about 36 for the top condensates, reaching 30 around the paleocontact, the predictions of the thermal-diffusion model prefitted in Example 6.1 enhanced the grading obtained by the original parameters ($\tau_i = 4$, for $i = 1, \ldots, nc$) and also the isothermal model. This has improved the agreement with experimental data. This observation suggests that the presence of heavier aromatic fractions, which lead to API decrease, really

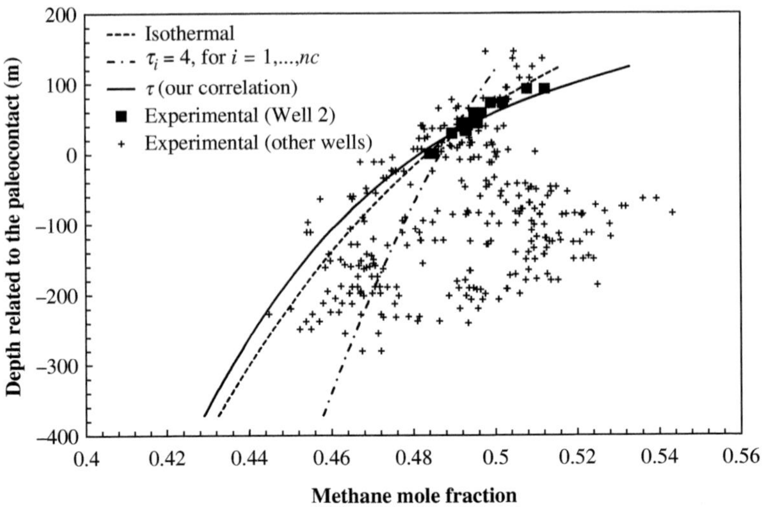

FIGURE 6.36 Distribution of CH_4 along depth for Well-2.

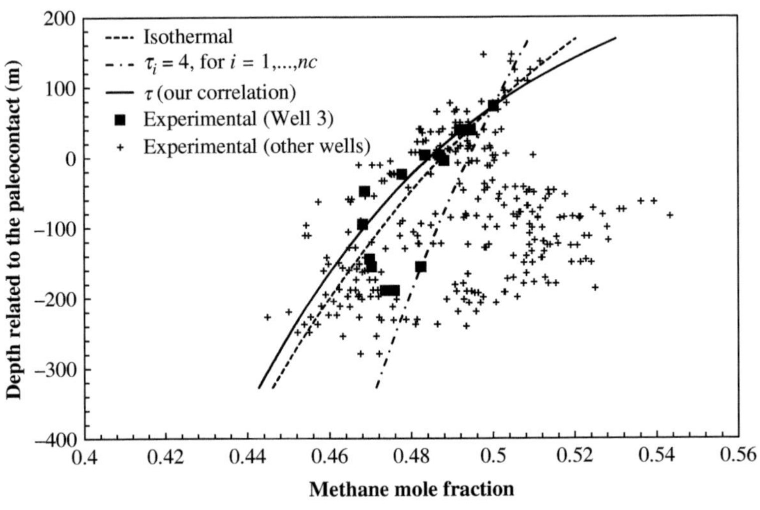

FIGURE 6.37 Distribution of CH_4 along depth for Well-3.

make thermal diffusion to start acting toward the gravitational field, enhancing segregation.

Other important point is the observation of different profiles of CO_2 mole fractions with depth *even above the paleocontact* in different wells, not observed for methane, for example. This is probably due to the solubility of CO_2 in the so-called "connate water" (the water covering the rock surface by

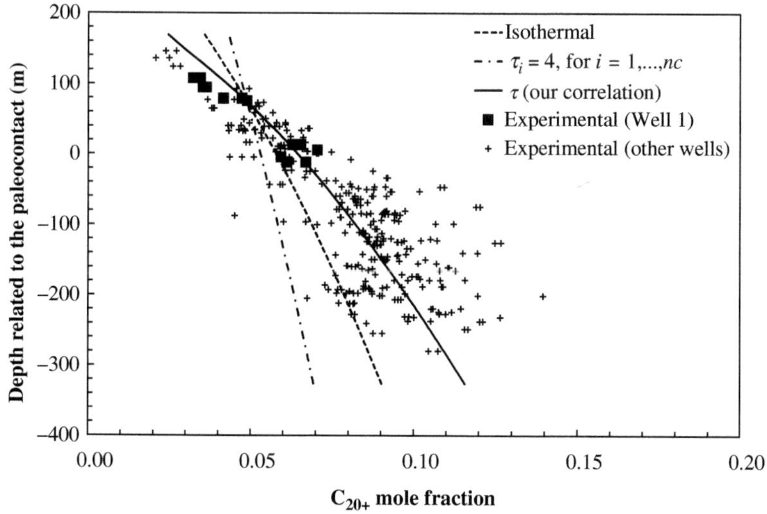

FIGURE 6.38 Distribution of C_{20+} along depth for Well-1.

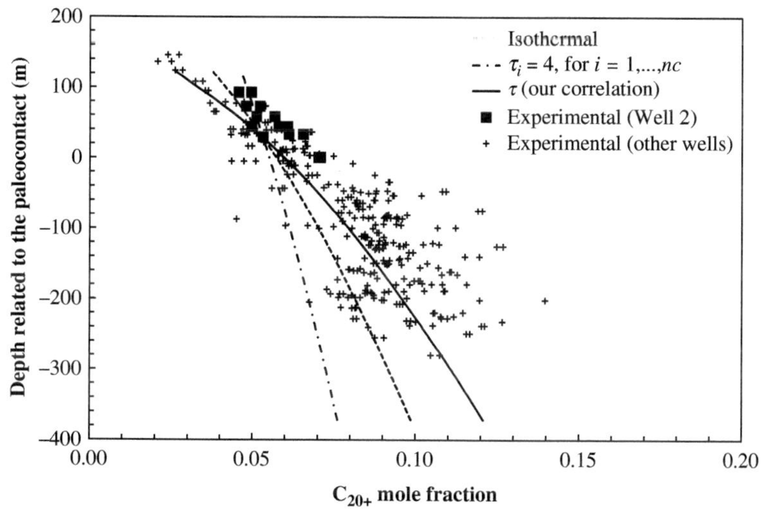

FIGURE 6.39 Distribution of C_{20+} along depth for Well-2.

adsorption and also trapped within the tightest porous of the oil zone by capillarity), enhanced, in this case, by the low temperature of the reservoir and also influenced by water salinity. The areal variation of the salinity impacts the distribution coefficient of CO_2 in an oil–water system. Such a phenomenon is negligible for most hydrocarbons, their solubility in water being around two orders of magnitude lower than the one of carbon dioxide. It can

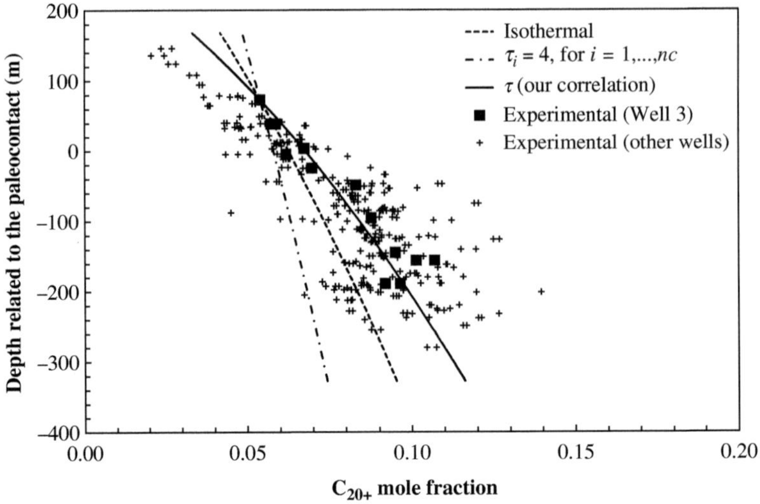

FIGURE 6.40 Distribution of C_{20+} along depth for Well-3.

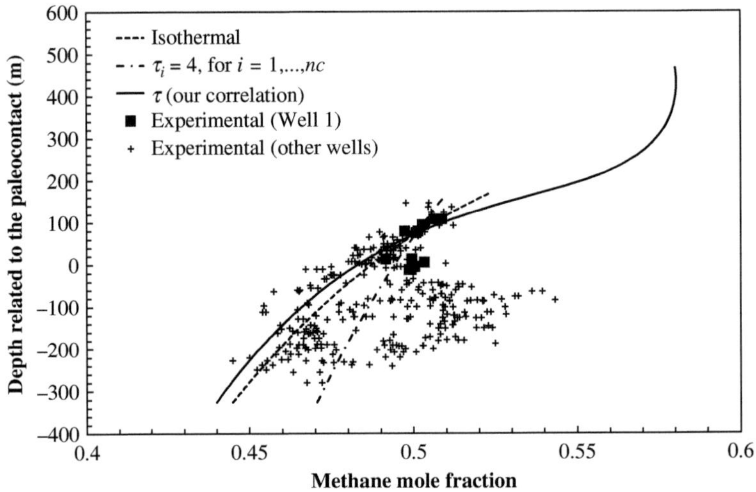

FIGURE 6.41 Complete critical transition calculated above the reservoir top, for example, in Well-1. The first dew point occurs still in porous formation, 170 m above the paleocontact. The first experimental gas condensate sample was detected at 108 m.

change their distribution profiles by increasing the dispersion. Including the connate water in the calculations of compositional grading needs the consideration of capillary forces and is beyond the scope of this book, but is stated as a future-work suggestion in Chapter 8.

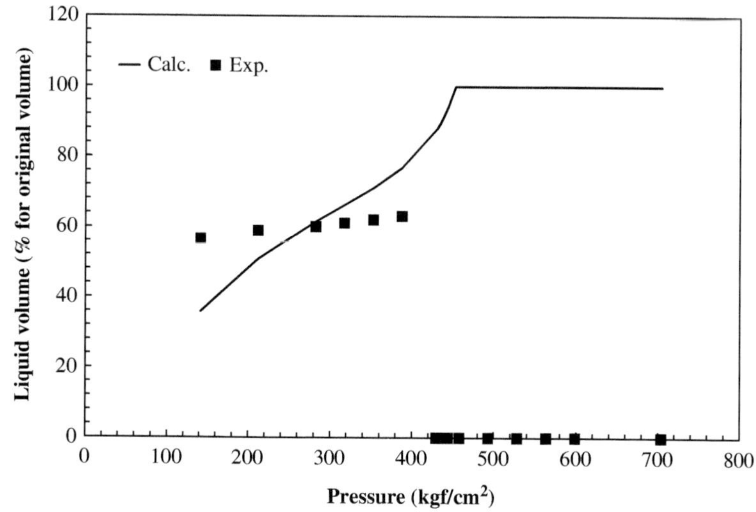

FIGURE 6.42 Comparison between the calculated and experimental liquid volumes in the CCE of Well-1, at 54°C. Calculations predict bubble point, but with good agreement on the saturation pressure value.

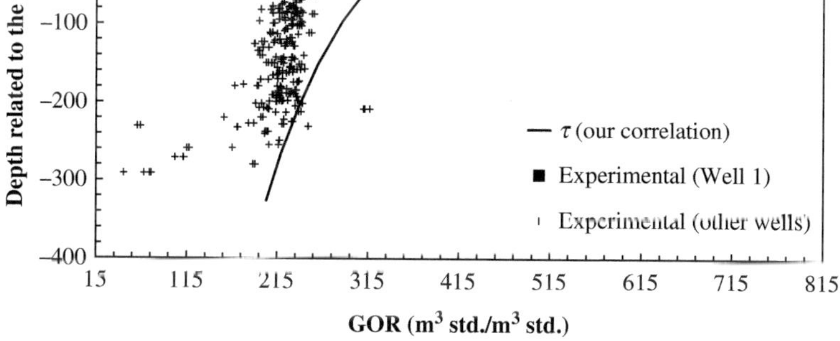

FIGURE 6.43 GOR variation along depth for Well-1, calculated with Peng–Robinson EoS. Heat-of-transport model with τ_i obtained by the proposed correlation.

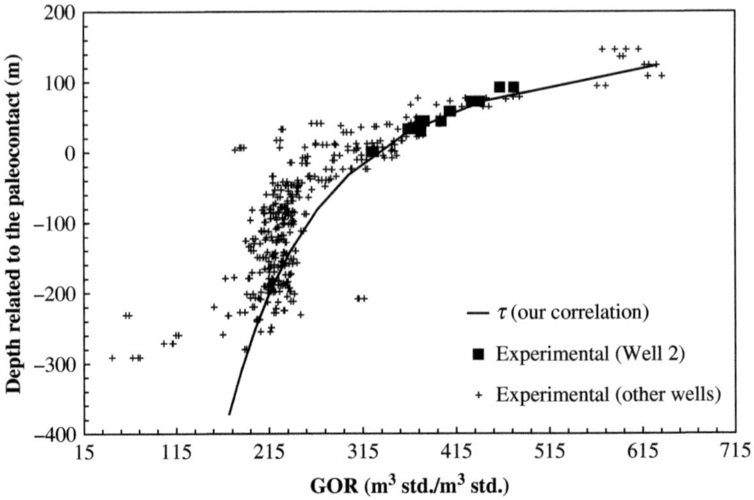

FIGURE 6.44 GOR variation along depth for Well-2, calculated with Peng–Robinson EoS. Heat-of-transport model with τ_i obtained by the proposed correlation.

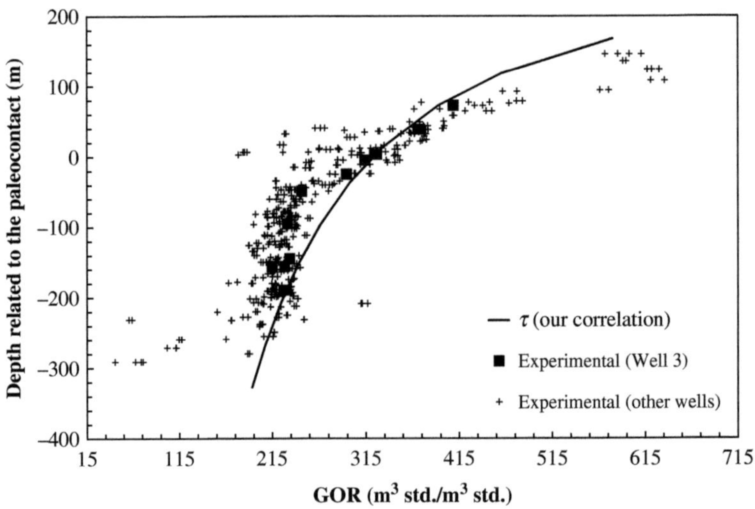

FIGURE 6.45 GOR variation along depth for Well-3, calculated with Peng–Robinson EoS. Heat-of-transport model with τ_i obtained by the proposed correlation.

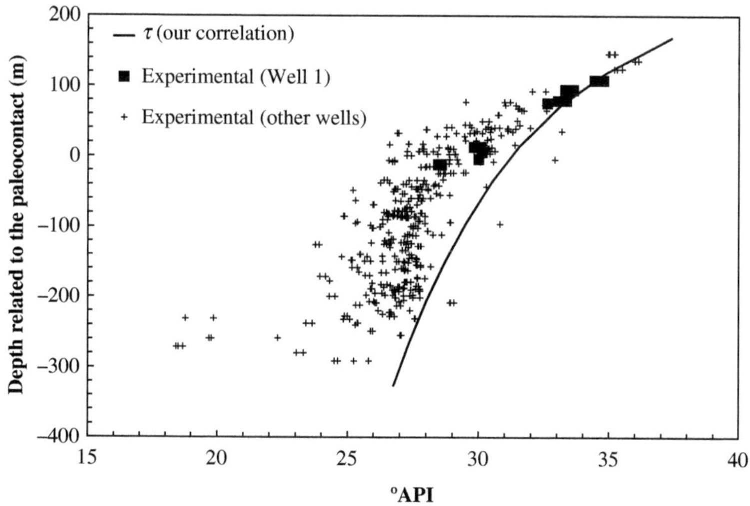

FIGURE 6.46 API variation along depth for Well-1, calculated with Peng–Robinson EoS. Heat-of-transport model with τ_i obtained by the proposed correlation.

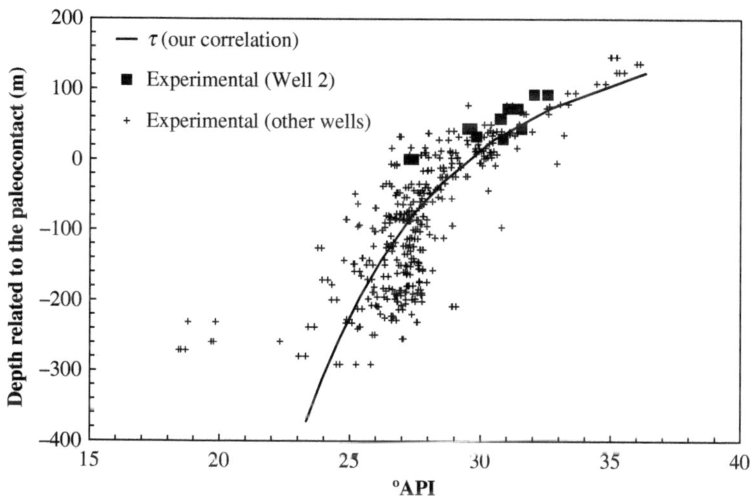

FIGURE 6.47 API variation along depth for Well-2, calculated with Peng–Robinson EoS. Heat-of-transport model with τ_i obtained by the proposed correlation.

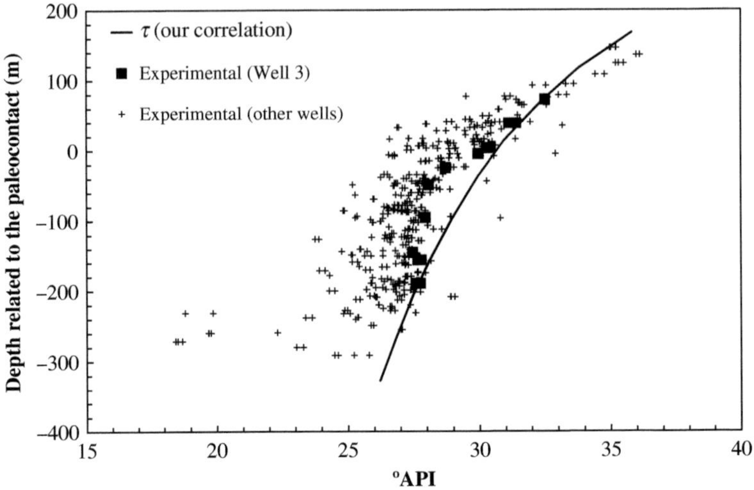

FIGURE 6.48 API variation along depth for Well-3, calculated with Peng–Robinson EoS. Heat-of-transport model with τ_i obtained by the proposed correlation.

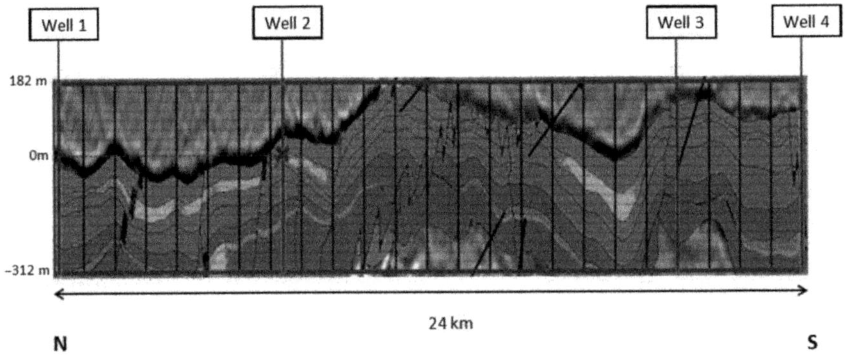

FIGURE 6.49 Section showing sediment layers and the 2D numerical domain (500 m × 24 km). Reference level on Well-2 (pointed as an "X"). Cells in the cap rock or in the basement have both porosity and permeability set to zero. The simulation is carried out only in the oil zone; that is why the domain bottom is the water–oil contact (WOC) line.

6.4 2D STEADY-STATE RESERVOIR

We now proceed to analyzing the influence of natural convection on compositional grading of oil and gas reservoirs. Consider therefore the reservoir which 2D section to be considered is shown in Fig. 6.49. It contains about 18% CO_2 and also presents a lateral temperature gradient of −1°C/km due to the increase in the cap-rock thickness from North to South. The section is around 24 km long and involves experimental data from four testing wells.

TABLE 6.5 Molar Masses and Composition Ranges for the Fluid Adjusted with Two Quadrature Points

Components	*M* (g/mol)	Composition Range (mol %)
CO_2	44	17–19
N_2–C_1	17	43–46
C_2–C_3	36	10–12
iC_4–C_7	59	5–8
C_8–C_{12}	151	5–8
C_{13}–C_{19}	231	5–7
$QC_{29.2}$	360	4–7
$QC_{63.2}$	815	2–4

Temperature and pressure at the reference level: 94°C and 617.8 bar.

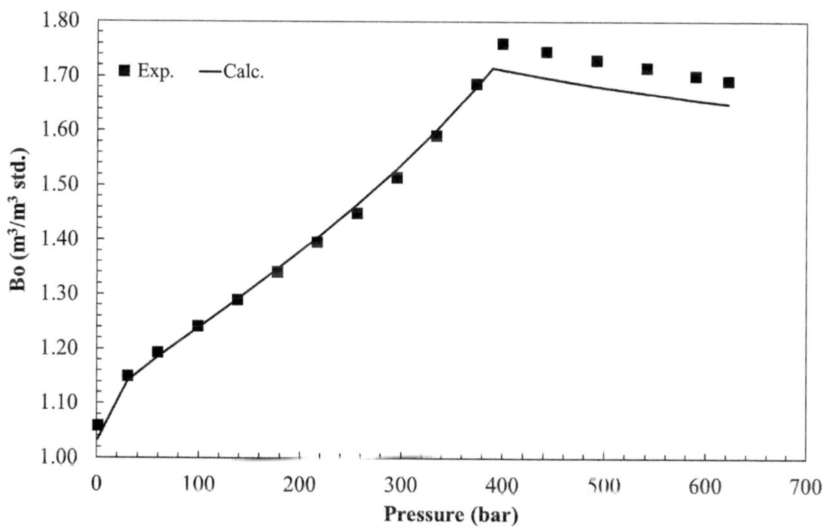

FIGURE 6.50 Calculated and experimental Bo for the reference sample at Well-2.

Porosity and permeability distributions are not available. Average values around 10% and 100 mD can be adjusted in order to reproduce the grading data on the testing wells. Table 6.5 shows molar masses and composition-variation ranges along the reservoir for the fluid adjusted with two quadrature points. Figs. 6.50–6.53 show, respectively, the experimental and

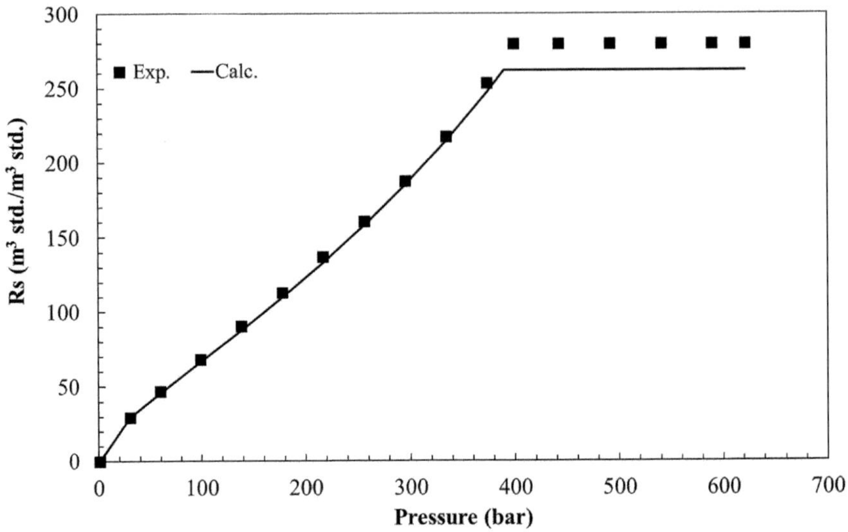

FIGURE 6.51 Calculated and experimental Rs for the reference sample at Well-2.

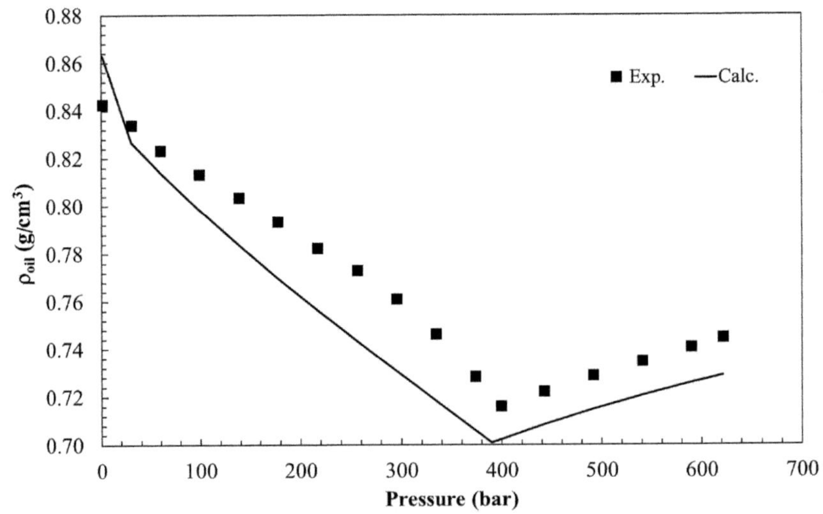

FIGURE 6.52 Calculated and experimental density for the reference sample at Well-2.

calculated values of Bo, Rs, density and viscosity for the reference sample in Well-2, which location in the domain is represented by an "X" in Fig. 6.49 (reference level at Well-2, 0 m). Results are very similar to the other wells.

Setting pressure and temperature conditions for the reference level in Well-2 at 94°C and 617.8 bar, we proceed initially to the 1D isothermal

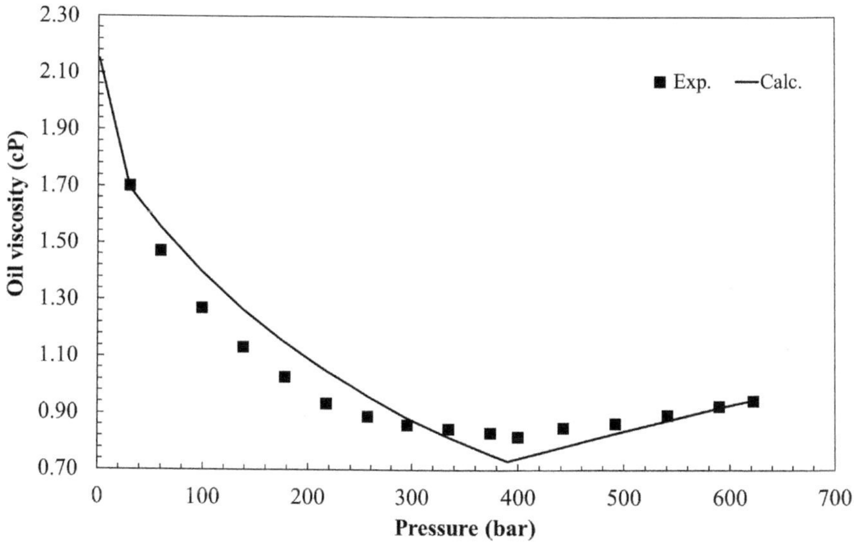

FIGURE 6.53 Calculated and experimental oil viscosity for the reference sample at Well-2.

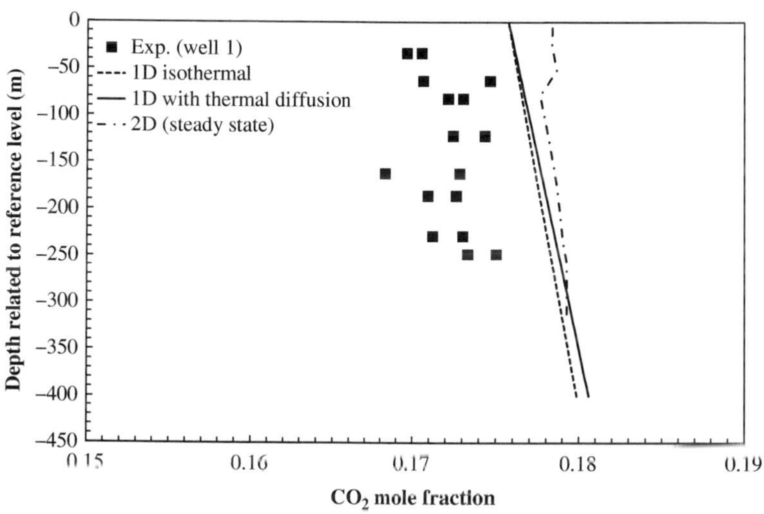

FIGURE 6.54 CO_2 distribution along depth in Well-1.

calculations. Then, neglecting the lateral temperature gradient, we repeat the 1D calculations using the ordinary vertical one (−0.025°C/m) and original values ($\tau_i = 4$, for $i = 1, \ldots, nc$) for the Heat of Transport model. Keeping the same reference, we proceed then to the simulation of fluid moving in 2D and geological time till one reaches steady state, sticking to the same original thermal-diffusion model. Finally, we compare the results obtained by the three types of calculation with the experimental data at the wells' coordinates.

The objective of this example is therefore to verify the influence of natural convection (2D simulation with lateral temperature gradient) on the fluid

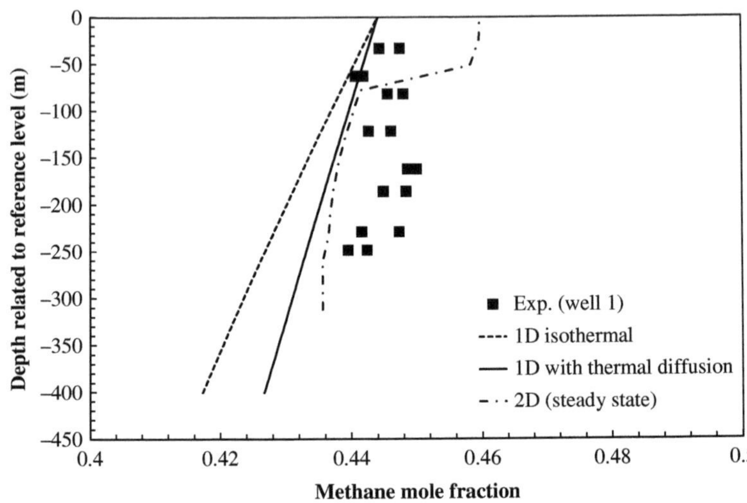

FIGURE 6.55 Methane distribution along depth in Well-1.

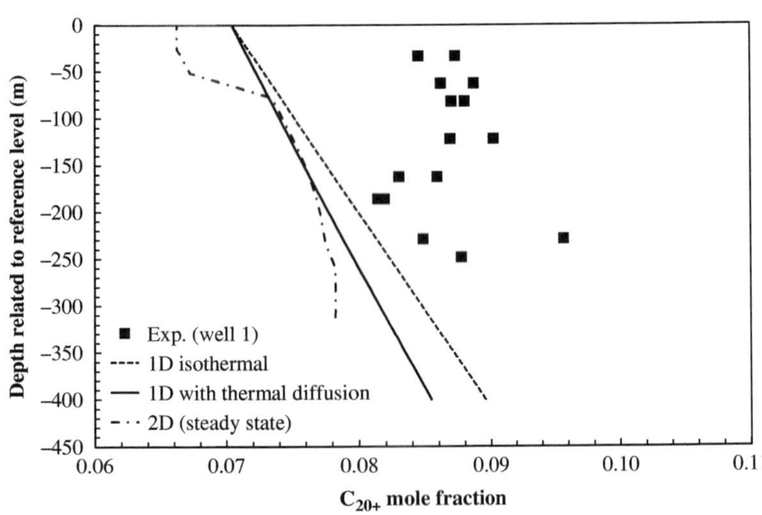

FIGURE 6.56 C_{20+} distribution along depth in Well-1.

distribution originally predicted by the 1D calculation, using only the vertical geothermal gradient. It is important to remark the premise that global connection is assured, low-permeability (nonreservoir) regions in the domain not having full lateral extension, i.e., do not propagate throughout the section.

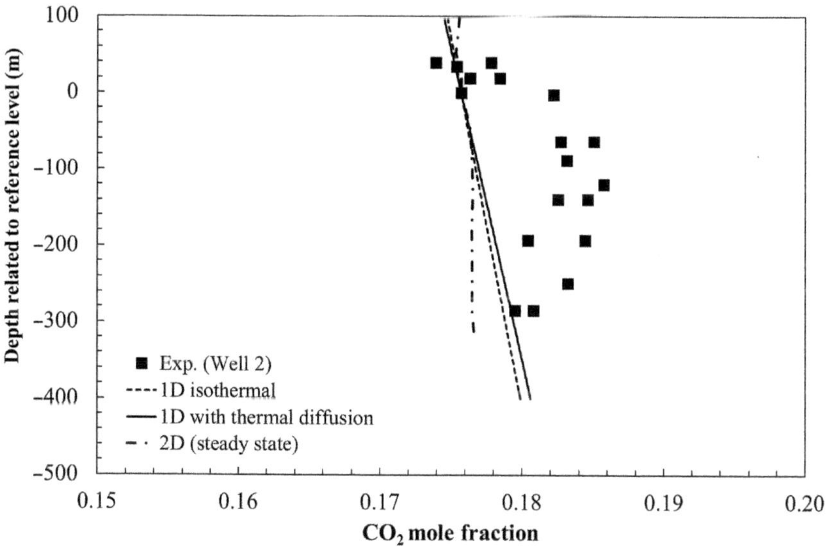

FIGURE 6.57 CO_2 distribution along depth in Well-2 (ref).

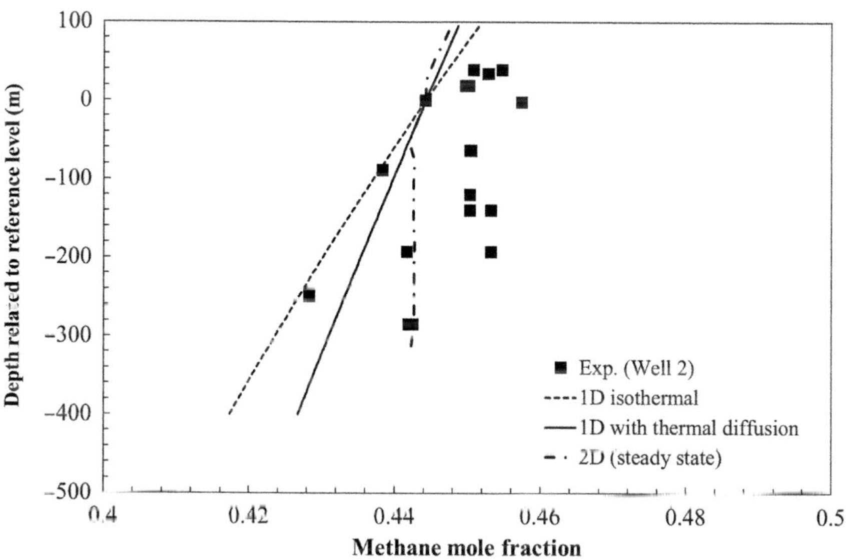

FIGURE 6.58 Methane's distribution along depth in Well-2 (ref).

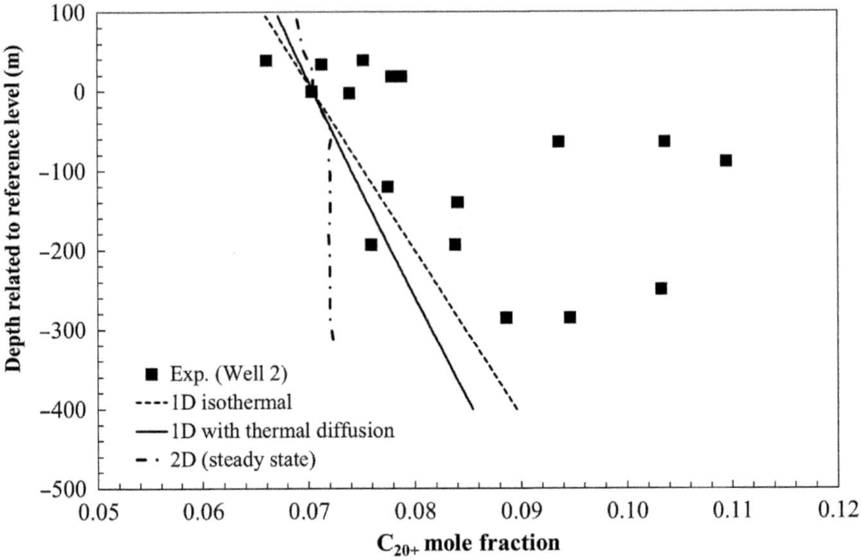

FIGURE 6.59 C_{20+} distribution along depth in Well-2 (ref).

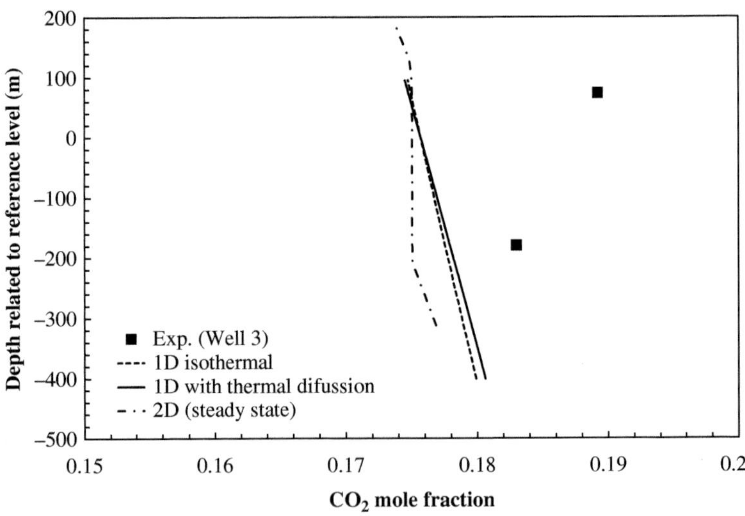

FIGURE 6.60 CO_2 distribution along depth in Well-3.

Simulations show that the reservoir is practically homogeneous after around 10 million years, suggesting that the steady state is achieved much before the current age of the reservoir (approximately 100 million years).

Figs. 6.54–6.56 show the comparison between experimental and simulated mole fractions of CO_2, CH_4, and C_{20+}, respectively, for isothermal,

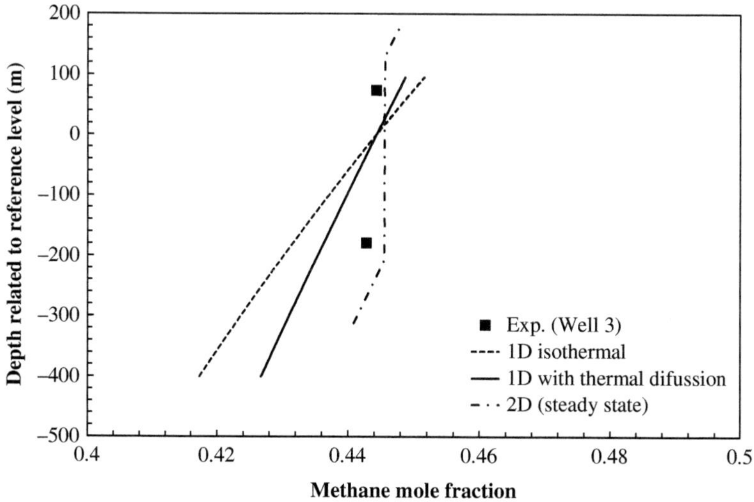

FIGURE 6.61 Methane distribution along depth in Well-3.

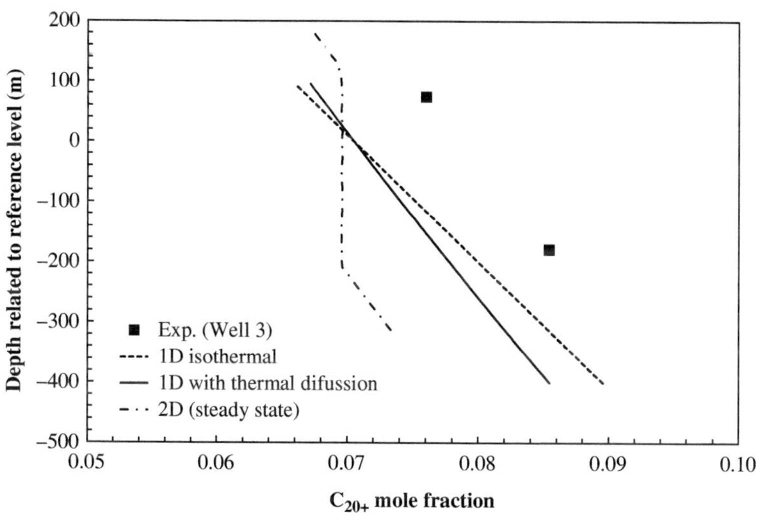

FIGURE 6.62 C_{20+} distribution along depth in Well-3.

nonisothermal 1D and nonisothermal 2D along the depth of Well-1. Figs. 6.57–6.59 show the same kind of comparison for Well-2. Figs. 6.60–6.62 refer to Well-3. Figs. 6.63–6.65 refer to Well-4. In general, one can conclude that natural convection tends, with the same thermal-diffusion parameters, to homogenize the reservoir, providing results in reasonable agreement with experimental data. Anomalies like the decrease in

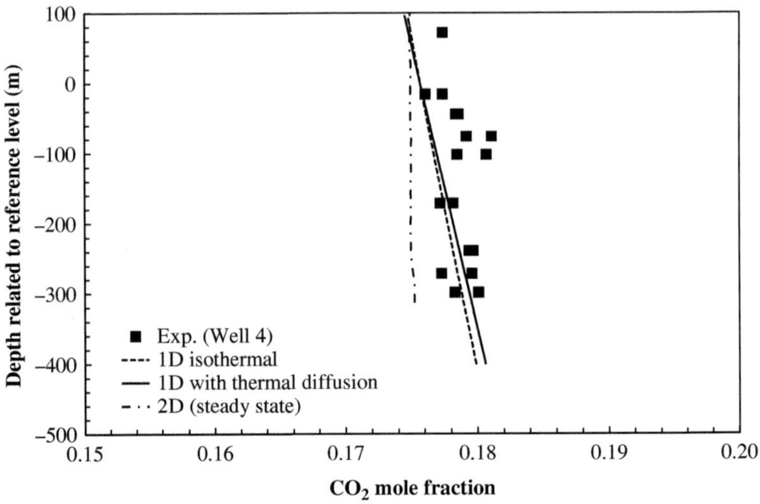

FIGURE 6.63 CO_2 distribution along depth in Well-4.

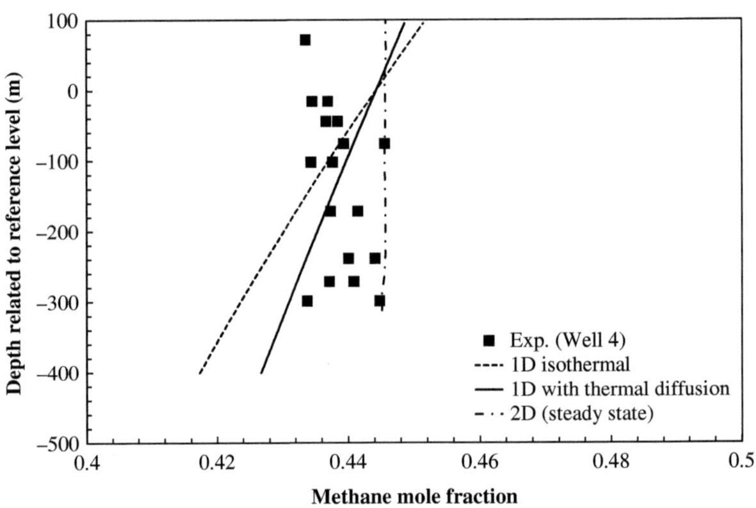

FIGURE 6.64 Methane distribution along depth in Well-4.

C_{20+} with depth in Well 4 could not be reproduced. Simulated mole fractions of this heavy fraction kept always below 8% mol, whereas the experimental data, in spite of dispersion, showed concentrations always above this value, except in the high zone of Well-2, where reference was arbitrarily set. It is worthy to report that asphaltene precipitation was observed in some further samples of this field by depletion (laboratory onset curves). Also, thermal-diffusion parameters of CO_2 might behave differently from the ones

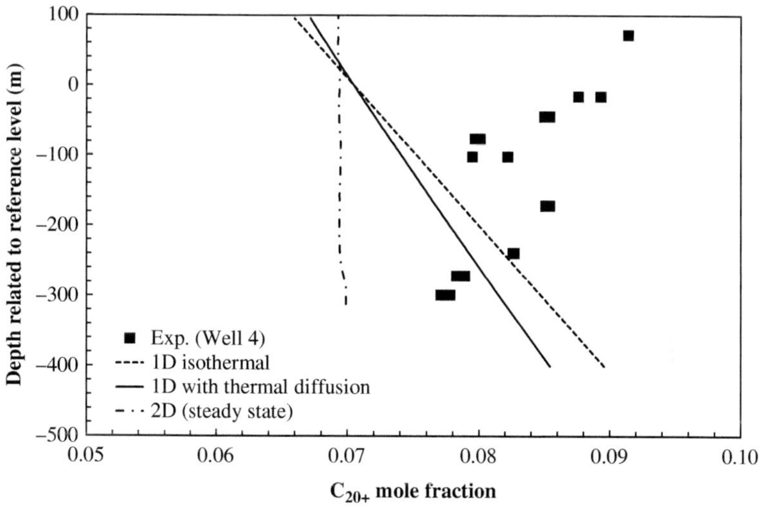

FIGURE 6.65 C_{20+} distribution along depth in Well-4.

of hydrocarbons. Reformulation of both thermodynamic and heat-of-transport models, specifically for this field, is required.

6.5 CONNECTIVITY IN A 2D RESERVOIR

During the development phase of a field, when one can count on some drilled wells, the hydraulic connection between reservoir fluids might be detected through the so-called interference tests: with one of the wells undergoing a formation test, that is under controlled flow for data acquiring, one must install a pressure meter in the adjacent wells and check whether or not one has depletion, which means local pressure drop as a consequence of mass withdrawing somewhere else. However, when one is still in exploratory phase, connectivity among different areas of a reservoir can only be inferred by geological data. At this moment, informations from fluid modeling can contribute valuably, corroborating other tools for a suitable prediction of compartmentalization or connection throughout the reservoir. Abrupt variations in grading tendencies, together with a poor agreement of PVT analysis around reference, might suggest barriers, filling/leakage on the go or even oils from different supplies coexisting in the trap. Unfortunately, these statements are not always clear and straight. We saw that thermal diffusion can change the fluid distribution with higher or lower intensity, but also natural convection, caused by lateral temperature gradients, tends generally to homogenize the fluid by its round moving. In some situations (still not fully comprehended in the literature), the fluid movement in steady state can enhance the thermal diffusion effects, turning the distribution of the components more heterogeneous, which cannot be predicted by isothermal or 1D calculations. Let us see if that happens in the reservoir sketched in Fig. 6.66. It is a deposition of a more recent sandstone layer (B) on a west portion of an older, and sandstone as well, reservoir (A). One wants to know whether the interface between both sands is permeable, so to ensure the connection between the fluids of Wells-2 (volatile oil, reference) and 1 (gas condensate). Although belonging to the same reservoir as the reference fluid, Well-3 (volatile oil) is at west of a permeability barrier which one also intends to classify like sealing or nonsealing. Due to the differences in both water depth and burial among the three considered wells, there is a lateral temperature gradient of −0.0013 K/m from West to East. The vertical thermal gradient in the whole area is −0.034 K/m upwards. The zooming of the cut region emphasizing reservoirs A and B in the simulation domain is shown in Figs. 6.67 and 6.68.

Assuming total connectivity throughout the numerical domain, one of the premises is a mean constant porosity (16.5%). Permeability estimated for the region around Well-3 is 265 mD. Around Well-2, one has 110 mD. For reservoir B around Well-1, we set 41 mD. Permeability of reservoir A, below B, in the left-hand side region of the domain is 94 mD. The reference level is in the midpoint of Well-2's perforated interval (which depth is arbitrarily set to zero). The temperature and pressure of this level are, respectively, 106°C and 569 bar.

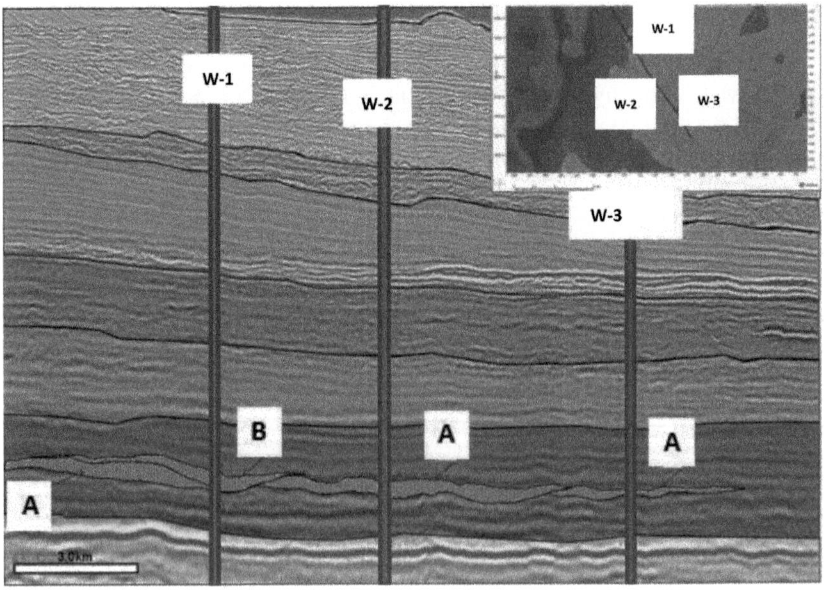

FIGURE 6.66 Top map of Reservoir A and 2D-section throughout depth from sea level, with sediment layers on our 2D domain. One wants to know whether there is connectivity over all the extension of Reservoirs A and B.

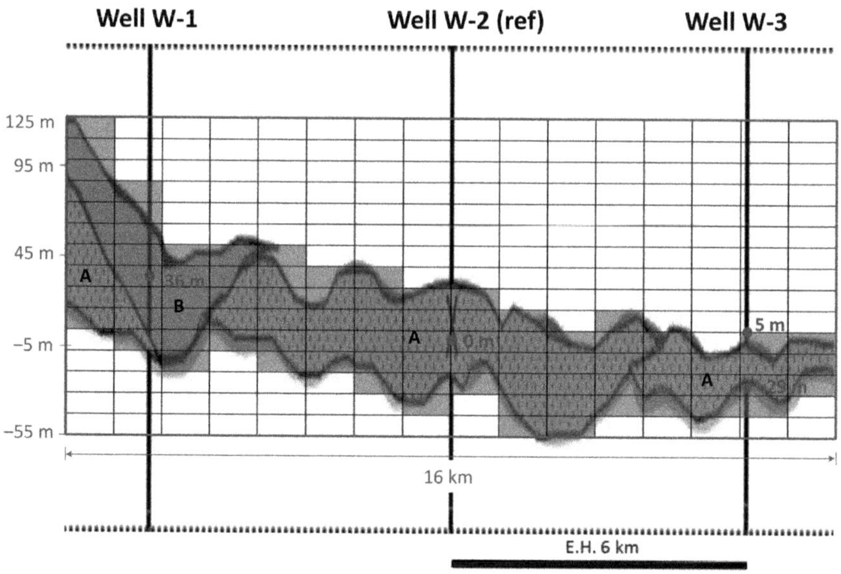

FIGURE 6.67 Zoom of the numerical simulation domain in Reservoirs A and B, locating the reference level in Well-2 (0 m) and testing sampled levels in Well-1 (36 m) and Well-3 (−29 and 5 m). Simulation will assume global connectivity throughout the domain. Red and magenta lines make up seismic interpretations, with uncertainties associated to the limits of the reservoirs, which are propagated by the zoom. Note that the samples of Well-3 refer to both top and bottom of the reservoir in that region. The blue cells will be given the constant value of 16.5% for porosity. Permeability distribution is detailed in Fig. 6.68. The white cells (off reservoir limits) will have both permeability and porosity set to zero.

Table 6.6 shows the experimental mole fractions of methane and C_{20+} of the reference and the other three testing levels, for which one intends to compare PVT properties calculated by grading with the ones measured in samples of Wells-1 (at 36 m) and 3 (−29 and 5 m), suggesting connection or compartmentalization according to the quality of the simulated results. One reinforce that, naturally, these reservoir fluid contains other fractions, which must be considered for its suitable characterization. However, we preferred to show in Table 6.6 only methane and C_{20+} compositions, since these are the most relevant components to support the discussions showed in this example.

Note that the sample of Well-1, besides being higher than the reference level, is also located in a higher temperature zone, which would support its classification as a gas condensate in a critical transition with the volatile oil of Well-2. As to the fluid of Well-3, being a volatile oil coherently located in a colder portion of the reservoir suggests higher concentration of heavies in that region; see its C_{20+} mole fraction, considerably greater than the other wells'. Curiously, the molar mass of this C_{20+} is also considerably less, API

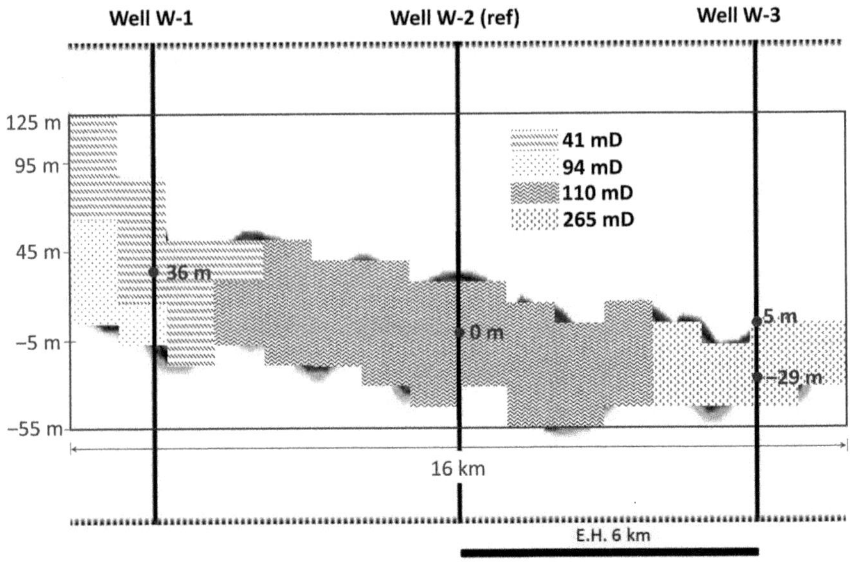

FIGURE 6.68 Permeability distribution along the reservoir: well-3, 265 mD; Well-2, 110 mD; Well-1, 41 mD; Reservoir A (below Reservoir B, not drilled), 94 mD. Note also the location of both reference and testing levels.

TABLE 6.6 Mole Fractions (in mol %) of Methane and C_{20+} and Properties of *Flash* Liberation at Reference and Testing Levels. Although Only C_1 and C_{20+} Compositions are Presented, The Fluid Contains also Other Typical Petroleum Fractions

Property	W-2 (0 m, Reference)	W-1 (−4 m)	W-3 (−44 m)	W-3 (−35 m)
C_1 composition (mol %)	63	71	60	60
C_{20+} composition (mol %)	3.4	3.4	5.4	5.3
GOR (m^3 std/m^3 std)	468	621	431	463
API	36	35	37	37
M of C_{20+} (g/mol)	481	444	364	359
ρ of C_{20+} (g/cm)	0.922	0.904	0.891	0.889
Classification	Volatile oil	Gas condensate	Volatile oil	Volatile oil

increasing consequently. Once confirming connection throughout the areas, would the lighter fractions of C_{20+} migrate to the colder region rather than the heavier ones? Would the fluid of Well-3 be a gas condensate if not for the temperature, relatively low? Would the fluid of Well-1 be a volatile oil if not for the temperature, relatively high? In the following, the results of the simulations assuming total connectivity are shown, using the thermal-diffusion model with parameters adjusted in the example presented in Section 6.1.

Figs. 6.69–6.72 show, respectively, the comparison between experimental and calculated (using three quadrature points) values of Bo, Rs, density and viscosity, all from PVT analysis of the reference sample in Well-2, which was carried out at 111°C, a little bit above the reference-level temperature. The agreement is excellent, minimizing the uncertainty with respect to the thermodynamic modeling.

The geological age of Reservoirs A and B is around 75 million years. The simulations reached steady state (that is one observed no more property variations with time) before 30 million years. So, any deviation relative to composition and properties calculated at the testing levels is due to barriers or filling/leaking in the trap by other reservoir fluids. Fig. 6.73 shows methane's distribution along the simulation domain at 30 million years (A) and 100 million years (B). Practically, there is no difference. Figs. 6.74–6.77 show methane and C_{20+} profiles along depth at the horizontal coordinates of Wells-3 and 1, respectively. Regarding methane content, we can state that it is not likely that Reservoir B is connected to Reservoir A, because Well-1 shows more than 70 moles% methane in the reservoir fluid, while simulation results indicate

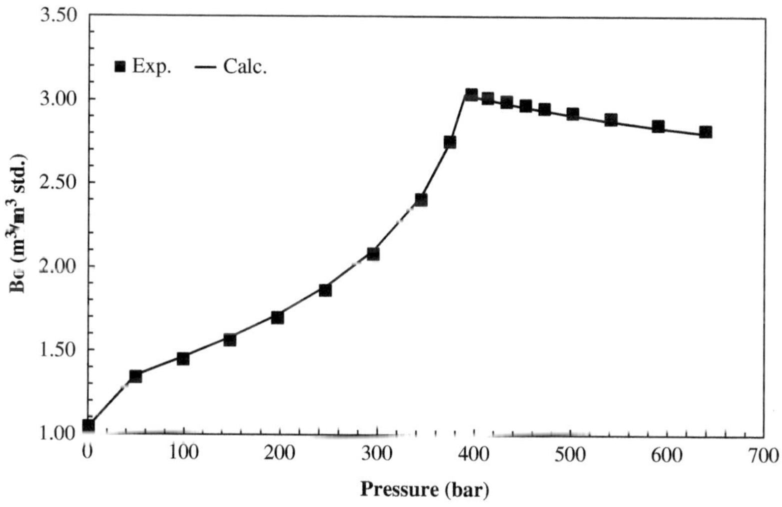

FIGURE 6.69 Bo in differential liberation at 111°C for the fluid of Well-2 at 0 m.

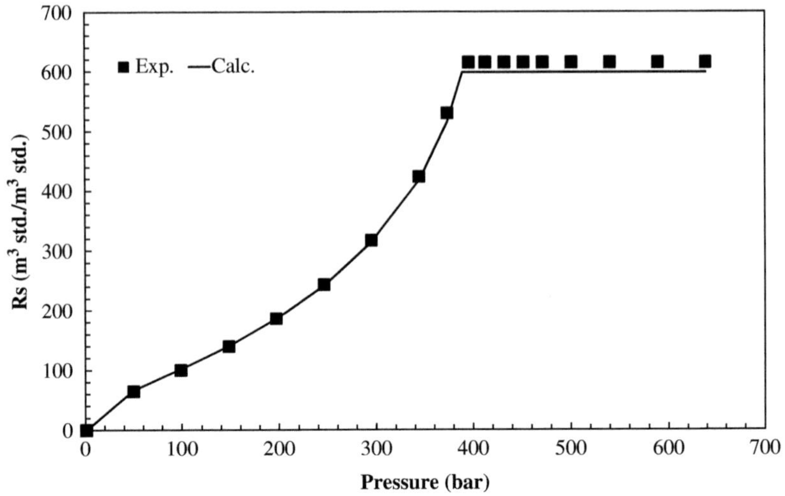

FIGURE 6.70 Rs in differential liberation at 111°C for the fluid of Well-2 at 0 m.

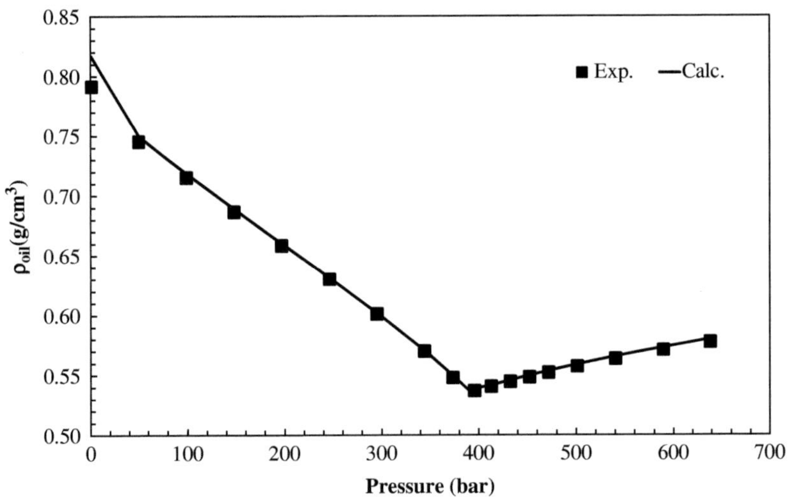

FIGURE 6.71 Oil density in differential liberation at 111°C for the fluid f Well-2 at 0 m.

that CH_4 mole fraction does not exceed 65% at any point of the domain. This statement is not so evident for C_{20+}, if we consider only Well-1 in the context of sandstone B. However, analyzing the plots of Well-3, one can note that the discrepancy relative to C_{20+} content is greater than that observed for methane. In other words, methane content suggests separation between Reservoir A (central portion) and B. C_{20+} content suggests separation of the right portion

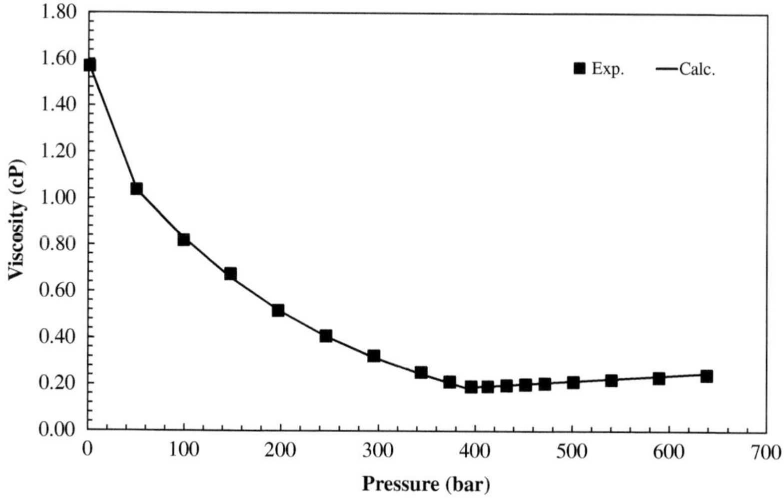

FIGURE 6.72 Oil viscosity in differential liberation at 111°C for the fluid of Well-2 at 0 m.

of Reservoir A in relation to the rest of the domain. Resuming, Wells 1, 2, and 3 would have been drilled in three disconnected reservoirs. These are: a (right portion), A (central portion), and B. One also notices, in Fig. 6.78, that the molar masses of C_{20+} observed in Wells 3 and 1 show distinct tendencies themselves, and also disagree with those obtained by simulation, which keeps them between 450 and 500 g/mol all along the domain. Such values are on the limit of the experimental error (5% to 10%) for the testing level in Well-1 (444 g/mol), but far beyond it when compared with the testing levels of Well-3 (around 360 g/mol).

The conclusions about barriers occurring among the wells suggested on the basis of the simulation results must be validated with other characterization methods, for example, with the isotopic signature of methane in the three oils. Depending on the thermal maturity of the oil (or gas) in the source rock, the relation between C^{13}/C^{12} or C^{14}/C^{12} isotopes varies within the range of parts per mil, allowing the detection of fluids that reached the trap at different eras. Therefore, besides the hypothesis of total separation, there is also the possibility of late filling of methane-rich fluid in Reservoir B. If this is the case, the isotopic signal in Well-1 must indicate higher thermal maturity in relation to the other well samples, independent of the sealing character of the permeability barrier (if this gas is still flowing into the reservoir). In other words, either the late filling of methane might have been retained within Reservoir B by some permeability barrier or it is still occurring, even if there is connection. Fig. 6.79 shows both calculated and experimental retrograde liquid volume in the CCE of Well-1 sample. Note that the composition simulated at the testing level suggests volatile oil due to the lower methane content, whereas the experimental data confirm rich,

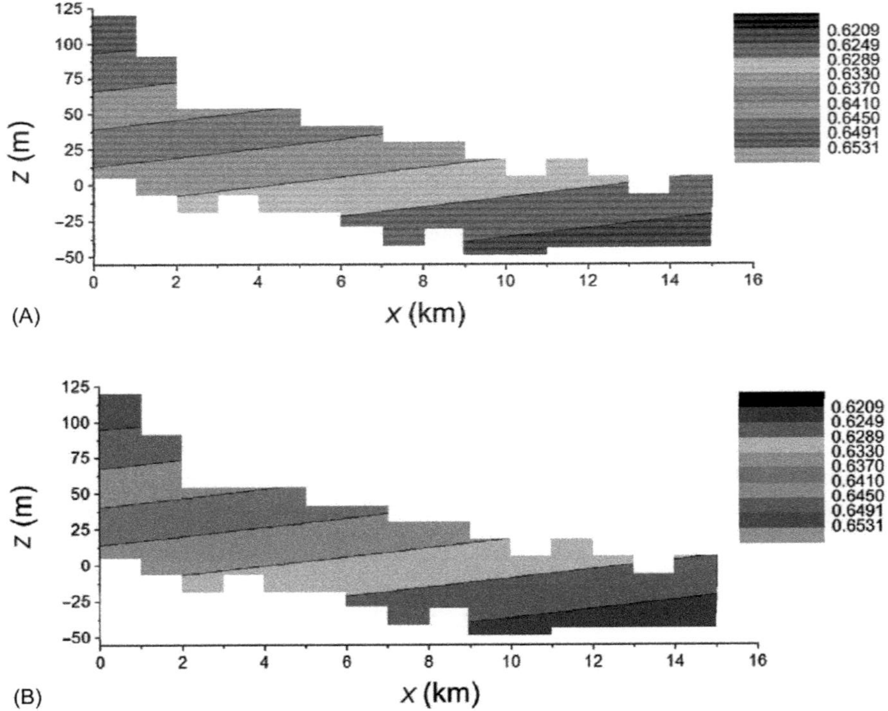

FIGURE 6.73 Methane distribution in the numerical domain after 30 million years (A) and 100 million years (B). Practically, there is no difference, suggesting that the steady state was reached much before the reservoir age.

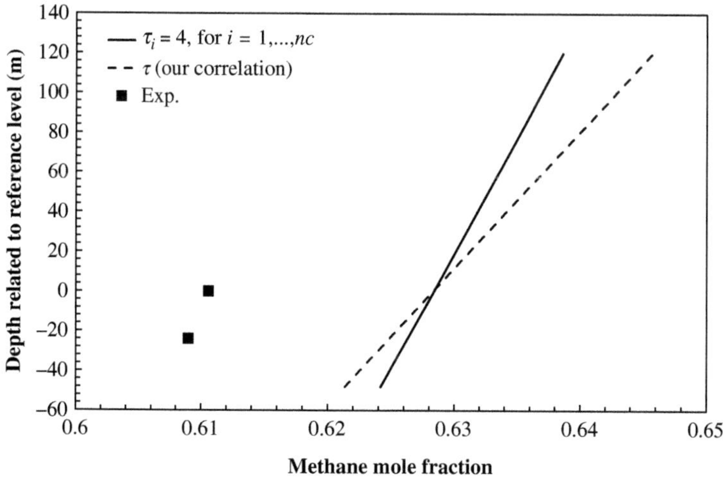

FIGURE 6.74 Methane distribution with depth at the horizontal coordinate of Well-3 (14 km from origin).

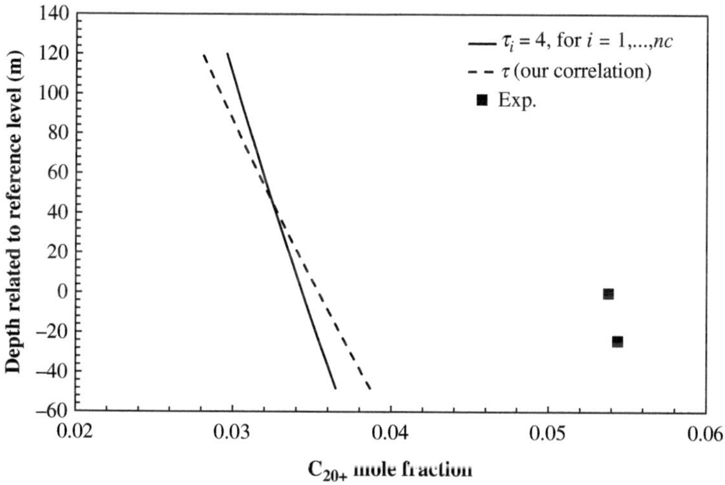

FIGURE 6.75 C_{20+} distribution with depth at the horizontal coordinate of Well-3 (14 km from origin).

near-critical gas condensate. The error in saturation pressure is greater than 10%. Fig. 6.80 shows the density of the reservoir fluid in the same CCE. Due to the underpredicted methane content, the calculated density (volatile oil) is considerably higher than the experimental one (gas condensate).

As to the oil of Well-3, one suggests another migrating stream due to the lighter character of its C_{20+}, requiring the permeability barrier between the

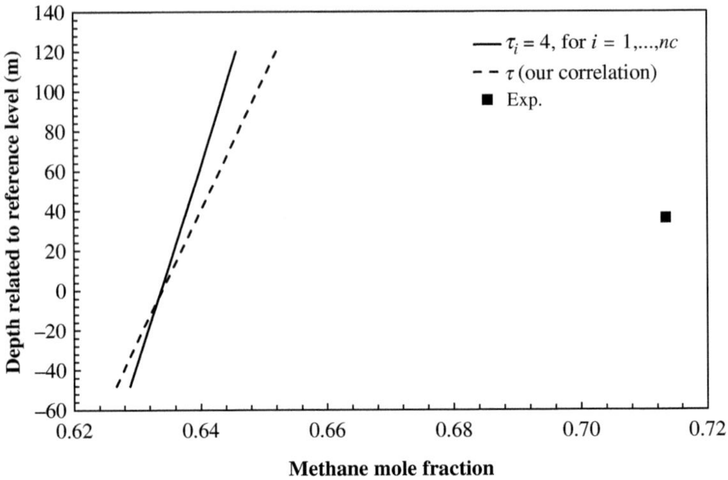

FIGURE 6.76 Methane distribution with depth at the horizontal coordinate of Well-1 (2 km from origin).

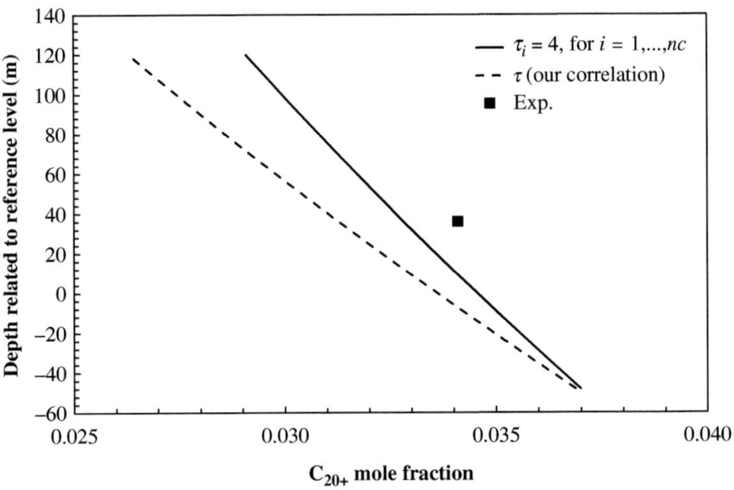

FIGURE 6.77 C_{20+} distribution with depth at the horizontal coordinate of Well-1 (2 km from origin).

central and the right portion of the domain to be necessarily sealing to avoid mixture, which would already have reached steady state, in case there was connection. Hydraulically, all reservoirs might be connected by a regional aquifer and, once this is the case, the production of one of them might, in principle, cause depletion on any other sector, even without direct connection.

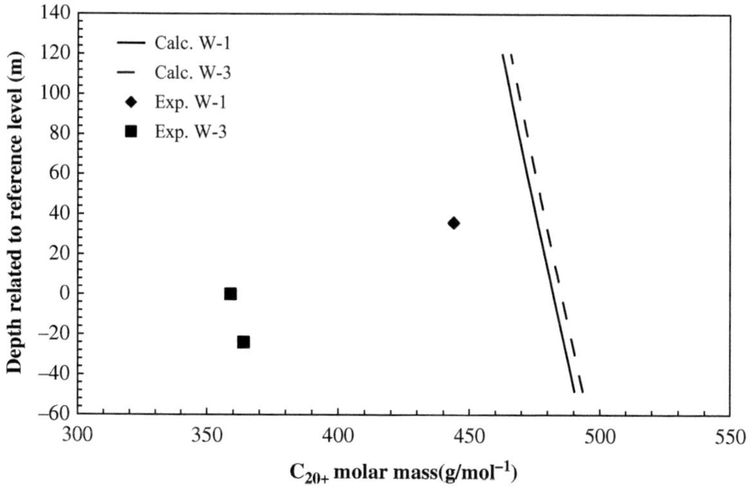

FIGURE 6.78 Molar mass of C_{20+} along depth at the coordinates of Wells-1 and 3.

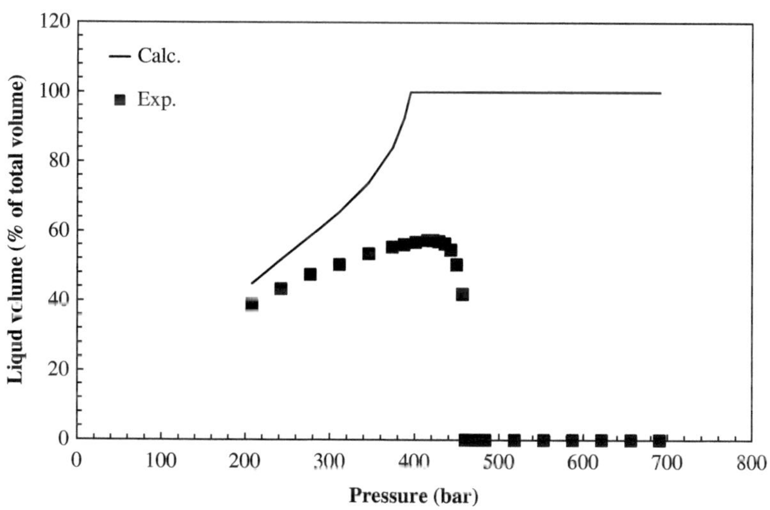

FIGURE 6.79 Retrograde liquid volume in the CCE of the testing-level sample of Well-1 at 116°C.

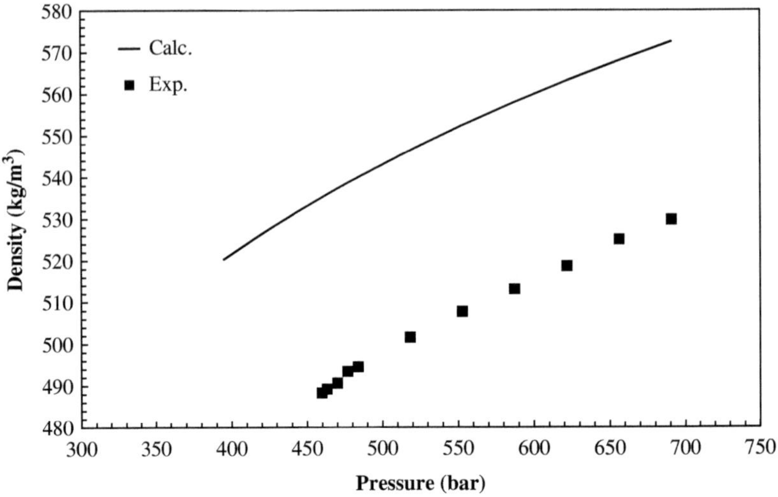

FIGURE 6.80 Density of the reservoir fluid in the CCE of the testing-level sample of Well-1 at 116°C.

6.6 FINAL REMARKS

The correlation readjusted in Eq. (6.1) for the parameter τ_i of the heat-of-transport model was applied successfully to the examples of this chapter. In general, one verified that the influence of thermal diffusion in the cases studied here is not so pronounced as it was in Yufutsu (see Section 5.2), due to either relatively low temperatures or heavier components occurrence, which impact the dead oil API. Corroborating the studies carried out with literature examples, one can state that, for fluids with API less than 33, thermal diffusion tends to enhance the gravitational field action, whereas, for lighter oils, its effect is of homogenizing the 1D distribution. As to natural convection, it tends to homogenize the 1D grading in 2D reservoirs independently of the heterogeneity of the geological model, at least for the default parameters of thermal diffusion. In the example presented in Section 6.5, the correlation for τ_i with the parameters given by Eq. (6.1) did not present great difference in relation to the original parameters. The technique of natural-convection simulation could be used successfully in the connectivity verification, at least under the assumed premises. For more complex fluids, new studies must be carried out, analyzing the performance of new thermodynamic models, using, perhaps, new parameters for the heat-of-transport. In Chapter 7, we proceed to the calculation of the 1D grading of a CO_2-rich reservoir, in liquid–liquid–vapor equilibrium. The thermal diffusion parameter estimated for carbon dioxide in Chapter 7 was used successfully in the 1D case of the example discussed in Section 6.3. As for 2D simulations (oil filling to the trap of the example of Section 6.3 through the paleocontact and the example reported on Section 6.4 with asphaltene precipitation), the performance of the proposed correlation must be evaluated together with new-EoS fittings.

REFERENCES

Firoozabadi, A., Ghorayeb, K., Shukla, K., 2000. Theoretical model of thermal diffusion factors in multicomponent mixtures. AIChE J. 46 (5), 892–900.

Moortgat, J., Firoozabadi, A., Li, Z., Espósito, R.O., 2013. *CO_2 injection in vertical and horizontal cores: measurements and numerical simulation*, SPE 135563. SPE J. 1–14.

Pedersen, K.S., Hjermstad, H.P., Modeling of Large Hydrocarbon Compositional Gradient. SPE 101275, Abu Dhabi International Petroleum Exhibition and Conference, November/2006.

Pedersen, K.S., Lindeloff, N., Simulations of Compositional Gradients in Hydrocarbon Reservoirs Under the Influence of a Temperature Gradient. SPE 84364, SPE Annual Technical Conference and Exhibition, Denver, Colorado, October/2003.

Shibata, S.K., Sandler, S.I., Behrens, R.A., 1987. *Phase equilibrium calculations for continuous and semicontinuous mixtures*. Chem. Eng. Sci. 42 (8), 1977–1988.

Chapter 7

The Influence of Molecular Association

In this chapter, we discuss the influence of molecular association on compositional grading of oil and gas reservoirs. This phenomenon, of which the hydrogen bonding is one of the main examples, is characterized by strong short-range attractive interactions and is intimately related to the polarity of fluid molecules. The angular orientation of the polar sites in a group of molecules and the short distance between them are crucial for the attractive potential around these areas to allow its association, which presents characteristics very similar to a strong (but not permanent) chemical bond.

The resulting molecular aggregates have considerable effect on the thermodynamic properties of the species. For example, water, which has four association sites per molecule, has a boiling point which is much higher than neon, which has practically the same size and mass but does not form hydrogen bonds (Economou and Donohue, 1991). As for ordinary fluids, made up only by nonpolar hydrocarbons, the mid-range interactions[1], based on van der Waals (dispersion) forces, are enough to describe the thermodynamic behavior.

In this case, the water, connate or injected in the reservoirs, is the only effectively polar component of the fluid system and may then be considered as a separate element. Its properties might be calculated, for example, by empirical correlations based on the salinity originated by the solution ions. The solubility of hydrocarbons in the water phase is neglected, and the only relation between these two phases is in the relative-permeability curve. However, for oils containing CO_2, high contents of asphaltenes or showing precipitation of the latter even at low contents, the heteroatom polar sites present in these components can change significantly the conditions for thermodynamic equilibrium due to the molecular associations.

Such associations might be *direct*, as in the case of hydrogen bonding between the water molecules and the agglomerations of asphaltene monomers,

1. Although interactions by van der Waals dispersion forces may be considered as of long-range in some contexts, when compared to Coulomb interactions (the latter effectively of long-range type), it becomes more suitable to classify them as mid-range ones.

Compositional Grading in Oil and Gas Reservoirs. DOI: http://dx.doi.org/10.1016/B978-0-12-812452-9.00007-1

or *cross*, as in the case of CO_2 and water molecules or between resins and asphaltenes, the latter being responsible for suspension of asphaltenes in the oil phase.

In this chapter, we do not discuss the direct associations between water molecules (hydrogen bonding), nor the cross associations arising from the solubility of CO_2 in the aqueous phase. We emphasize basically the direct associations of asphaltene–asphaltene molecules and the resin–asphaltene cross association. To model this kind of interaction, still via equations of state, one adopts the so-called CPA-EoS (*Cubic Plus Association Equation of State*), proposed initially by Kontogeorgis et al. (1996), which incorporates only the association term (short-range strong attraction) belonging to SAFT-EoS (*Statistical Association Fluid Theory Equation of State*) (Chapman et al., 1990) to a traditional two-parameters cubic equation of state [e.g., Peng–Robinson (PR) or Soave–Redlich–Kwong (SRK)], which contains only the repulsive and attractive mid-range terms. The chapter is organized as follows: first, we describe the association term from SAFT and additional parameters that it adds to the model used so far. Then we proceed to a brief literature review, discussing the main papers that proposed parameter fitting based on experimental data of asphaltene precipitation or volumetric properties measured in liquid–vapor or liquid–liquid–vapor equilibria. Finally, we present a practical example of compositional grading in a three-phase reservoir, where the occurrence of the second liquid phase is due to the excess of CO_2 in the reservoir fluid.

7.1 THE ASSOCIATION TERM

One of the main contributions in the literature for modeling associative polar systems was the SAFT equation of state, proposed by Chapman et al. (1990) and Huang and Radosz (1990), and analyzed afterwards by various authors, for example, Economou and Donohue (1991). Among the different approaches for deriving the equation, the one using the Perturbation Theory from hard spheres (PC-SAFT) is the most widely used. It adds independent terms to the expression of the hard sphere (HS) Helmholtz energy, A^{HS}—for example the one of Carnahan and Starling (1969)—to consider the effects of dispersion (mid-range or dispersion interactions, A^{disp}), formation of polymeric chains to the generation of nonspherical molecules, A^{chain}, and, mainly, the molecular association among these chains, A^{assoc}. Then, the expression that generates the equation of state by its derivative with respect to volume may be written as

$$A^{res} = A^{HS} + A^{disp} + A^{chain} + A^{assoc} \tag{7.1}$$

Fig. 7.1 shows the perturbation effects of each term to the purely repulsive interaction of the HS. The traditional cubic equations present only the repulsive and attractive (dispersion) terms, with assumptions simpler than those used by SAFT in the terms (A) and (B) of Fig. 7.1.

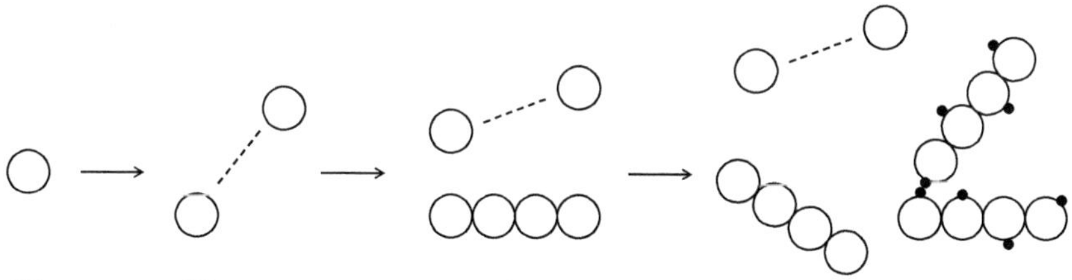

FIGURE 7.1 Additional effects of dispersion (B); chain formation (C); and associations among them through polar sites (D) to the purely repulsive potential of the hard spheres (A) modeled by SAFT. The position of the sites is random, and the association takes place when both orientation and distance are favorable.

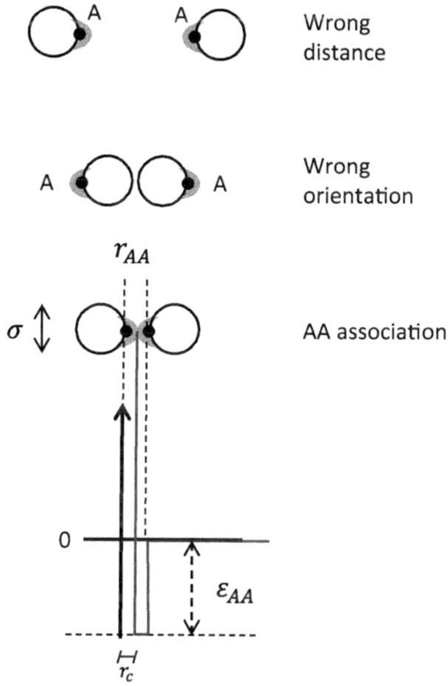

FIGURE 7.2 Hard sphere (of diameter σ) model for a pure component with a single type of association site (A). This is a simple example of molecular association due to a strong short-range attraction modeled by the square-well potential and also influenced by the sites orientation. *Adapted from Chapman, W.G., Gubbins, K.E., Jackson, G. and Radosz, M., New reference equation of state for associating liquids, Ind. Eng. Chem. Res., 29, pp. 1709–1721, 1990.*

Fig. 7.2 details the association term for a pair of HS with a single type of site (A). These spheres (whose diameter is σ) can form only a dimer of A–A association when both distance and orientation are favorable. Placing one of the sites on the origin of the potential axis, the radius of the shaded region is called *critical radius of association*, r_c. When the distance between two sites r_{AA} is between r_c and $2r_c$, the association forms with energy ε_{AA}.

The degree of dimerization depends on this association energy. One represents this energy by a square-well potential (deeper and narrower than the one of van der Waals dispersion interactions discussed in Appendix F), which origin is on site A.

Chapman et al. (1990) do not restrict the number of association sites on an individual molecule and nominate these sites by capital letters, *A*, *B*, *C*, etc. One assumes that each association site has a different attraction with the other sites of other molecules. So, the symbol A_iB_j means interaction between site *A* of molecule *i* with site *B* of molecule *j*. Details about the terms of Eq. (7.1) may be found in Kontogeorgis and Folas (2010) and Firoozabadi (2015), as well as in the original papers of the SAFT-EoS (Chapman et al., 1990, Huang and Radosz, 1990). Here, we present the A^{assoc} term, which was developed by Wertheim (1984a,b) and adopted by Jackson et al. (1988) and Chapman et al. (1990). For a pure component, the variation of Helmholtz energy due to association is given by:

$$\frac{A^{\text{assoc}}}{NkT} = \sum_{A}\left(\ln \chi^{A} - \frac{\chi^{A}}{2}\right) + \frac{1}{2}M \tag{7.2}$$

where N is the number of moles, k is the Boltzmann constant, T is the absolute temperature (K), χ^A is the fraction of molecules not associated to the site A, and M is the total number of sites. χ^A is given by

$$\chi^{A} = \frac{1}{1 + \sum_{B}\rho\chi^{B}\Delta_{AB}} \tag{7.3}$$

where $\rho = 1/\overline{V}$ is the molar density (inverse of the molar volume, $\overline{V}$), and Δ_{AB} is the so-called *association strength* between sites A and B. It can be written as the probability of a site of a certain molecule having both distance and angle orientation compatible with the site of other molecule to form the association. For HS, that means to apply the radial distribution function $g^{\text{HS}}(r_{12})$ to the central molecule (1) plus a probability factor of a site B belonging to other molecule around (2) being at both orientation and distance (given by r_{12}) compatible with a site A of the central molecule. This factor is the so-called Mayer Function, $f_{A_1B_2}$:

$$f_{A_1B_2} = \exp\left(-\frac{\phi_{A_1B_2}(r_{A_1B_2})}{kT}\right) - 1 \tag{7.4}$$

where $\phi_{A_1B_2}(r_{A_1B_2})$ is an interaction potential; for example, the narrow and deep square-well potential of Fig. 7.2.

Appendix F shows a brief discussion on the concept of radial distribution function, using the square-well potential to compute the dispersion interactions that will generate the attractive terms of the cubic equations of state via Perturbation Theory.

Analog to the Configuration Energy described in Appendix F, the integral over the spherical-coordinate angles gives the factor 4π, resting only the integral over the radial direction in relation to the center of molecule 1. So, dropping the subscripts 1 and 2 of indexes A and B, one has

$$\Delta_{AB} = 4\pi \int g^{\mathrm{HS}}(r_{12}) f_{AB}(r_{AB}) r_{12}^2 dr_{12} \tag{7.5}$$

Due to the short-range characteristics of the association potential, we can simplify the integral of Eq. (7.5), replacing $g^{\mathrm{HS}}(r_{12})$ and r_{12}^2 by their respective values evaluated at the contact ($r_{12} = \sigma$), because it is in a very restrictive region around the contact that the association will take place:

$$\Delta_{AB} = 4\pi\, g^{\mathrm{HS}}(\sigma)\sigma^2 \int f_{AB}(r_{AB}) dr_{12} \tag{7.6}$$

Due to numerous orientation possibilities as a function of the distances and angles between centers of molecules and their sites, the Mayer Function cannot be withdrawn from the integral. However, Jackson et al. (1988) proposed one more simplification on defining the so-called *A*–*B association volume* (K_{AB}):

$$K_{AB} \equiv \frac{\sigma^2}{F_{AB}} \int f_{AB}(r_{AB}) dr_{12} \tag{7.7}$$

where

$$F_{AB} = \exp\left(\frac{\varepsilon_{AB}}{kT}\right) - 1 \tag{7.8}$$

Therefore, the final expression of the association strength Δ_{AB} reduces to

$$\Delta_{AB} = 4\pi\, g^{\mathrm{HS}}(\sigma) K_{AB} F_{AB} \tag{7.9}$$

Later, Chapman et al. (1990) redefined this volume as

$$\kappa_{AB} \equiv \frac{4\pi}{\sigma^3} K_{AB} \tag{7.10}$$

Replacing Eq. (7.10) in Eq. (7.9), one gets

$$\Delta_{AB} = \sigma^3 g^{\mathrm{HS}}(\sigma) \kappa_{AB} F_{AB} \tag{7.11}$$

Once we are given the radial distribution function for HS $g^{\mathrm{HS}}(\sigma)$, which can be given, for example, by the equation of Carnahan and Starling (1969):

$$g^{\mathrm{HS}}(\sigma) = \frac{2-\eta}{2(1-\eta)^3} \tag{7.12}$$

where η is the so-called reduced density, given by

$$\eta = \rho \frac{\pi\sigma^3}{6} \tag{7.13}$$

Eq. (7.12) may be replaced in Eq. (7.11) for the calculation of Δ_{AB}, with κ_{AB} and ε_{AB} being the parameters of this association model. It is interesting to note that the functional form of Eq. (7.3) needs an iterative procedure for the computation of χ^A, which depends on T and ρ.

Extension of Eq. (7.2) to mixtures is straightforward (Kontogeorgis and Folas, 2010; Firoozabadi, 2015):

$$\frac{A^{\text{assoc}}}{NkT} = \frac{1}{N}\sum_i N_i \left[\sum_{A_i} \left(\ln \chi^{A_i} - \frac{\chi^{A_i}}{2} \right) + \frac{1}{2} M_i \right] \tag{7.14}$$

where χ^{A_i}, that is, the mole fraction of molecules of type i *not* associated to the site A, in a mixture with other components, is given by

$$\chi^{A_i} = \frac{1}{1 + \sum_j \sum_{B_j} \rho_j \chi^{B_j} \Delta_{A_i B_j}} \tag{7.15}$$

It is also interesting to note that χ^{A_i} depends on the molar density of molecules j:

$$\rho_j = \frac{N_j}{N}\rho = x_j \rho \tag{7.16}$$

And also on the association strength:

$$\Delta_{A_i B_j} = \sigma_{ij}^3 g_{ij}^{\text{HS}}(\sigma_{ij}) \kappa_{A_i B_j} \left[\exp\left(\frac{\varepsilon_{A_i B_j}}{kT} \right) - 1 \right] \tag{7.17}$$

which depends on the temperature and on some mixing rules for σ_{ij}^3 and $g_{ij}^{\text{HS}}(\sigma_{ij})$, presented later on this chapter.

7.2 CPA EQUATION OF STATE

The conventional cubic equations already show both repulsive and dispersion (mid-range van der Waals-type attraction) terms, related to the HS reference fluid. These equations work reasonably well for nonpolar components, where these two types of interactions are enough to provide a sufficiently accurate calculation of the thermodynamic properties of the fluids. Due to the mathematical complexity of the four terms in SAFT, and still to the high cost/benefit ratio of its application in terms of computational effort, Kontogeorgis et al. (1996) suggested to combine the terms already existent in the cubic with the association term of the SAFT EoS described in Eq. (7.14), neglecting also the polymeric chain formation. The resulting equation of state was called *CPA-EoS* (*Cubic Plus Association Equation of State*). So, all molecules remain individually spherical, but able to associate. Fig. 7.3 shows the Perturbation Theory under the point of view of the *CPA-EoS* proposal. The repulsive potential (Fig. 7.3A) is based on the free volume of van der Waals, that is $RT/(\overline{V} - b)$ (see Appendix F). The attractive one (dispersion, Fig. 7.3B),

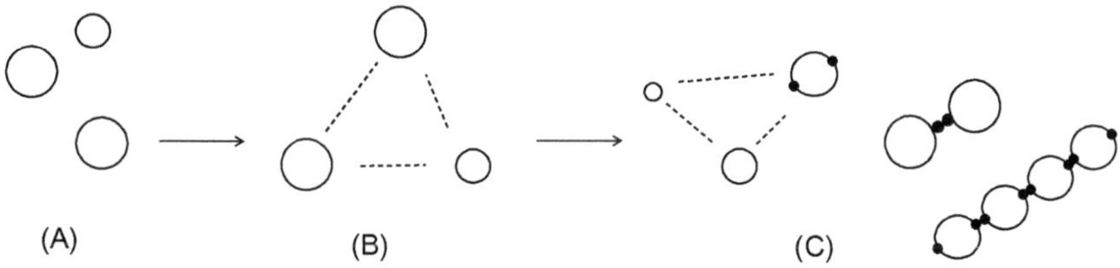

FIGURE 7.3 Perturbation theory under CPA-EoS proposal. Sketch (A) refers to the repulsive potential $RT/(\overline{V}-b)$. Sketch (B) is responsible for the attraction term. Sketch (C) considers association of individually spherical molecules, for there is no chain formation. Molecules with only one association site form dimers. Molecules with two or more sites may form two- or three-dimensional structures.

discussed in Appendix F, is compatible with van der Waals type equations, according to the complexity of the expression of the coordination number. The addition of the SAFT association term is sketched in Fig. 7.3C. Molecules with only one association site form only dimers. Molecules with two or more sites may form two- or three-dimensional structures. Molecules with sites also interact with nonsite molecules by ordinary dispersion. The absence of association sites makes the component to be modeled exclusively by the conventional cubic part. This is an advantage of CPA-EoS, since it uses all the experience of the petroleum industry in modeling reservoir fluids with the equations of PR and/or SRK.

The derivative of A^{assoc} with respect to volume provides the additional term in pressure units, the final expression of CPA-EoS, considering PR-EoS in the cubic part, being (Kontogeorgis and Folas, 2010)

$$P=\frac{RT}{\overline{V}-b}-\frac{a}{\overline{V}(\overline{V}+b)+b(\overline{V}-b)}-\frac{1}{2}RT\overline{V}\left(1+\rho\frac{\partial g^{\text{HS}}}{\partial\rho}\right)\sum_i x_i\sum_{A_i}(1-\chi^{A_i}) \tag{7.18}$$

where R is the universal gas constant, $\overline{V}$ is the molar volume, a and b are the attractive and repulsive parameters of PR-EoS, respectively (see Chapter 2 and Appendix F).

There are many studies regarding CPA-EoS performance, as reviewed by Kontogeorgis et al. (2006a,b) and Kontogeorgis and Folas (2010). Here, we cite only some of those related to petroleum industry, which will help the reader acquire sensitivity about the new parameters introduced by the association term. Li and Firoozabadi (2009) studied the influence of cross association of contaminants like H_2S and CO_2 with water in the vapor–liquid equilibrium of such binary systems. Hydrocarbons were also considered, for according to the authors, although lights as methane and ethane are saturated and nonpolar, the strong dipole moments of water can induce the temporary separation of both positive and negative charges of these molecules. This premise, we mean the possibility of cross association between nonpolar hydrocarbons and water, improved the prediction of the phase envelopes of these binary systems, their densities being also better predicted using temperature-dependent volume translations.

Later, the same authors studied the induction of asphaltene precipitation by adding *n*-alkanes to bitumens (Li and Firoozabadi, 2010a) and also by depletion in live oils (Li and Firoozabadi, 2010b). Jindrová et al. (2016) tested both PR and CPA equations of state in CO_2-bitumen systems over a wide range of temperatures and pressures. The authors concluded that PR-EoS describes well the mixtures of bitumens with hydrocarbons and CO_2 when there is no second liquid phase or, even when this phase exists, its asphaltene content is not very high. As to CPA-EoS, it is a natural choice for asphaltene high-content systems, since they show considerable degrees of direct and cross associations, leading also to the split out of a second liquid-phase.

The authors consider the C_{20+} fraction to be made up of only asphaltenes (A) and a second lump of heavy components (R), generalizing the concept of *resins*. Concentrations of these two pseudos come from the SARA analysis of the dead oil. They also propose that each one of these molecules has four identical association sites. Only direct associations of A–A type and cross associations of A–R type are allowed, which simplifies the expression of the association Helmholtz energy to:

$$\frac{A^{\text{assoc}}}{NRT} = M_A x_A \left[\left(\ln \chi^A - \frac{1-\chi^A}{2} \right) \right] + M_R x_R \left[\left(\ln \chi^R - \frac{1-\chi^R}{2} \right) \right] \tag{7.19}$$

in which

$$\chi^A = \frac{1}{1 + \rho x_A M_A \chi^A \Delta_{AA} + \rho x_R M_R \chi^R \Delta_{AR}} \tag{7.20}$$

$$\chi^R = \frac{1}{1 + \rho x_A M_A \chi^A \Delta_{AR}} \tag{7.21}$$

The association strength Δ_{ij}, with $i = A$ and $j = A$ or R, is given by

$$\Delta_{ij} = g^{\text{HS}} \kappa_{ij} b_{ij} \left[\exp\left(\frac{\varepsilon_{ij}}{kT} \right) - 1 \right] \tag{7.22}$$

in which

$$b_{ij} = \frac{b_i + b_j}{2} \tag{7.23}$$

and g^{HS} keeps calculated by Eq. (7.12), but with reduced density obtained by the covolume, instead of the HS volume of Eq. (7.13):

$$\eta = \frac{b\rho}{4} \tag{7.24}$$

Jindrová et al. (2016) consider $M_A = M_R = 4$; $\kappa_{AA} = \kappa_{AR} = 0.01$ and $\varepsilon_{AA}/k = 2000$ K. So, the only adjustable parameter is the cross-association energy ε_{AR}.

Now, we proceed to modeling the compositional grading in a three-phase reservoir with high CO_2 content, applying the CPA-EoS according to the methodology proposed by Jindrová et al. (2016).

It is important to mention that, apart from this research line, other groups, like, for example, Gonzalez et al. (2008), modeled the asphaltene precipitation in live oils (by both depletion and CO_2 injection) using SAFT EoS *without considering the association term.* The so-called asphaltene cluster is modeled by the chain polymerization term and the prediction of the onset curves (i.e., the limit of the asphaltene precipitation as a second liquid-phase) is very reasonable when compared to experimental data. Also, the Yen-Mullins model for asphaltene colloids characterization in dead oils (Mullins, 2010), combined with the Flory–Huggins–Zuo equation of state (Zuo et al., 2013) have been used to determine both connectivity and steady state reaching. Using the methodologies of Gonzalez et al. (2008) and Zuo et al. (2013) in compositional grading calculations is out of the scope of this book.

7.3 RESERVOIR WITH HIGH CO_2 CONTENT

Fig. 7.4 shows, qualitatively, the fluid distribution to be reproduced by the one-dimensional (1D) compositional grading of a reservoir with high CO_2 contents. The reservoir contains a dense-gas cap with about 75 mol % of

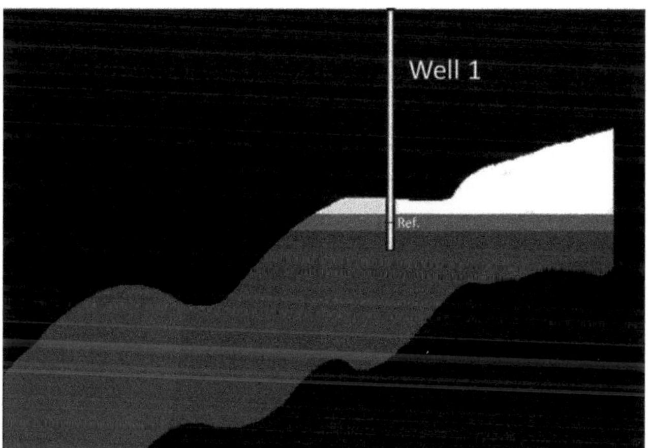

FIGURE 7.4 Qualitative description of the fluid distribution in the high CO_2 content reservoir. Gas cap (*white* zone) on a heavy-oil zone (*brown* zone, or *light gray* in print versions), where one also sees a second liquid-phase with high asphaltene content (*dark gray* zone). Detail of Well-1 with the reference sample-taking depth.

CO_2 above a viscous oil region with CO_2 content over 50%. The excess of CO_2 has done a light-component extraction to the top, generating a gas cap and a more viscous oil with a high saturation pressure gradient, about -1 bar/m, toward the bottom of the structure. The pronounced increase of C_{20+}'s molar mass up to 700 g/mol in local samples collected at the deepest levels in the oil zone and the consequent increase in viscosity suggest asphaltene segregation in such a way that one could argue about an ultraviscous second liquid-phase split up just above the water-oil contact (WOC). One wonders how much this phase could, in a certain way, limit the aquifer action as a production drive.

The objective of this example is to reproduce the local compositional data (basically CO_2, methane, and C_{20+}) measured along the depth of Well-1 through the 1D nonisothermal compositional grading calculation, pointing the gas–oil contact and/or the second denser liquid phase, inferring its properties between the deepest sampling point and the water–oil contact. We emphasize that there are not enough experimental data to confirm the liquid–liquid interface at reservoir conditions. The slope change in the observed distribution profiles may induce to this interpretation.

The premise of liquid–liquid interface occurrence in the oil zone implies necessarily in the existence of liquid–liquid and liquid–liquid–vapor regions in the phase envelopes of the sampled fluids. Such equilibria cannot be described with the conventional cubic equations of state, like PR-EoS. To do so, we use CPA-EoS, following the methodology of Jindrová et al. (2016) for the characterization of the associative components of the reservoir fluid. After the SARA analysis of the reference level sample in Well-1, which indicated around 3.5% w/w of asphaltenes in the dead oil, the rest of the C_{20+} fraction was discretized in two more pseudocomponents: one of them representing the lighter components and the other (predominant), representing the resins. This procedure was necessary due to the huge difference in the molar masses of C_{20+} observed in the local samples of both gas ($M_{C_{20+}} \sim 345$ g/mol) and oil ($M_{C_{20+}} \sim 550$ g/mol) zones.

In other words, the lightest of the pseudocomponents must have necessarily a molar mass lower than the one reported for the gas-zone heavy fraction, which would not occur if one used only resin and asphaltene to make up the C_{20+}. Once this is done, we have assumed that resin is the only component able to undergo a cross association with asphaltene. It is important to mention that other splitting and lumping strategies could be used with equivalent results.

Other interesting observation is that the asphaltene molar mass is also an adjustable parameter of the model. When adopting PR-EoS to represent one of the fluids studied in their paper, Jindrová et al. (2016) proposed a value greater than 3000 g/mol, suggesting this number to represent an aggregate previously set in a single big molecule to interact only by dispersion forces with the other components. On adopting CPA-EoS, the adjusted values for asphaltene monomer in the cases studied by the authors stayed between 1000 and 1500 g/mol.

TABLE 7.1 Molar Masses and Pseudocomponent Composition Variation Along Depth

Components	*M* (g/mol)	Composition Range (mol %)
CO_2	44	47–80
N_2–C_1	16	14–18
C_2–C_5	43	2–3
C_6–C_{12}	145	1–3
C_{13}–C_{19}	241	0–3
$QC_{24.62}$	340	0–4
Resin	540	0–18
Asphaltene	1500	0–8

Back to our case study, Table 7.1 shows the molar masses and the variation ranges of the pseudocomponent compositions with depth. Once the conventional Pressure, Volume and Temperature (PVT) analysis considers only vapor–liquid equilibrium, initially only the cubic-part parameters were adjusted to reproduce these properties. With this adjustment, it is possible to reproduce the gas–oil contact depth (confirmed experimentally at 40 m above the reference) changing only the thermal-diffusion parameter of CO_2 (τ_{CO_2}), as hydrocarbons will follow the correlation of Eqs. (5.3)–(5.5), with the parameters presented in Eq. (6.1).

In the absence of experimental data of asphaltene precipitation onset, assuming that the liquid–liquid interface exists at approximately 20 m below the reference level, we have adjusted the association parameters intending to reproduce this contact, keeping the same procedure described in the previous paragraph for the gas–oil contact location. In other words, given a set of parameters for the cubic part that can reproduce the PVT analysis of the oil, the association parameters are then regressed to reproduce the distribution profiles and properties with depth, turning on the 1D compositional grading algorithm.

Once this is done, one must evaluate the impact of the adjusted association parameters on the reproduction of the PVT analysis. If calculated properties were disturbed, it is necessary to refit the cubic-part parameters, keeping the association ones obtained on the previous step. The latter are, on their turn, readjusted (together with the thermal-diffusion parameter of CO_2) based on the distribution profiles of the composition grading, repeating the procedure sketched in Fig. 7.5 until one reaches convergence.

The τ_i parameters for the hydrocarbons were obtained along depth by the correlation developed in Eqs. (5.3)–(5.5), with the parameters presented in Eq. (6.1). For CO_2, adjusted values (of τ_{CO_2}) range from 0.1 to 2.0.

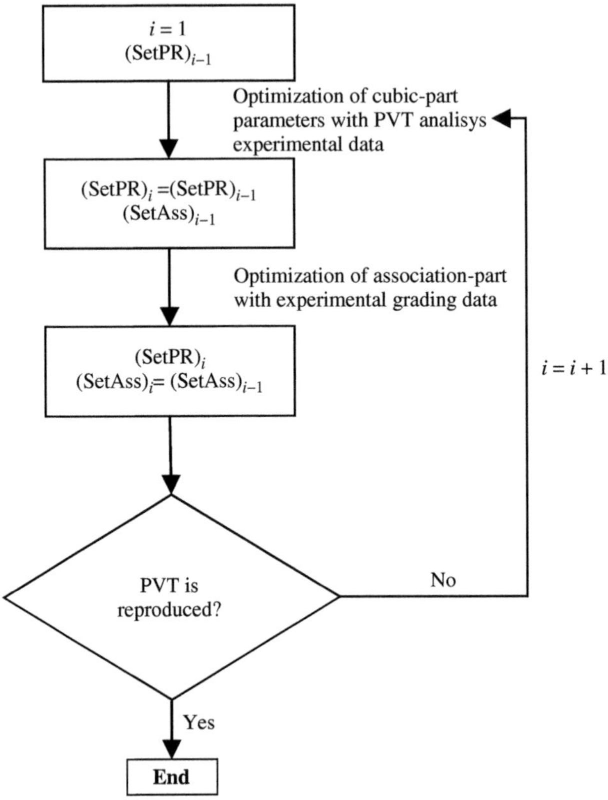

FIGURE 7.5 Methodology for the calculation of the cubic-part (SetPR) and association-part (SetAss) parameters of CPA-EoS.

The predictions of PVT properties (Bo, Rs, density, and oil viscosity—as shown in previous chapters) calculated by PR and CPA equations of state were similar and in satisfactory agreement with the experimental data. As will be seen later in this text, the similarity on the performance of the two models is due to the small influence of associative effects in the prediction of bubble-point curve, which is not the case for the asphaltene onset curves. In other words, keeping all the cubic-part parameters values in conventional trends, considering the association effects is the only way to predict a second asphaltene-rich liquid-phase. With the conventional instrumentation used in PVT analysis of oil and gas industry, it is still not possible to detect the second liquid-phase experimentally. Therefore, in fluids with features similar to the one presented in this chapter, additional information should be provided in order to confirm the existence of two liquid phases, allowing an appropriate parameterization and choice of the model. As previously mentioned, in this case study the lack of information in the PVT analysis was fulfilled by reservoir-condition evidences, like the abrupt slope changes on the distribution profiles of the compositional grading data.

Fig. 7.6 shows the liquid volumes in the Constant Composition Expansion of the gas sample at 91 m above the reference level. Values are calculated by CPA-EoS using the composition obtained at this level by calculations (dashed line) and the experimental composition of the gas PVT analysis (continuous line). The former presents a considerable underestimation of this property, most likely due to the high sensitivity of the retrograde condensate to the heavy fractions obtained by the compositional grading calculation.

Figs. 7.7–7.10 show, respectively, the comparisons between the experimental and calculated distribution profiles of CO_2, methane, and C_{20+} along depth. The four curves refer to PR and CPA equations of state for both isothermal and nonisothermal calculations. Starting from the reference level toward the bottom of the structure, one observes that the compositional grading is dominated by molecular association and that the geothermal gradient (−0.025°C/m) does not have a major influence on the location of the interface between the two liquid phases. In fact, the interface was also reasonably well pointed by the isothermal calculation using the fitted CPA-EoS. Fig. 7.11A presents the phase envelope calculated from the experimental data of the reference point. The well conditions at this depth are very close to the liquid–liquid equilibrium curve (upper onset). That explains why a small pressure drop leads to precipitation of liquid asphaltene-rich phase. Fig. 7.11B presents the phase envelope calculated by PR-EoS for the same composition. The bubble-point curve remains unchanged, liquid–liquid and liquid–liquid–vapor regions not being observed, which demonstrates the importance of the association term on predicting the liquid–liquid interfaces.

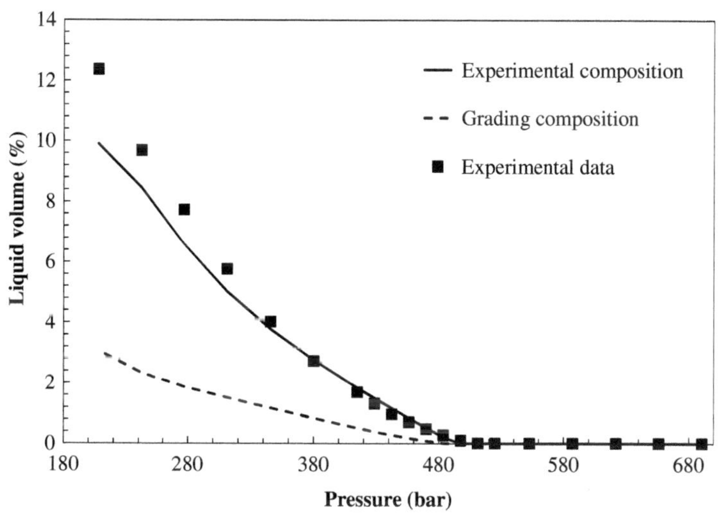

FIGURE 7.6 Liquid volumes in the Constant Composition Expansion at 62.8°C of the sample taken in the gas zone, 91 m above the reference level. Calculated values refer to CPA using the composition obtained at this level by grading (dashed line) and the experimental composition of the PVT analysis (continuous line).

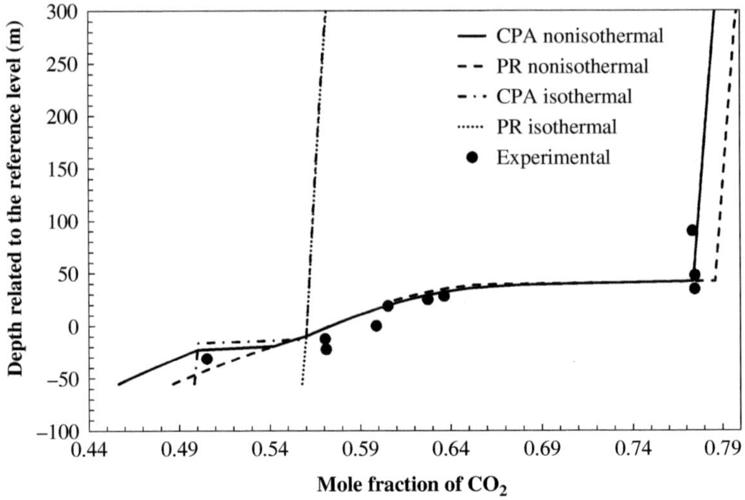

FIGURE 7.7 CO_2 distribution along depth. Comparison between calculated and experimental data using both PR-EoS and CPA-EoS in isothermal and nonisothermal runs.

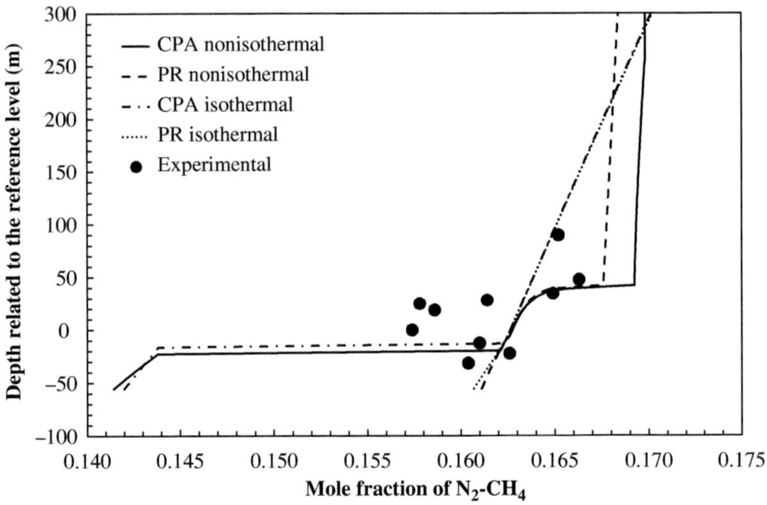

FIGURE 7.8 Methane distribution along depth. Comparison between calculated and experimental data using both PR-EoS and CPA-EoS in isothermal and nonisothermal runs.

The combined effect is important from both thermal diffusion (due to the presence of CO_2) and molecular association (resins and asphaltenes) on the liquid–liquid–vapor "*equilibrium*" of this reservoir fluid. We emphasize the word "*equilibrium*" in quotes due to the entropy generation arising from the heat flux caused by the geothermal gradient, observing, however, that the system is already static, at steady state.

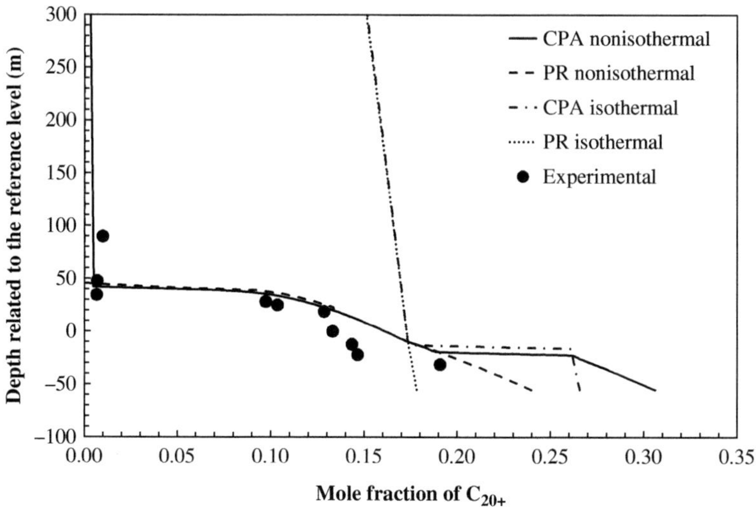

FIGURE 7.9 C_{20+} distribution along depth. Comparison between calculated and experimental data using both PR-EoS and CPA-EoS in isothermal and nonisothermal runs.

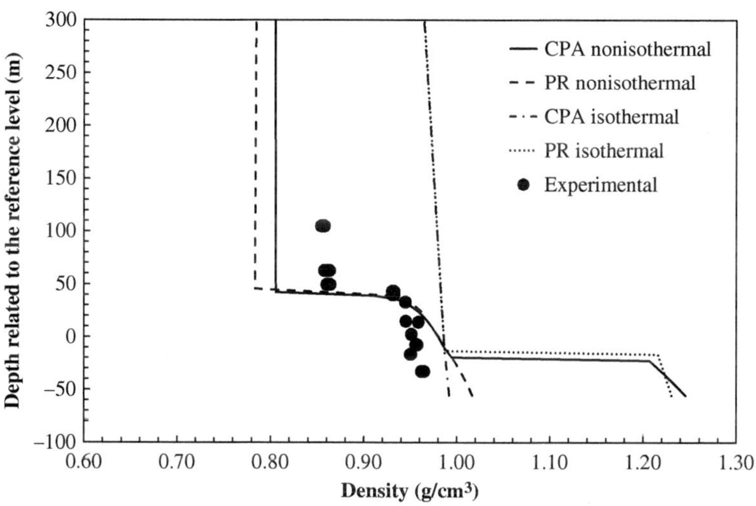

FIGURE 7.10 Density distribution along depth. Comparison between calculated and experimental data using both PR-EoS and CPA-EoS in isothermal and nonisothermal runs.

Figs. 7.12 and 7.13 show gas–oil ratio and saturation pressure (P_{sat}) profiles along depth, respectively. The calculated values were obtained using fitted CPA-EoS on the nonisothermal run. PR-EoS provides similar results. The huge P_{sat} gradient is not due to asphaltene's grading, but to CO_2

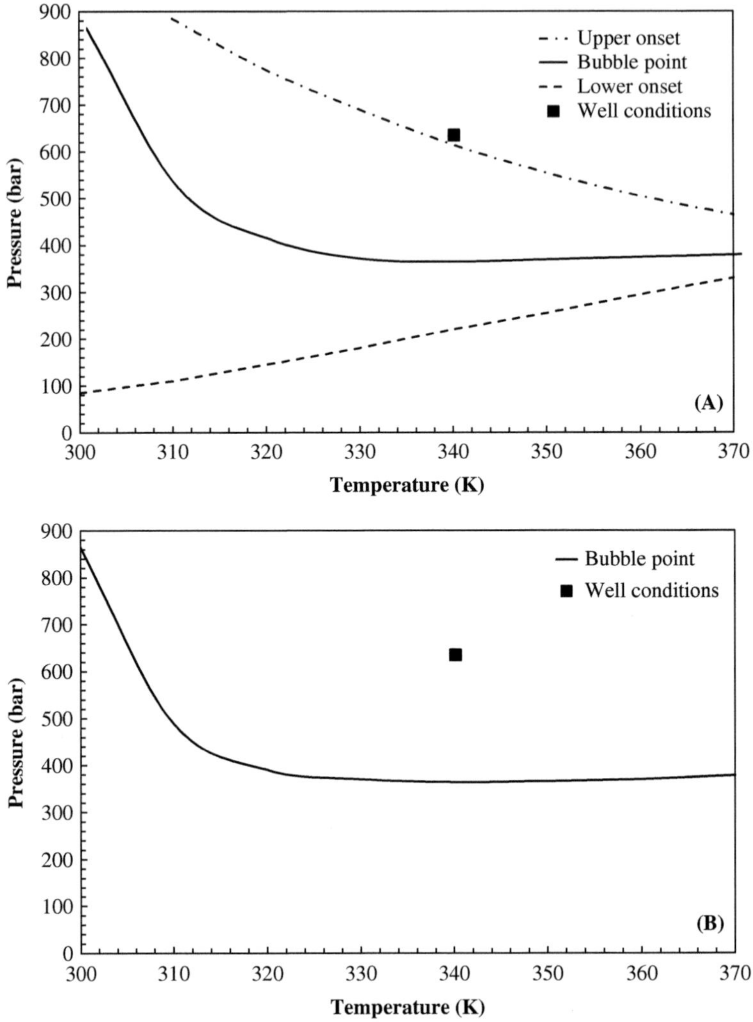

FIGURE 7.11 (A) Phase envelopes calculated for the composition of the reference level (reservoir conditions also shown) using CPA-EoS. (B) Phase envelopes calculated for the composition of the reference level (reservoir conditions also shown) using PR-EoS.

distribution profile in the oil zone. Fig. 7.14 shows the phase envelopes for compositions calculated at some depths above the reference level. As the vertical (high-derivative) region of the bubble-point curve (which only exists this way because of CO_2) approaches reservoir conditions, the derivative of P_{sat} increases rapidly, simultaneously to the reduction of the asphaltene phase quantity. Finally, given the composition calculated below the interface between both liquid phases, it is possible to estimate its viscosity by the fitted JST correlation (Jossi et al., 1962), as shown in Fig. 7.15.

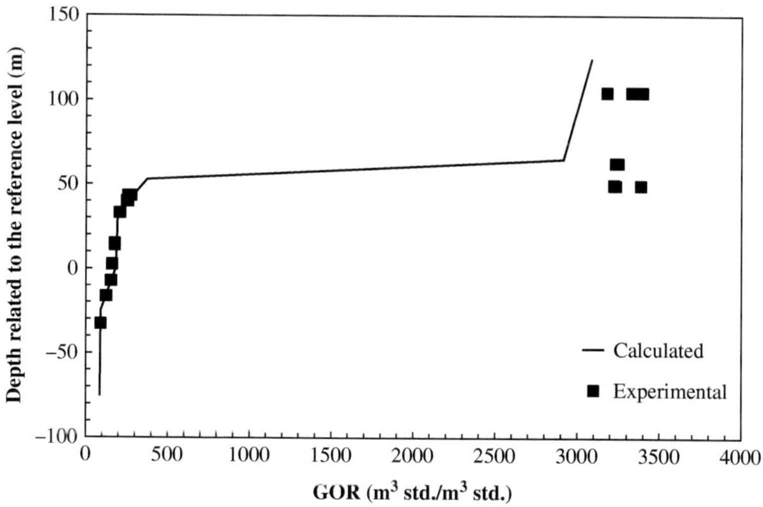

FIGURE 7.12 Variation of gas–oil ratio (GOR) along depth in oil and gas zones. Comparison of calculated and experimental data using fitted CPA-EoS in the nonisothermal case. PR-EoS provides similar results.

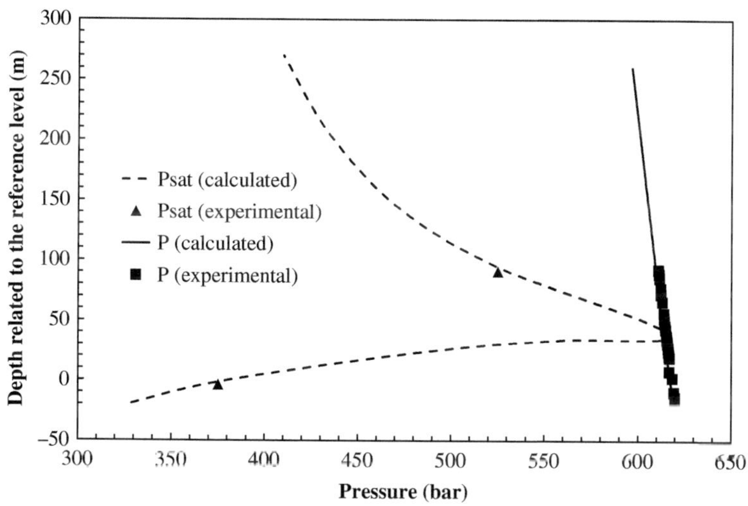

FIGURE 7.13 Static (P) and saturation pressure (P_{sat}) variation along depth for fitted CPA-EoS in the nonisothermal run. The high variation of P_{sat} profile along depth is due to CO_2 grading, not to asphaltene's one.

7.4 FINAL REMARKS

A high CO_2 content reservoir, showing also considerable asphaltene concentrations (those evidenced by atypical depth-increasing values of C_{20+} molar mass) was selected for increasing the complexity of phase behavior in the porous media at high pressure. The possible interpretation of a second more viscous

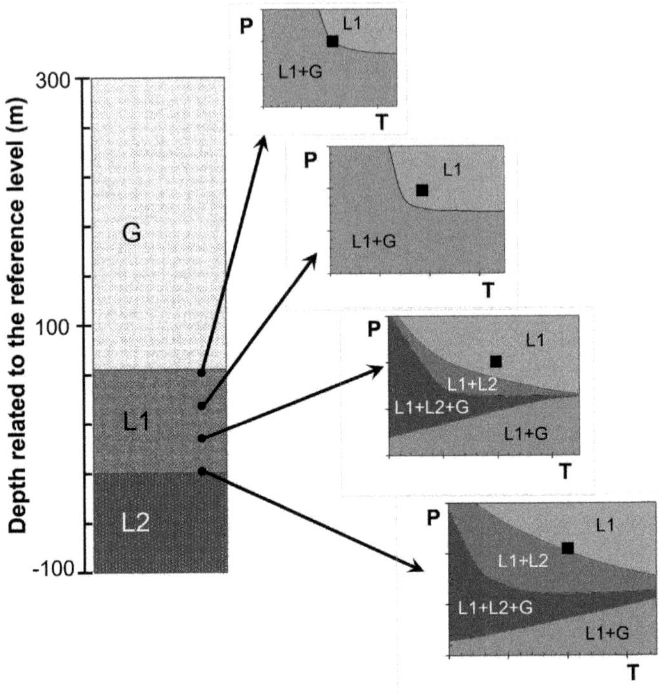

FIGURE 7.14 Phase envelopes changing above the reference level using fitted CPA-EoS: as the vertical region of the bubble-point curve approaches the reservoir conditions, the variation of P_{sat} increases rapidly, while the second liquid asphaltene-rich phase vanishes.

liquid-phase occurrence at reservoir conditions brought up the necessity of a more appropriate thermodynamic model to describe the molecular association of asphaltenes with both themselves and resins. This model still needs to be combined with the effect of thermal diffusion, by adjusting the Heat-of-Transport parameter of CO_2 (not considered so far) to reproduce the gas–oil contact, already confirmed experimentally by sampling the dense gas phase.

One observes that the intense variation on the P_{sat} profile, mainly in the oil zone, is not due to heavy components, nor to the closeness to the critical point but to the excess of CO_2, which leads the phase envelope to show a high-derivative of P_{sat} with temperature around the reservoir conditions.

The parameters of the adjusted model indicate that the isolate cubic part of CPA-EoS is able to reproduce the properties measured along depth, although it cannot predict the second liquid-phase in the compositional grading calculations. In commercial reservoir simulators that work only with vapor–liquid equilibrium, a conventional well-fitted model can compensate the lack of information arising from such unavailability, providing production curves still reliable. In the case of this reservoir specifically, new studies must be done, comparing the performance of both models for the same numerical domain and drainage plan. Model parameters might be readjusted based on new compositional grading data to be achieved.

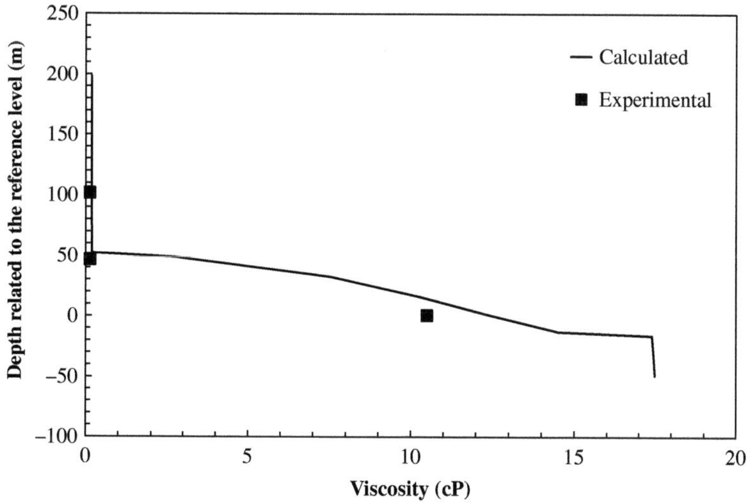

FIGURE 7.15 Viscosity variation along depth using JST correlation (Jossi et al., 1962), reduced densities coming from CPA-EoS in the nonisothermal run (PR-EoS shows similar results). The heaviest liquid-phase, for which there are no experimental data, can provide useful information regarding formation-test matching and reservoir simulation initialization.

One emphasizes that the solubility of CO_2 in water (to be modeled with new cross-association parameters) was not considered and may influence the distribution profiles from the aquifer up.

REFERENCES

Carnahan, N.F., Starling, K.E., 1969. Equation of state for nonattracting rigid spheres. J. Chem. Phys. 51, 635–636.

Chapman, W.G., Gubbins, K.E., Jackson, G., Radosz, M., 1990. New reference equation of state for associating liquids. Ind. Eng. Chem. Res. 29, 1709–1721.

Economou, I.G., Donohue, M.D., 1991. Chemical, quasi-chemical and perturbation theories for associating fluids. AIChE J. vol. 37 (no 12), 1875–1894.

Firoozabadi, A., 2015. Thermodynamics and Applications of Hydrocarbons Energy Production. McGraw Hill Professional, New York.

Gonzalez, D.L., Vargas, F.M., Hirasaki, G.J., Chapman, W.G., 2008. *Modeling study of CO_2-induced asphaltene precipitation*. Energy Fuels 22, 757–762.

Huang, S.H., Radosz, M., 1990. Equation of state for small, large, polydisperse and associationg molecules: extension to fluid mixtures. Ind. Eng. Chem. Res. 29, 2284.

Jackson, G., Chapman, W.G., Gubbins, K.E., 1988. Phase equilibria of associating fluids. Spherical molecules with multiple bonding sites. Mol. Phys. Vol. 65 (No 1), 1–31.

Jindrová, T., Mikyska, J., Firoozabadi, A., 2016. Phase behavior modeling of bitumen and light normal alkanes and CO_2 by PR-EOS and CPA-EOS. Energy Fuels 30, 515–525.

Jossi, J.A., Stiel, L.I., Thodos, G., 1962. The viscosity of pure substances in the dense gaseous and liquid phases. AIChE J. 8 (1), 59–63.

Kontogeorgis, G.M., Folas, G.K., 2010. Thermodynamic Models for Industrial Applications: From Classical to Advanced Mixing Rules to Association Theories. John Wiley & Sons Ltd, Chichester, UK.

Kontogeorgis, G.M., Michelsen, M.L., Folas, G.K., Derawl, S., von Solms, N., Stenby, E.H., 2006a. Ten years with the CPA (Cubic-Plus-Association) equation of state. Part 1. Pure compounds and self-associating systems. Ind. Eng. Chem. Res. 45 (14), 4855–4868.

Kontogeorgis, G.M., Michelsen, M.L., Folas, G.K., Derawl, S., von Solms, N., Stenby, E.H., 2006b. Ten years with the CPA (Cubic-Plus-Association) equation of state. Part 2. Cross-associating and multicomponent systems. , Ind. Eng. Chem. Res. 45 (14), 4869–4878.

Kontogeorgis, G.M., Voutsas, E.C., Yacoumis, I.V., Tassios, D.P., 1996. An equation of state for associating fluids. Ind. Eng. Chem. Res. 35, 4310–4318.

Li, Z., Firoozabadi, A., 2009. Cubic-plus-association equation of state for water-containing mixtures: is "cross association" necessary? AIChE J. 55 (7), 1803–1813.

Li, Z., Firoozabadi, A., 2010a. Modeling asphaltene precipitation by *n*-alkanes from heavy oils and bitumens using cubic-plus-association equation of state. Energy Fuels 24, 1106–1113.

Li, Z., Firoozabadi, A., 2010b. Cubic plus association equation of state for asphaltene precipitation in live oils. Energy Fuels 24, 2956–2963.

Mullins, O.C., 2010. The modified Yen model. Energy Fuels 24 (4), 2179–2207.

Wertheim, M.S., 1984a. Fluids with highly directional attractive forces. I—Statistical thermodynamics. J. Stat. Phys. 35, 19–34.

Wertheim, M.S., 1984b. Fluids with highly directional attractive forces. II—Thermodynamic perturbation theory and integral equations. J. Stat. Phys. 35, 35–47.

Zuo, J.Y., Mullins, O.C., Freed, D., Elshahawy, H., Dong, C., Seifert, D.J., 2013. Advances in the Flory–Huggins–Zuo equation of state for asphaltene gradients and formation evaluation. Energy Fuels 27 (4), 1722–1735.

Chapter 8

General Comments and Perspectives

The results presented along this textbook do not lead us to definitive conclusions. But do provide a basis for new discussions and elaboration of new hypothesis, premises, and ideas that would certainly be of help to increase the reliability of the compositional grading predictions. We reinforce that this discipline, although fascinating from the point of view of aggregating value from theory formulation to its final application, must be considered as one more important tool on predicting fluid properties and settling diagnostics of connection/disconnection in oil and gas reservoirs. We strongly recommend the interaction between fluid-modeling engineers with geologists, geochemists, and geophysicists that may bring up more information to corroborate the conclusions generated by grading calculations. In many occasions, the acquired information by these correlate-area professionals serve as inputs for the formulation or reformulation of problems. In this chapter, we enumerate some suggestions for future works that can help improve current predictions; some of them already going on in new joint venturing projects between academy and industry.

8.1 NEW THERMAL DIFFUSION MODELS

The semiempirical character of the Heat-of-Transport model described by Glasstone et al. (1941) and Dougherty and Drickamer (1955) points itself to the need of a more consistent approach from a theoretical point of view for the energy transport at microscopic level. The inherent difficulty to obtain reliable heat transfer experimental data (or Soret coefficient) for mixtures of interest in different conditions of temperature, pressure, and composition hampers the development of a more reliable model. A recent development of an experimental equipment to work in this problem, at least related to diffusion coefficient and Soret coefficient experimental data of binary mixtures, was proposed by Hashmi et al. (2016). Another important (and complementary) strategy is the development of molecular simulation methods (Furtado et al., 2015a,b) to obtain these coefficients. In the same way Firoozabadi et al. (2000) generalized the original expression for multicomponent mixtures, we

Compositional Grading in Oil and Gas Reservoirs. DOI: http://dx.doi.org/10.1016/B978-0-12-812452-9.00008-3

expect to generate new and more robust models, with adjustable parameters that show more consistency compared to τ_i. Parallel to this direction, which results are expected to be available in the medium to long term, we also suggest a consistency check between both main semiempirical models reviewed in the literature and presented in Chapter 4 (Firoozabadi et al., 2000; Pedersen and Lindeloff, 2003). So, we propose a third (maybe more robust) procedure, equating both expressions. The expression generalized by Firoozabadi et al. (2000) for the heat-of-transport in multicomponent mixtures is given by

$$Q_i^* = -\frac{1}{\tau_i}\overline{U}_i + \frac{\overline{V}_i}{\sum_{j=1}^{nc} x_j \overline{V}_j} \sum_{k=1}^{nc} x_k \frac{1}{\tau_k}\overline{U}_k, \quad \text{for } i = 1, \ldots, nc \tag{8.1}$$

For the nc^{th} component, we have

$$Q_{nc}^* = -\frac{1}{\tau_{nc}}\overline{U}_{nc} + \frac{\overline{V}_{nc}}{\sum_{j=1}^{nc} x_j \overline{V}_j} \sum_{k=1}^{nc} x_k \frac{1}{\tau_k}\overline{U}_k \tag{8.2}$$

Pedersen and Lindeloff (2003) have proposed:

$$Q_i^* - Q_{nc}^* = \overline{h}_{nc} - \overline{h}_i, \quad \text{for } i = 1, \ldots, nc - 1 \tag{8.3}$$

Replacing Eqs. (8.1) and (8.2) into Eq. (8.3), we have

$$\begin{aligned} &-\frac{1}{\tau_i}\overline{U}_i + \frac{\overline{V}_i}{\sum_{j=1}^{nc} x_j \overline{V}_j} \sum_{k=1}^{nc} x_k \frac{1}{\tau_k}\overline{U}_k + \frac{1}{\tau_{nc}}\overline{U}_{nc} \\ &-\frac{\overline{V}_{nc}}{\sum_{j=1}^{nc} x_j \overline{V}_j} \sum_{k=1}^{nc} x_k \frac{1}{\tau_k}\overline{U}_k = \overline{h}_{nc} - \overline{h}_i, \quad \text{for } i = 1, \ldots, nc - 1 \end{aligned} \tag{8.4}$$

where

$$\overline{h}_i = h_i^{\text{IG}}(T) + \overline{h}_i^R, \quad \text{for } i = 1, \ldots, nc \tag{8.5}$$

where $\overline{h}_i^R$ is the partial molar residual enthalpy of the component i, and h_i^{IG} is the enthalpy of the component i in a hypothetical ideal-gas condition (hence the superscript "IG").

And also,

$$h_i^{\text{IG}}(T) = h_i^{\text{IG}}(273.15\ \text{K}) + \int_{273.15}^{T} Cp_i dT, \quad \text{for } i = 1, \ldots, nc \tag{8.6}$$

For which, Pedersen and Hjermstad (2006) propose

$$h_i^{\text{IG}}(273.15\ \text{K}) = (-1342 + 8.367 M_i)R, \quad \text{for } i = 1, \ldots, nc \tag{8.7}$$

which is not convenient from the point of view of the chemical nature of hydrocarbons, as they should not have their ideal-gas enthalpies reestimated case to case, depending on the reservoir.

Therefore, given a methodology suitable to replace Eq. (8.7) on the calculation of the ideal-gas enthalpies, mainly for the nontabled pseudocomponents, parameters τ_i might be reestimated by Eq. (8.4) over a wide database; built up, for example, with the case studies we went through in this text.

8.2 COMPOSITIONAL GRADING CONSIDERING SOLUBILITY OF CO_2 IN WATER

For fields with a considerable CO_2 content, the solubility of this component in both connate oil-zone and aquifer water can influence the distribution of the other components with impacts still unknown on the prediction of composition of nondrilled areas. Illustratively, we present that, for isothermal systems, the gravity action prevents water in the oil-zone without considering the capillary curve. Besides, the utilization of a thermodynamic model suitable for polar multiphase associative systems (e.g., CPA-EoS) is indispensable. Fig. 8.1 shows our proposal to modeling this type of problem. Beyond the aquifer, kept below the oil zone by gravity, the water, due to capillarity, manages to occupy the tightest porous of the oil zone, which can be modeled as a second phase separated from the oil in a same depth level by rigid, diathermic, and permeable membranes.

Admitting the system of Fig. 8.1 to be isothermal, one can write the fugacity-equality equations for each component for the levels where both oil and water coexist:

$$f_i^{\text{oil}} = f_i^{\text{oil,ref}} \exp\left(-\frac{M_i g(z - z_{\text{ref}})}{RT} \right), \quad \text{for } i = 1, \ldots, nc \tag{8.8}$$

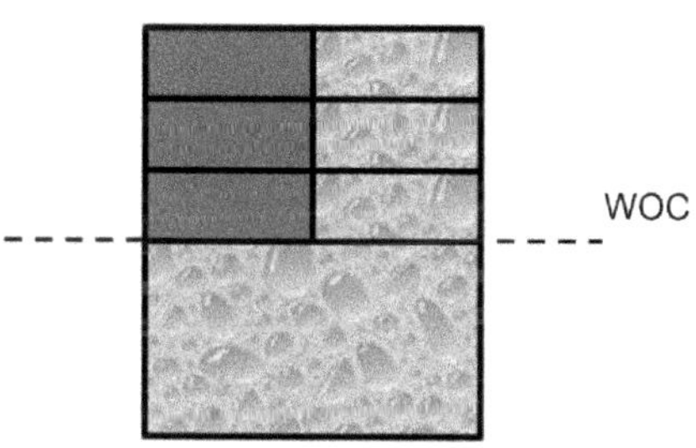

FIGURE 8.1 Modeling scheme of the compositional grading problem considering the solubility of CO_2 (and hydrocarbons) in water: above the water–oil contact, the transition and oil zones are modeled by discrete depth levels which separate the oil (brown zone, or *dark gray* in print versions) from the aqueous phase (*blue* zone, or *light gray* in print versions) by rigid, diathermic and permeable membranes. The capillary-pressure curve and an equation-of-state suitable for polar systems are required.

$$f_i^{\text{oil}} = f_i^{aq}, \quad \text{for } i = 1, \ldots, nc \tag{8.9}$$

where z_{ref} is the reference level; f_i^{aq}, f_i^{oil}, and $f_i^{\text{oil,ref}}$ are the fugacities of component i in the aqueous, oil and oil at reference phases, respectively.

Besides, the pressures of both oil (P^{oil}) and aqueous (P^{aq}) phases are related by the capillary-pressure (P_c) curve, measured in the laboratory:

$$P^{\text{oil}} - P^{aq} = P_c(S_w) \tag{8.10}$$

where S_w is the water saturation, i.e., the water volume fraction in the porous medium.

The problem might then be formulated as follows: given T, $P^{\text{oil}}_{\text{ref}}$, $\underline{x}^{\text{oil}}_{\text{ref}}$ and the capillary-pressure curve, obtain P^{oil}, $\underline{x}^{\text{oil}}$, P^{aq}, $\underline{x}^{aq}$, and S_w along depth.

There are still two additional decoupled relations:

$$P^{\text{oil}} - P^{aq} = \frac{2\sigma}{r} \tag{8.11}$$

where σ is the surface tension, obtained, for example, by the *Parachor correlation* (Danesh, 1998), and r is the pore radius as a function of depth.

There is also the restriction between the pressure gradients from the WOC depth (z^{WOC}):

$$P^{\text{oil}} - P^{aq} = \int_{z^{WOC}}^{z} \left(\rho^{aq} - \rho^{\text{oil}}\right) g dz \tag{8.12}$$

Electrolytes (which modeling will not be discussed in this work) might also be considered, influencing the solubility of the components in water directly or via salinity, which can be estimated in connate water to fit the experimental data. There are empirical correlations that predict the solubility of CO_2 in water as a function of salinity (Gonçalves, 2014), which reduces the complexity of the equation-of-state used.

8.3 SYSTEMS WITH CHEMICAL REACTIONS

Water-injection projects (or water alternate gas, WAG) in carbonate reservoirs might present rock-fluid interactions, like the acid attack to the reactive porous media, generating solution ions which influence the CO_2 distribution coefficient (Nghiem et al., 2011). Coreflooding experiments in reactive-rock plugs might be useful for estimation of kinetic parameters in the chemical reactions involved (Pegoraro, 2012). In this type of reservoir simulation, both permeability and porosity change along time due to dissolution and precipitation of calcium and magnesium carbonates, among other salts, on the rock matrix. Other inert ions such as Na^+ and Cl^- also influence indirectly the equilibrium of the reaction, described in Eq. (8.13a), and also the kinetics of the reactions, Eqs. (8.13b) and (8.13c), due to their high concentration and, consequently, strong effect in the ionic strength of the solution (Nghiem et al., 2011).

$$CO_2 + H_2O \leftrightarrow H^+ + HCO_3^- \tag{8.13a}$$

$$CaCO_3 + H^+ \rightarrow Ca^{2+} + HCO_3^- \tag{8.13b}$$

$$CaMg(CO_3)_2 + 2H^+ \rightarrow Ca^{2+} + Mg^{2+} + 2HCO_3^- \tag{8.13c}$$

For calculations of static or steady-state compositional grading, besides the "fugacity-equality" and capillary-pressure equations (Eqs. (8.8)–(8.10)), one has to consider the chemical equilibrium previously assumed along geological time in the water–oil–rock system. The simulation of 2D or 3D domain must include source terms by chemical reactions in the differential mass-balance equations showed in Table 4.1. The set of composition at the reference level in the oil zone is no longer enough to get the problem well posed, as the ions do not distribute, remaining only in the water phase. It is worthy to recall that the ionic concentration is measured at laboratory conditions after flashing the acquifer samples. The calculation of the ionic concentration to be set in the water zone and the parameter fitting of the unified thermodynamic model for the whole system remains a challenge for the resolution of multiphase compositional grading. For rocks containing anhydrite (calcium sulfate) in their framework, contaminants like H_2S might be generated, for example, by thermochemical reduction of sulfate, a reaction that involves hydrocarbons present in the oil, mainly methane:

$$CH_4 + CaSO_4 \rightarrow H_2S + H_2O + CaCO_3 \tag{8.14}$$

This reaction (Eq. (8.14)) is favored at temperatures above 130°C. The field of Ghawar, in Saudi Arabia, shows continuous production of H_2S, observed since its discovery till nowadays (Temeng et al., 1998). The sulfide is generated only from a certain depth, for which, due to the geothermal gradient, this temperature limit is achieved. The slow diffusion process makes lower concentrations to be observed at shallower depths, where the temperature is still impeditive for the reaction. According to Temeng et al. (1998), regardless the dispersion of the data, the concentrations of all hydrocarbons decrease with depth, including C_{7+}. All of them are well correlated with H_2S contents. An interesting work proposal would calculate the compositional grading setting the reference level at a depth high enough for the sulfide concentration to be zero. The difference between the predictions of such calculations for the other depths and the values observed in practice by Temeng et al. (1998) is related to the production rate of H_2S, which, if determined, might be useful to predict the sulfide contents along the productive life of the field. Another proposal would be to include both H_2S and methane consumption rate in the mass-balance equations of these components and reestimate kinetic parameters to reproduce experimental data. The inclusion of similar terms for heavier hydrocarbons can be done gradually, in order to fine tune the compositions observed around the generation depth.

8.4 OTHER RECOMMENDATIONS

We used the CPA-EoS to describe a reservoir behavior containing asphaltenes and high contents of CO_2. However, there are other proposals in the literature, such as SAFT-EoS. One suggests to evaluate the performance of SAFT-EoS (till the perturbation chain term, A^{chain}) on reproducing the asphaltene onset curves (Gonzalez et al., 2008) in other case studies, in comparison to CPA-EoS. With respect to reservoir simulation and 2D and 3D compositional grading calculations, there may be convergence problems in complex systems (miscible-gas injection, asphaltene precipitation, WAG, etc.) due to limitations of the traditional flash algorithms implemented so far. Similar to those already shown in Chapter 2, these algorithms consider only the vapor–liquid equilibrium with cubic equations of state, water being modeled separately considered as immiscible in the condensed phase. Besides, they present robustness limitations in the vicinity of the critical region, certainly reached in the swelling test as one sees the switching from bubble to dew point in saturation pressure. In the case of compositional grading calculations in multiphase systems, more robust algorithms, suitable to deal with molecular associations, must be used. The simulation of the CO_2–water–oil system in the example described in Section 6.3 (Chapter 6), which could confirm the steady-state achieving above the paleocontact, would be made possible by the simultaneous description of all three liquid phases (water, light, and heavy oil, and also the gas cap) that surely would show when the oil entered the trap. We are developing a PVT simulator containing a similar optimization algorithm as described in Appendix E, to carry out the CPA-EoS parameters regression based mainly on the liquid–liquid–vapor equilibrium data of systems containing both CO_2 and asphaltenes, like the one described in the example of Section 7.3. Briefly speaking, unifying the modeling of all fluids present in the reservoir, including water and electrolytes and their interactions with the rock, is a real necessity due to the more and more complex phase behavior of the recently found reservoirs, especially those rich in CO_2. One expects this challenge to unfold in opportunities for young investigators, increasing the comprehension on the subject and, consequently, the recovery efficiencies of the new research and development projects.

REFERENCES

Danesh, A., 1998. PVT and Phase Behavior of Petroleum Reservoir Fluids. Elsevier Science, Amsterdam (Netherlands).

Dougherty Jr, E.L., Drickamer, H.G., 1955. Thermal diffusion and molecular motion in liquids. J. Phys. Chem. 59 (5), 443–449.

Firoozabadi, A., Ghorayeb, K., Shukla, K., 2000. Theoretical model of thermal diffusion factors in multicomponent mixtures. AIChE J. 46 (5), 892–900.

Furtado, F.A., Silveira, A.J., Abreu, C.R.A., Tavares, F.W., 2015a. Non-equilibrium molecular dynamics used to obtain Soret coefficients of binary hydrocarbon mixtures. Braz. J. Chem. Eng. 32 (3), 683–698.

Furtado, F.A., Abreu, C.R., Tavares, F.W., 2015b. A low-disturbance nonequilibrium molecular dynamics algorithm applied to the determination of thermal conductivities. AIChE J. 61 (9), 2881–2890.

Glasstone, S., Laidler, K.J., Eyring, H., 1941. The Theory of Rate Processes. McGraw Hill Book Co, New York (USA).

Gonçalves, N.P.F., "*Solubilidade do CO_2 em aqüíferos salinos de reservatórios de petróleo*", Tese de Mestrado, Universidade de Aveiro, Departamento de Química, 2014.

Gonzalez, D.L., Vargas, F.M., Hirasaki, G.J., Chapman, W.G., 2008. *Modeling study of CO_2-induced asphaltene precipitation*. Energy Fuels 22, 757–762.

Hashmi, S.M., Senthilnathan, S., Firoozabadi, A., 2016. Thermodiffusion of polycyclic aromatic hydrocarbons in binary mixtures. J. Chem. Phys. 145 (18), 184503. Available from: http://dx.doi.org/10.1063/1.4966191.

Nghiem, L., Shrivastava, V., Kohse, B., "*Modeling aqueous phase behavior and chemical reactions in compositional simulation*", SPE 141417, Reservoir Simulation Symposium, The Woodlands, Texas, EUA, February, 2011.

Pedersen, K.S., Hjermstad, H.P., 2006. Modeling of large hydrocarbon compositional gradient. Abu Dhabi International Petroleum Exhibition and Conference. Society of Petroleum Engineers, SPE-101275-MS, November, 2006.

Pedersen, K.S., Lindeloff, N., "*Simulations of compositional gradients in hydrocarbon reservoirs under the influence of a temperature gradient*", SPE 84364, SPE Annual Technical Conference and Exhibition, Colorado, USA, 2003.

Pegoraro, R.T., 2012. *Escoamento Trifásico em Meios Porosos: Permeabilidade Relativa Óleo-Gás-Água*", MSc. Thesis. Federal University of Rio de Janeiro, Brazil.

Temeng, K.O., Al-Sadeg, M.J., Al-Mulhim, W.A., "*Compositional grading in the Ghawar Khuff reservoirs*", SPE 49270, Annual Technical Conference and Exhibition, New Orleans, Lousiana, 1998.

Appendix A

Transport Equations Derived from Irreversible Thermodynamics

In this appendix, we will present the complete derivation of the entropy-generation terms based on principles of irreversible thermodynamics, obtaining the expressions necessary to the calculation of the mass diffusive fluxes as a function of temperature (T), pressure (P), and composition ($\underline{x}$), to be used in Eq. (4.10) (see Chapter 4, Irreversible Thermodynamics Applied to Reservoir Engineering). We will also derive the expressions necessary to transform these fluxes to the molar-averaged velocity axis, so they can be used in Eq. (4.13) (see Chapter 4, Irreversible Thermodynamics Applied to Reservoir Engineering).

A.1 ENERGY TRANSPORT EQUATION

The works of De Groot and Mazur (1962), Fitts (1962), and Haase (1969) are the main references in Irreversible Thermodynamics. Among the general-approach derivations presented by these authors, we will proceed now to the ones that most apply to the modeling of petroleum reservoirs with entropy production. Following the nomenclature of Fitts (1962), associated with each mass element ρdV of a fluid is its extensive internal energy $\rho e dV$, where e is the specific internal energy of the fluid. Additionaly, the element of volume has a kinetic energy equal to $(1/2)\rho|\underline{u}|^2 dV$, where $|\underline{u}|^2 = \underline{u} \cdot \underline{u}$; $\underline{u}$ being the velocity vector of the fluid. The total energy of an arbitrary volume V of the fluid referent to an external fixed reference axis is therefore:

$$E_{\text{total},V} = \int_V \rho\left(e + (1/2)|\underline{u}|^2\right) dV \tag{A.1}$$

Its variation rate is

$$\frac{dE_{\text{total},V}}{dt} = \int_V \frac{\partial}{\partial t}\rho\left(e + (1/2)|\underline{u}|^2\right)dV \tag{A.2}$$

The total energetic flux (liquid rate that crosses volume V) through its surface A consists of two parts, a convective and a conductive one. Being $\underline{u} \cdot \underline{n}\, dA$ (with positive normal vector pointing outwards), the volume of fluid that crosses V per unit time through surface A, the energy rate flowing by convection is given by $\int_A \rho(e + (1/2)|\underline{u}|^2)\underline{u} \cdot \underline{n}\, dA$. The conduction stream is the energy flow due to both diffusion and heat conduction, henceforth named $\underline{j}_E$.

So, the liquid rate of energy ("inlets" – "outlets") that crosses V by surface A is given by

$$\dot{E}_{\text{cross}} = -\int_A \left[\underline{j}_E + \rho\left(e + (1/2)|\underline{u}|^2\right)\underline{u}\right] \cdot \underline{n}dA \tag{A.3}$$

The work W exerted on the fluid in V also consists of two parts: the one realized by body forces ($\underline{X}$), and the one realized by surface forces $\underline{\underline{\sigma}} \cdot \underline{n}\, dA$, where $\underline{\underline{\sigma}}$ is the stress tensor.

For the great majority of the cases studied here, the dominant body force is gravity ($\underline{X} = \underline{g}$). Therefore, one has

$$\frac{dW}{dt} = \int_V \sum_{i=1}^{nc} \underline{u}_i \cdot \rho_i \underline{X}_i dV + \int_A \underline{u} \cdot \underline{\underline{\sigma}} \cdot \underline{n}\, dA \tag{A.4}$$

Defining the diffusive flux of component i in relation to the bulk velocity of the fluid:

$$\underline{j}_i = \rho_i(\underline{u}_i - \underline{u}) \rightarrow \rho_i \underline{u}_i = \underline{j}_i + \rho_i \underline{u}, \quad \text{for } i = 1, 2, ..., nc \tag{A.5}$$

we have

$$\sum_{i=1}^{nc} \underline{u}_i \cdot \rho_i \underline{X}_i = \sum_{i=1}^{nc} (\underline{j}_i + \rho_i \underline{u}) \cdot \underline{X}_i = \sum_{i=1}^{nc} \underline{j}_i \cdot \underline{X}_i + \sum_{i=1}^{nc} \rho_i \underline{X}_i \cdot \underline{u} \tag{A.6}$$

So, we can write

$$\frac{dW}{dt} = \int_V \left[\sum_{i=1}^{nc} (\underline{j}_i \cdot \underline{X}_i + \rho_i \underline{X}_i \cdot \underline{u})\right] dV + \int_A \underline{u} \cdot \underline{\underline{\sigma}} \cdot d\underline{A} \tag{A.7}$$

Then, the energy variation rate in V and the rate transported across surface A (which already includes the heat) must be equated to the work to give

$$\int_V \frac{\partial}{\partial t}\rho(e+(1/2)|\underline{u}|^2)\,dV + \int_A \left[\underline{j}_E + \rho(e+(1/2)|\underline{u}|^2)\underline{u}\right]\underline{n}\,dA = \int_V \left[\sum_{i=1}^{nc}(\underline{j}_i\cdot\underline{X}_i + \rho_i\underline{X}_i\cdot\underline{u})\right]dV + \int_A \underline{u}\cdot\underline{\underline{\sigma}}\cdot\underline{n}\,dA \tag{A.8}$$

Applying Gauss theorem on the surface integrals, one gets

$$\int_A \underline{u}\cdot\underline{\underline{\sigma}}\cdot\underline{n}\,dA = \int_V \nabla\cdot(\underline{u}\cdot\underline{\underline{\sigma}})dV \tag{A.9}$$

$$\int_A \left[\underline{j}_E + \rho\left(e+(1/2)|\underline{u}|^2\right)\underline{u}\right]\underline{n}\,dA = \int_V \nabla\cdot\left[\underline{j}_E + \rho\left(e+(1/2)|\underline{u}|^2\right)\underline{u}\right]dV \tag{A.10}$$

Replace Eqs. (A.9) and (A.10) in Eq. (A.8) to get

$$\int_V \left\{\left[\frac{\partial}{\partial t}\rho\left(e+(1/2)|\underline{u}|^2\right)\right] + \nabla\cdot\left[\underline{j}_E + \rho\left(e+(1/2)|\underline{u}|^2\right)\underline{u}\right] - \nabla\cdot(\underline{u}\cdot\underline{\underline{\sigma}}) - \left[\sum_{i=1}^{nc}(\underline{j}_i\cdot\underline{X}_i + \rho_i\underline{X}_i\cdot\underline{u})\right]\right\}dV = 0 \tag{A.11}$$

Once volume V is arbitrary, the integrand of Eq. (A.11) must be null:

$$\frac{\partial}{\partial t}\rho\left(e+(1/2)|\underline{u}|^2\right) + \nabla\cdot\left[\rho\left(e+(1/2)|\underline{u}|^2\right)\underline{u}\right] = \sum_{i=1}^{nc}\underline{j}_i\cdot\underline{X}_i + \rho\underline{u}\cdot\underline{X} + \nabla\cdot(\underline{u}\cdot\underline{\underline{\sigma}}) - \nabla\cdot\underline{j}_E \tag{A.12}$$

Recognizing that $\frac{\partial\rho}{\partial t} + \nabla\cdot(\rho\underline{u}) = 0$ (continuity equation) and $\frac{\partial}{\partial t}\left(e+(1/2)|\underline{u}|^2\right) + \underline{u}\cdot\nabla\left(e+(1/2)|\underline{u}|^2\right) = \frac{D}{Dt}\left(e+(1/2)|\underline{u}|^2\right)$, expanding the term on the left-hand side of the equality in Eq. (A.12), we have

$$\begin{aligned}
&\frac{\partial}{\partial t}\rho\left(e+(1/2)|\underline{u}|^2\right) + \nabla\cdot\left[\rho(e+(1/2)|\underline{u}|^2)\underline{u}\right] = \frac{\partial\rho}{\partial t}(e+(1/2)|\underline{u}|^2) + \rho\frac{\partial}{\partial t}(e+(1/2)|\underline{u}|^2)\\
&+ \rho\underline{u}\cdot\nabla(e+(1/2)|\underline{u}|^2) + (e+(1/2)|\underline{u}|^2)\nabla\cdot(\rho\underline{u})\\
&= (e+(1/2)|\underline{u}|^2)\underbrace{\left[\frac{\partial\rho}{\partial t} + \nabla\cdot(\rho\underline{u})\right]}_{\text{zero}} + \rho\underbrace{\left[\frac{\partial}{\partial t}(e+(1/2)|\underline{u}|^2) + \underline{u}\cdot\nabla(e+(1/2)|\underline{u}|^2)\right]}_{\frac{D}{Dt}(e+(1/2)|\underline{u}|^2)}\\
&= \rho\frac{D}{Dt}(e+(1/2)|\underline{u}|^2) = \rho\frac{De}{Dt} + \rho 1/2\frac{D}{Dt}(\underline{u}\cdot\underline{u}) = \rho\frac{De}{Dt} + \rho\underline{u}\cdot\frac{D\underline{u}}{Dt}
\end{aligned} \tag{A.13}$$

Now, let us expand the term $\nabla \cdot (\underline{u} \cdot \underline{\underline{\sigma}})$ using index notation and knowing that $\underline{\underline{\sigma}}$ is symmetric:

$$\nabla \cdot (\underline{u} \cdot \underline{\underline{\sigma}}) = \sum_{k=1}^{3} \frac{\partial}{\partial x_k} \sum_{l=1}^{3} u_l \sigma_{lk} = \sum_{k=1}^{3} \sum_{l=1}^{3} \frac{\partial}{\partial x_k} u_l \sigma_{lk} = \sum_{k=1}^{3} \sum_{l=1}^{3} u_l \frac{\partial \sigma_{kl}}{\partial x_k} + \sum_{k=1}^{3} \sum_{l=1}^{3} \sigma_{lk} \frac{\partial u_l}{\partial x_k}$$
$$= \underline{u} \cdot (\nabla \cdot \underline{\underline{\sigma}}) + \underline{\underline{\sigma}} : \nabla \underline{u} \tag{A.14}$$

Substituting Eqs. (A.13) and (A.14) into Eq. (A.12), we have

$$\sum_{i=1}^{nc} \underline{j}_i \cdot \underline{X}_i + \rho \underline{u} \cdot \underline{X} + \underline{u} \cdot (\nabla \cdot \underline{\underline{\sigma}}) + \underline{\underline{\sigma}} : \nabla \underline{u} - \nabla \cdot \underline{j}_E = \rho \frac{De}{Dt} + \rho \underline{u} \cdot \frac{D\underline{u}}{Dt} \tag{A.15}$$

Rearranging the terms in Eq. (A.15), one can write

$$\rho \frac{De}{Dt} = \underline{\underline{\sigma}} : \nabla \underline{u} + \sum_{i=1}^{nc} \underline{j}_i \cdot \underline{X}_i - \nabla \cdot \underline{j}_E + \underline{u} \left(\frac{\rho \underline{X} + \nabla \cdot \underline{\underline{\sigma}} - \rho D\underline{u}}{Dt} \right) \tag{A.16}$$

The last term on the right-hand side of Eq. (A.16) vanishes because of momentum equation. Then, one finally gets to the energy transport equation:

$$\rho \frac{De}{Dt} = \underline{\underline{\sigma}} : \nabla \underline{u} + \sum_{i=1}^{nc} \underline{j}_i \cdot \underline{X}_i - \nabla \cdot \underline{j}_E \tag{A.17}$$

A.2 ENERGY TRANSPORT EQUATION BASED ON LOCAL EQUILIBRIUM

Let us derive Eq. (A.17) using the local equilibrium hypothesis. According to this hypothesis, each infinitesimal volume element would be in equilibrium, for the time of molecular relaxation is much lower than the characteristic time of local macroscopic phenomena and then the fundamental relations of Thermodynamics hold punctually. Knowing that the specific internal energy is given by the sum of mass partial internal energies $\tilde{e}_i$, one can write

$$\frac{dE_{\text{total},V}}{dt} = \int_V \frac{\partial}{\partial t} \left[\sum_{i=1}^{nc} \rho_i \left(\tilde{e}_i + (1/2) |\underline{u}_i|^2 \right) \right] dV \tag{A.18}$$

Using $\underline{j}_i = \rho_i (\underline{u}_i - \underline{u}) \rightarrow \underline{u}_i = \frac{1}{\rho_i} \underline{j}_i + \underline{u}$, one has

$$|\underline{u}_i|^2 = \underline{u}_i \cdot \underline{u}_i = \left(\frac{1}{\rho_i} \underline{j}_i + \underline{u} \right) \cdot \left(\frac{1}{\rho_i} \underline{j}_i + \underline{u} \right) = \frac{1}{\rho_i^2} |\underline{j}_i|^2 + \frac{2}{\rho_i} \underline{j}_i \cdot \underline{u} + \underline{u} \cdot \underline{u} \tag{A.19}$$

Diffusion is a slow process, mainly when the local-equilibrium postulate holds. Then, let us neglect the term $\frac{1}{\rho_i^2}\left|\underline{j}_i\right|^2$ and write

$$\sum_{i=1}^{nc}\rho_i\left(\tilde{e}_i+(1/2)\left|\underline{u}_i\right|^2\right)=\sum_{i=1}^{nc}\rho_i\left[\tilde{e}_i+(1/2)\left(\frac{2}{\rho_i}\underline{j}_i\cdot\underline{u}+\underline{u}\cdot\underline{u}\right)\right]=\rho e+(1/2)\rho\left|\underline{u}\right|^2 \tag{A.20}$$

So, the total rate of energy variation is identical to Eq. (A.2), which was derived without the assumption of local equilibrium.

The energy stream that crosses surface A of volume V due to material flux (convection) is given by

$$\dot{E}\Big|_{\substack{\text{through } A \text{ due to}\\ \text{mass flow}}}=\int_A\sum_{i=1}^{nc}\rho_i(\tilde{e}_i+(1/2)\left|\underline{u}_i\right|^2)\underline{u}_i\cdot\underline{n}\,dA \tag{A.21}$$

It is very important to mention that Eq. (A.21) differs from Eq. (A.3)—derived without the local equilibrium hypothesis—because it already contains both diffusive and convective fluxes but does not contain the heat, which makes now the totality of the conduction stream and must be added separately. We will call it "Pure Heat Flux" q' (not associated to the mass flow). So, Eq. (A.21) reads

$$\dot{E}\Big|_{\text{through } A}=\int_A\left[q'+\sum_{i=1}^{nc}+\rho_i\left(\tilde{e}_i+(1/2)\left|\underline{u}_i\right|^2\right)\underline{u}_i\right]\cdot\underline{n}\,dA \tag{A.22}$$

Giving that $\underline{u}_i=\frac{1}{\rho_i}\underline{j}_i+\underline{u}$ and $\left|\underline{u}_i\right|^2=\frac{2}{\rho_i}\underline{j}_i\cdot\underline{u}+\left|\underline{u}\right|^2$, we can write

$$\begin{aligned}\sum_{i=1}^{nc}\rho_i\left(\tilde{e}_i+(1/2)\left|\underline{u}_i\right|^2\right)\underline{u}_i&=\sum_{i=1}^{nc}\rho_i\left[\tilde{e}_i+(1/2)\left(\frac{2}{\rho_i}\underline{j}_i\cdot\underline{u}+\left|\underline{u}\right|^2\right)\right]\left[\frac{1}{\rho_i}\underline{j}_i+\underline{u}\right]\\&=\sum_{i=1}^{nc}\rho_i\tilde{e}_i\left[\frac{1}{\rho_i}\underline{j}_i+\underline{u}\right]+\sum_{i=1}^{nc}\rho_i(1/2)\left|\underline{u}\right|^2\left[\frac{1}{\rho_i}\underline{j}_i+\underline{u}\right]+\underbrace{\sum_{i=1}^{nc}\underline{j}_i\cdot\underline{u}\left[\frac{1}{\rho_i}\underline{j}_i+\underline{u}\right]}_{\text{zero}}\\&=\sum_{i=1}^{nc}\tilde{e}_i\underline{j}_i+\sum_{i=1}^{nc}\rho_i\tilde{e}_i\underline{u}+\underbrace{\sum_{i=1}^{nc}(1/2)\left|\underline{u}\right|^2\underline{j}_i}_{\text{zero}}+\sum_{i=1}^{nc}\rho_i(1/2)\left|\underline{u}\right|^2\underline{u}=\sum_{i=1}^{nc}\tilde{e}_i\underline{j}_i+\rho\left(e+(1/2)\left|\underline{u}\right|^2\right)\underline{u}\end{aligned} \tag{A.23}$$

So,

$$\int_A\left[q'+\sum_{l-1}^{nc}+\rho_i\left(\tilde{e}_i+(1/2)\left|\underline{u}_i\right|^2\right)\underline{u}_i\right]\cdot\underline{n}\,dA=\int_A\left[q'+\sum_{i-1}^{nc}\tilde{e}_i\underline{j}_i+\rho\left(e+(1/2)\left|\underline{u}\right|^2\right)\underline{u}\right]\cdot\underline{n}\,dA \tag{A.24}$$

Comparing Eq. (A.24) with Eq. (A.3)—which was derived without the local equilibrium hypothesis—we can write

$$\underline{j}_E = \underline{q}' + \sum_{i=1}^{nc} \underline{j}_i \tilde{e}_i \tag{A.25}$$

So we can recover the energy transport equation:

$$\rho \frac{De}{Dt} = \underline{\underline{\sigma}}:\nabla \underline{u} - \nabla \cdot \underline{q}' - \nabla \cdot \sum_{i=1}^{nc} \underline{j}_i \tilde{e}_i + \sum_{i=1}^{nc} \underline{j}_i \cdot \underline{X}_i \tag{A.26}$$

Now let us define another important heat-transfer rate, q:

$$\underline{q} \equiv \underline{q}' - \sum_{i=1}^{nc} \underline{j}_i P\tilde{v}_i \tag{A.27}$$

Replacing Eq. (A.25) into Eq. (A.27) one gets

$$\underline{q} \equiv \underline{j}_E - \sum_{i=1}^{nc} \underline{j}_i (\tilde{e}_i + P\tilde{v}_i) \tag{A.28}$$

or

$$\underline{j}_E = \underline{q} + \sum_{i=1}^{nc} \underline{j}_i \tilde{h}_i \tag{A.29}$$

Therefore, the energy transport equation in terms of $\underline{q}$ reads

$$\rho \frac{De}{Dt} = \underline{\underline{\sigma}}:\nabla \underline{u} - \nabla \cdot \underline{q} - \nabla \cdot \sum_{i=1}^{nc} \underline{j}_i \tilde{h}_i + \sum_{i=1}^{nc} \underline{j}_i \cdot \underline{X}_i \tag{A.30}$$

At this point, it might be important to the interested reader to relate the heat quantities, here defined according to Fitts (1962), with those of De Groot and Mazur (1962), another classic reference on the subject. De Groot and Mazur (1962) have defined Eq. (A.29) as follows:

$$J'_q = J_q - \sum_{i=1}^{nc} \underline{j}_i \tilde{h}_i \tag{A.31}$$

Comparing Eqs. (A.31) and (A.29), we may state that $\underline{q} = J'_q$ and $\underline{j}_E = J_q$.

A.3 ENTROPY BALANCE

The entropy-balance equation, coupled with the energy transport one assuming local equilibrium everywhere in the reservoir will provide the theoretical basis for the species distribution under entropy production due to irreversibilities in a globally nonequilibrium system. In this section, we will derive the entropy-balance equation and use the fundamental relation of

Thermodynamics, replacing Eq. (A.30) in it. Once this is done, we will have computed all sources of entropy production that govern the motion of the reservoir fluid inside the porous medium.

Let us start by deriving the entropy equation based on the conservation laws of any thermodynamic property Y. For the arbitrary control volume V, the general balance equation for Y is

$$\int_V \frac{\partial}{\partial t}(\rho y)dV = -\int_A \underline{N}_Y \cdot d\underline{A} + \int_V r_Y dV \tag{A.32}$$

where y is the property Y per unit mass; $\underline{N}_Y$ is the flux of Y across surface A of V; and r_Y is the generation term of Y per unit volume in the time unit. Applying Gauss theorem on the surface integral and regrouping the terms, we have

$$\int_V \left[\frac{\partial}{\partial t}(\rho y) + \nabla \cdot N_Y - r_Y\right] dV = 0 \tag{A.33}$$

Once V is arbitrary, the only way of satisfying Eq. (A.33) is nullifying the integrand, so

$$\left[\frac{\partial}{\partial t}(\rho y) + \nabla \cdot N_Y - r_Y\right] = 0 \tag{A.34}$$

In the particular case of Y being the total mass, $y = 1$, $\underline{N}_Y = \rho \underline{u}$, and $r = 0$ (because there is no mass generation). So, Eq. (A.34) reads

$$\frac{\partial \rho}{\partial t} + \nabla \cdot (\rho \underline{u}) = 0 \tag{A.35}$$

If, however, Y is the mass of any component i, $y = w_i$ (mass fraction), $N_Y = \rho_i \underline{u}_i$, and r_i is the generation rate of i by chemical reactions. So, one can write,

$$\left[\frac{\partial}{\partial t}(\rho w_i) + \nabla \cdot (\rho_i u_i) - r_i\right] = 0 \tag{A.36}$$

Knowing that $w_i = \rho_i / \rho$ and $\underline{j}_i = \rho_i(\underline{u}_i - \underline{u})$, Eq. (A.36) might be rewritten as

$$\rho \frac{\partial w_i}{\partial t} + w_i \frac{\partial \rho}{\partial t} + \nabla \cdot \left[\rho_i \left(\frac{\underline{j}_i}{\rho_i} + \underline{u}\right)\right] = r_i \tag{A.37}$$

$$\rho \frac{\partial w_i}{\partial t} + w_i \frac{\partial \rho}{\partial t} + \nabla \cdot \underline{j}_i + \nabla \cdot (\rho_i \underline{u}) = r_i \tag{A.38}$$

But $\rho_i = \rho w_i$, so,

$$\rho \frac{\partial w_i}{\partial t} + w_i \frac{\partial \rho}{\partial t} + \nabla \cdot \underline{j}_i + w_i \nabla \cdot (\rho \underline{u}) + \rho \underline{u} \cdot \nabla w_i = r_i \tag{A.39}$$

Grouping the terms, one has

$$\rho\left[\frac{\partial w_i}{\partial t} + \underline{u}\cdot\nabla w_i\right] + w_i\left[\frac{\partial \rho}{\partial t} + \nabla\cdot(\rho\underline{u})\right] + \nabla\cdot\underline{j}_i = r_i \tag{A.40}$$

The second term of the left-hand side of Eq. (A.40) is zero because of the continuity equation (Eq. (A.35)). Applying the Substantial Derivative concept, one has

$$\rho\frac{Dw_i}{Dt} + \nabla\cdot\underline{j}_i = r_i \tag{A.41}$$

Now, finally, suppose that Y is the entropy, so $y = s$ and Eq. (A.34) reads

$$\left[\frac{\partial}{\partial t}(\rho s) + \nabla\cdot\underline{N}_S - r_S\right] = 0 \tag{A.42}$$

It is important to emphasize that in Eq. (A.42), the flux $\underline{N}_S$ is not given only by $\rho s\underline{u}$, since the diffusive individual fluxes also carry entropy. The substantial derivative for ρs gives

$$\frac{D(\rho s)}{Dt} = \frac{\partial(\rho s)}{\partial t} + \underline{u}\cdot\nabla(\rho s) \tag{A.43}$$

Replacing Eq. (A.43) in Eq. (A.42), one gets

$$\frac{D(\rho s)}{Dt} - \underline{u}\cdot\nabla(\rho s) + \nabla\cdot Ns - r_S = 0 \tag{A.44}$$

$$\rho\frac{Ds}{Dt} + \underbrace{s\frac{D\rho}{Dt} + \rho s\nabla\cdot\underline{u}}_{\text{zero (continuity eq.)}} - \nabla\cdot(\rho s\underline{u}) + \nabla\cdot Ns - r_S = 0 \tag{A.45}$$

$$\rho\frac{Ds}{Dt} + \nabla\cdot(\underline{N}_S - \rho s\underline{u}) - r_S = 0 \tag{A.46}$$

Now, let us define $\underline{j}_S \equiv \underline{N}_S - \rho s\underline{u}$, the entropy flux due only to the transport of heat by diffusion, for the convective flux was subtracted. Let us also redefine the term of entropy generation as $r_S \equiv \frac{\Phi}{T}$, so we can write

$$\rho\frac{Ds}{Dt} = \frac{\Phi}{T} - \nabla\cdot\underline{j}_S \tag{A.47}$$

Based on local equilibrium hypothesis, one can use the fundamental relation in the following way:

$$\frac{De}{Dt} = T\frac{Ds}{Dt} - P\frac{D(1/\rho)}{Dt} + \sum_{i=1}^{nc}\mu_i\frac{Dw_i}{Dt} \tag{A.48}$$

Rewriting Eq. (A.48), one has

$$\rho\frac{Ds}{Dt}=\frac{\rho}{T}\frac{De}{Dt}-\frac{P}{\rho T}\frac{D\rho}{Dt}-\frac{\rho}{T}\sum_{i=1}^{nc}\mu_i\frac{Dw_i}{Dt} \tag{A.49}$$

Our task now is to replace the expressions for $\frac{De}{Dt}$ (Eq. (A.30)), $\frac{D\rho}{Dt}$ (continuity equation) and $\frac{Dw_i}{Dt}$ (Eq. (A.41)) in Eq. (A.49) and then replace the resultant expression on the left-hand side of Eq. (A.47).

First, let us obtain the relation between enthalpy and entropy to use in Eq. (A.30):

$$\tilde{h}_i=\tilde{e}_i+P\tilde{v}_i=T\tilde{s}_i-P\tilde{v}_i+\tilde{\mu}_i+P\tilde{v}_i\rightarrow\tilde{h}_i=T\tilde{s}+\tilde{\mu}_i \tag{A.50}$$

where $\tilde{\mu}_i=\mu_i/M_i$ is the chemical potential per unit mass (as well as other thermodynamic properties that are in mass basis).

$$\rho\frac{De}{Dt}=\underline{\underline{\sigma}}:\nabla\underline{u}-\nabla\cdot\underline{q}-\nabla\cdot\sum_{i=1}^{nc}\underline{j}_i\left(T\tilde{s}_i+\tilde{\mu}_i\right)+\sum_{i=1}^{nc}\underline{j}_i\cdot\underline{X}_i \tag{A.51}$$

Note that,

$$T\nabla\cdot\left(\underline{q}/T\right)=\nabla\cdot\underline{q}+T\underline{q}\nabla\left(1/T\right)=\nabla\cdot\underline{q}+T\underline{q}\left(-\frac{1}{T^2}\right)\nabla T=\nabla\cdot\underline{q}-\underline{q}\nabla\ln T \tag{A.52}$$

Replacing $\nabla\cdot\underline{q}=T\nabla\cdot(\underline{q}/T)+\underline{q}\cdot\nabla\ln T$ into Eq. (A.48), we have

$$\rho\frac{De}{Dt}=\underline{\underline{\sigma}}:\nabla\underline{u}-T\nabla\cdot(\underline{q}/T)-\underline{q}\cdot\nabla\ln T-\nabla\cdot\sum_{i=1}^{nc}\underline{j}_iT\tilde{s}_i-\nabla\cdot\sum_{i=1}^{nc}\underline{j}_i\tilde{\mu}_i+\sum_{i=1}^{nc}\underline{j}_i\cdot\underline{X}_i \tag{A.53}$$

Note also that

$$\nabla\cdot\sum_{i=1}^{nc}T\underline{j}_i\tilde{s}_i=T\nabla\cdot\sum_{i=1}^{nc}\underline{j}_i\tilde{s}_i+\left(\sum_{i=1}^{nc}\underline{j}_i\tilde{s}_i\right)\cdot\nabla T \tag{A.54}$$

Redefining $\underline{j}_S$ in more detail,

$$\underline{j}_S=\frac{\underline{q}}{T}+\sum_{i=1}^{nc}\underline{j}_i\tilde{s}_i \tag{A.55}$$

we can write

$$\rho\frac{De}{Dt}=\underline{\underline{\sigma}}:\nabla\underline{u}-T\nabla\cdot\underline{j}_S-\underline{q}\cdot\nabla\ln T-\left(\sum_{i=1}^{nc}\underline{j}_i\tilde{s}_i\right)\cdot\nabla T-\nabla\cdot\sum_{i=1}^{nc}\underline{j}_i\tilde{\mu}_i+\sum_{i=1}^{nc}\underline{j}_i\cdot\underline{X}_i \tag{A.56}$$

Now, let us make

$$\nabla \cdot \sum_{i=1}^{nc} \underline{j}_i \tilde{\mu}_i = \sum_{i=1}^{nc} \underline{j}_i \cdot \nabla \tilde{\mu}_i + \sum_{i=1}^{nc} \tilde{\mu}_i \nabla \cdot \underline{j}_i \tag{A.57}$$

and also,

$$\nabla \tilde{\mu}_i = \nabla_T \tilde{\mu}_i - \tilde{s}_i \nabla T \tag{A.58}$$

to write

$$\rho \frac{De}{Dt} = \underline{\underline{\sigma}} : \nabla \underline{u} - T \nabla \cdot \underline{j}_S - \underline{q} \cdot \nabla \ln T - \sum_{i=1}^{nc} \underline{j}_i \nabla_T \tilde{\mu}_i - \sum_{i=1}^{nc} \tilde{\mu}_i \nabla \cdot \underline{j}_i + \sum_{i=1}^{nc} \underline{j}_i \cdot \underline{X}_i \tag{A.59}$$

Now, let us rewrite the continuity equation using the substantial derivative:

$$\frac{D\rho}{Dt} + \rho \nabla \cdot \underline{u} = 0 \tag{A.60}$$

We can also rewrite the divergent of velocity as

$$\nabla \cdot \underline{u} = \underline{\underline{I}} : \nabla \underline{u} \tag{A.61}$$

where $\underline{\underline{I}}$ is the identity tensor. So, one has

$$\frac{D\rho}{Dt} = -\rho \underline{\underline{I}} : \nabla \underline{u} \tag{A.62}$$

Replacing Eqs. (A.41), (A.59), and (A.62) in Eq. (A.49), one gets finally:

$$\begin{aligned} \rho \frac{Ds}{Dt} = {} & \frac{1}{T} \underline{\underline{\sigma}} : \nabla \underline{u} - \nabla \cdot \underline{j}_S - \frac{\underline{q}}{T} \cdot \nabla \ln T - \frac{1}{T} \sum_{i=1}^{nc} \underline{j}_i \cdot \nabla_T \tilde{\mu}_i \\ & - \sum_{i=1}^{nc} \frac{\tilde{\mu}_i}{T} \nabla \cdot \underline{j}_i + \frac{1}{T} \sum_{i=1}^{nc} \underline{j}_i \cdot \underline{X}_i - \frac{P}{\rho T} \left(-\rho \underline{\underline{I}} : \nabla \underline{u} \right) - \frac{1}{T} \sum_{i=1}^{nc} \tilde{\mu}_i \left(\varphi_i - \nabla \cdot \underline{j}_i \right) \end{aligned} \tag{A.63}$$

Comparing Eq. (A.63) to Eq. (A.47), one can conclude that

$$\Phi = (\underline{\underline{\sigma}} + P\underline{\underline{I}}) : \nabla \underline{u} - \underline{q} \cdot \nabla \ln T - \sum_{i=1}^{nc} \underline{j}_i \cdot (\nabla_T \tilde{\mu}_i - \underline{X}_i) - \sum_{i=1}^{nc} \mu_i r_i \tag{A.64}$$

where $\underline{X}_i$ is the body force acting in i (normally, $\underline{X}_i = -\underline{g}$).

Eq. (A.64) shows four sources of entropy generation:

1. $(\underline{\underline{\sigma}} + P\underline{\underline{I}}) : \nabla \underline{u}$ is the viscous dissipation; at equilibrium, $\underline{\underline{\sigma}} = -P\underline{\underline{I}}$, and this term vanishes;
2. $-\underline{q} \cdot \nabla \ln T$ is the heat flux; at equilibrium, $\nabla \ln T = 0$;
3. $-\sum_{i=1}^{nc} \underline{j}_i \cdot \left(\nabla_T \tilde{\mu}_i - \underline{X}_i \right)$ is diffusion in an equivalent isothermal system; at equilibrium, $\nabla_T \tilde{\mu}_i = \underline{X}_i$;
4. $-\sum_{i=1}^{nc} \tilde{\mu}_i r_i$ is the chemical reaction (or any other source of i); at equilibrium, $r_i = 0$.

Each entropy-generation term above might be interpreted as a *force* $\times$ *flux* product. For example, in the first one (viscous dissipation), $\underline{\underline{\sigma}} + P\underline{\underline{I}}$ is the flux and $\nabla \underline{u}$ is the force. Heat and diffusion fluxes are consequences of "temperature-gradient" and "chemical-potential-gradient" forces. Clearly, one can write,

$$\Phi = \sum_{\alpha} \varphi_{\alpha} \mathfrak{I}_{\alpha} \tag{A.65}$$

where φ_{α} are the fluxes, and $\mathfrak{I}_{\alpha}$ are the forces. When fluxes and forces obey the relation above, that is, have the same index α, they are called *conjugated* The treatment of the fluxes and forces involved in the entropy equation derivation is a particular case of the general formulation of Onsager (1931a, 1931b). Not all the experimental transport relations are simply a proportionality between a flux and a force. When two transport processes occur simultaneously, they can interfere on each other, producing cross phenomena. A classic example is the interference between diffusion and heat conduction, which generates the so-called *Soret* and *Dufour* effects.

The Soret effect, or thermal diffusion, is the establishment of a mass flow as a result of a temperature gradient. This effect might be mathematically described by adding a term proportional to the temperature gradient to the Fick Law. The Dufour effect is a heat flux that shows due to a concentration gradient. It is possible to describe it by adding terms proportional do concentration gradient to the Fourier's Law.

Generally, fluxes φ_{α} are homogenous and linear functions of forces $\mathfrak{I}_{\beta}$:

$$\varphi_{\alpha} = \sum_{\beta} L_{\alpha\beta} \mathfrak{I}_{\beta} \tag{A.66}$$

where the so-called phenomenological coefficients of Onsager, $L_{\alpha\beta}$, are independent of the forces. Coefficients $L_{\alpha\alpha}$ relate conjugated forces and fluxes, while $L_{\alpha\beta}$ $(\alpha \neq \beta)$ generate cross phenomena.

A.4 APPLYING THE ENTROPY EQUATION TO RESERVOIR SIMULATION

By Eq. (A.64), it can be easily shown that, for a vertical isothermal column in equilibrium, there will not be entropy production. If there is not temperature gradient, nor viscous dissipation or chemical reactions, and considering $-\underline{X}_i = \underline{g}$ (z pointing upwards), we will have $\frac{d\tilde{\mu}_i}{dz} + \underline{g} = 0$ or $d\mu_i = -M_i g dz$, as already derived in Chapter 3, for the compositional grading subject to the gravitational field.

For the general cases of nonequilibrium, studied by Ghorayeb and Firoozabadi (2000), let us now manipulate Eq. (A.64) to model the phenomena that govern the motion of fluids inside a reservoir. Dividing Eq. (A.64) by T and still neglecting viscous dissipation, one has

$$\frac{\Phi}{T} = r_S = -\frac{1}{T^2}\underline{q}\cdot\nabla T - \frac{1}{T}\sum_{i=1}^{nc}\underline{j}_i(\nabla_T\tilde{\mu}_i + \underline{g}) \tag{A.67}$$

Knowing that $\sum_{i=1}^{nc}\underline{j}_i = 0 \rightarrow \underline{j}_{nc} = -\sum_{i=1}^{nc-1}\underline{j}_i$, one can write

$$\sum_{i=1}^{nc}\underline{j}_i(\nabla_T\tilde{\mu}_i + \underline{g}) = \sum_{i=1}^{nc-1}\underline{j}_i(\nabla_T\tilde{\mu}_i + \underline{g}) + \underline{j}_{nc}(\nabla_T\tilde{\mu}_{nc} + \underline{g}) = \sum_{i=1}^{nc-1}\underline{j}_i(\nabla_T\tilde{\mu}_i - \nabla_T\tilde{\mu}_{nc}) \tag{A.68}$$

Replacing Eq. (A.68) in Eq. (A.67), we have

$$\frac{\Phi}{T} = r_S = -\frac{1}{T^2}\underline{q}\cdot\nabla T - \frac{1}{T}\sum_{i=1}^{nc-1}\underline{j}_i(\nabla_T\tilde{\mu}_i - \nabla_T\tilde{\mu}_{nc}) \tag{A.69}$$

Using the properties of operator "∇", we can rewrite Eq. (A.69) in the following way:

$$\frac{\Phi}{T} = r_S = \underline{q}\cdot\nabla(1/T) - \frac{1}{T}\sum_{i=1}^{nc-1}\underline{j}_i\,\nabla_T(\tilde{\mu}_i - \tilde{\mu}_{nc}) \tag{A.70}$$

The entropy-generation terms are still products of the fluxes (heat or diffusive) by their respective forces ($\nabla(1/T)$ and $\nabla_T(\tilde{\mu}_i - \tilde{\mu}_{nc})$), but Eq. (A.65) might be rewritten in a more convenient way:

$$\frac{\Phi}{T} = \sum_{m=0}^{nc-1}\underline{\varphi}_m\cdot\underline{\mathfrak{I}}_m \tag{A.71}$$

where $\underline{\varphi}_m$ are the following fluxes:

$$\underline{\varphi}_0 = \underline{q} \quad \text{and} \quad \underline{\varphi}_i = \underline{j}_i, \quad \text{for } i = 1, 2, \ldots, nc-1 \tag{A.72}$$

And $\underline{\mathfrak{I}}_m$ are the independent forces:

$$\underline{\mathfrak{I}}_0 = \nabla(1/T) \quad \text{and} \quad \underline{\mathfrak{I}}_i = -\frac{1}{T}\nabla_T(\tilde{\mu}_i - \tilde{\mu}_{nc}), \quad \text{for } i = 1, 2, \ldots, nc-1 \tag{A.73}$$

Now, let us rewrite the Onsager relations for $\underline{\varphi}_m$:

$$\underline{\varphi}_m = \sum_{k=0}^{nc-1}L_{mk}\underline{\mathfrak{I}}_k \tag{A.74}$$

Combining Eqs. (A.71) and (A.74), one can clearly note that the entropy production is influenced by cross phenomena.

$$\frac{\Phi}{T} = \sum_{m=0}^{nc-1}\sum_{k=0}^{nc-1}L_{mk}\underline{\mathfrak{I}}_k\cdot\underline{\mathfrak{I}}_m \tag{A.75}$$

Getting back to Eq. (A.74), if one takes off the first term of the summations, one will obtain

$$\underline{\varphi}_m = L_{m0}\underline{\mathfrak{J}}_0 + \sum_{i=1}^{nc-1} L_{mi}\underline{\mathfrak{J}}_i \tag{A.76}$$

Replacing Eq. (A.76) in Eq. (A.71):

$$\frac{\Phi}{T} = \sum_{m=0}^{nc-1}\left(L_{m0}\underline{\mathfrak{J}}_0 + \sum_{i=1}^{nc-1} L_{mi}\underline{\mathfrak{J}}_i\right)\cdot\underline{\mathfrak{J}}_m \tag{A.77}$$

Grouping the terms with the index "0," one has

$$\frac{\Phi}{T} = \left(L_{00}\underline{\mathfrak{J}}_0 + \sum_{i=1}^{nc-1} L_{0i}\underline{\mathfrak{J}}_i\right)\cdot\underline{\mathfrak{J}}_0 + \sum_{k=1}^{nc-1}\left(L_{k0}\underline{\mathfrak{J}}_0 + \sum_{i=1}^{nc-1} L_{ki}\underline{\mathfrak{J}}_i\right)\cdot\underline{\mathfrak{J}}_k \tag{A.78}$$

Using relations of Eq. (A.73), we can compare Eqs. (A.70) and (A.78) and note that

$$\underline{q} = L_{00}\nabla(1/T) + \sum_{i=1}^{nc-1} L_{0i}\left[-\frac{1}{T}\nabla_T(\tilde{\mu}_i - \tilde{\mu}_{nc})\right] \tag{A.79}$$

$$\underline{j}_i = L_{i0}\nabla(1/T) + \sum_{k=1}^{nc-1} L_{ik}\left[-\frac{1}{T}\nabla_T(\tilde{\mu}_i - \tilde{\mu}_{nc})\right] \tag{A.80}$$

Eqs. (A.79) and (A.80) are important results of diffusive and heat fluxes as functions of cross phenomena. Note that Fourier's Law is a particular case of Eq. (A.79), in which $L_{00}\nabla(1/T) = -K\nabla T \rightarrow -\dfrac{L_{00}}{T^2}\nabla T = -K\nabla T \rightarrow K = \dfrac{L_{00}}{T^2}$.

The other terms of Eq. (A.79) are due to the Dufour effect (heat flux due to concentration gradients). The first term of Eq. (A.80) is the Soret effect (molecular diffusion flux due to temperature gradiente, or simply "termal diffusion").

In a multicomponent reservoir, it is important to write the diffusive-flux expression in terms of temperature, pressure, and composition, the primary variables by which the chemical potential is dependent on. Starting by Eq. (A.80), let us derive such an expression, using some local equilibrium thermodynamic relations and the symmetrical properties of the phenomenological coefficients ($L_{ik} = L_{ki}$).

Knowing that $\tilde{\mu}_k = \tilde{\mu}_k(T, P, w_1, w_2, \ldots, w_{nc-1})$, one can write

$$\nabla_T\tilde{\mu}_k = \left(\frac{\partial\tilde{\mu}_k}{\partial P}\right)_{T,\underline{x}}\nabla P + \sum_{j=1}^{nc-1}\left(\frac{\partial\tilde{\mu}_k}{\partial w_j}\right)_{T,P,x_{i\neq j}}\nabla w_j \tag{A.81}$$

By Gibbs–Duhem equation at constant T:

$$\sum_{j=1}^{nc} w_j\nabla_T\tilde{\mu}_j = \tilde{v}\nabla P \tag{A.82}$$

where $\tilde{v}$ is the specific volume, that is $\tilde{v} = 1/\rho$, where ρ is the global mass concentration (kg/m^3, in SI). Let us rewrite Eq. (A.82) as

$$\sum_{j=1}^{nc-1} w_j \nabla_T \tilde{\mu}_j + w_{nc} \nabla_T \tilde{\mu}_{nc} = \frac{\nabla P}{\rho} \tag{A.83}$$

Isolating $\nabla_T \tilde{\mu}_{nc}$ on the left-hand side of Eq. (A.83), one has

$$\nabla_T \tilde{\mu}_{nc} = \frac{1}{w_{nc}} \left[-\sum_{j=1}^{nc-1} w_j \nabla_T \tilde{\mu}_j + \frac{\nabla P}{\rho} \right] \tag{A.84}$$

Now, let us subtract Eq. (A.84) from Eq. (A.81):

$$\begin{aligned} \nabla_T(\tilde{\mu}_k - \tilde{\mu}_{nc}) = {} & \left(\frac{\partial \tilde{\mu}_k}{\partial P} \right)_{T,\underline{x}} \nabla P + \sum_{j=1}^{nc-1} \left(\frac{\partial \tilde{\mu}_k}{\partial w_j} \right)_{T,P,x_{i \neq j}} \nabla w_j \\ & - \frac{1}{w_{nc}} \left[-\sum_{j=1}^{nc-1} w_j \nabla_T \tilde{\mu}_j + \frac{\nabla P}{\rho} \right] \end{aligned} \tag{A.85}$$

Using Eq. (A.81) to eliminate $\nabla_T \tilde{\mu}_j$ in Eq. (A.84), one has

$$\begin{aligned} \nabla_T(\tilde{\mu}_k - \tilde{\mu}_{nc}) = {} & \left(\frac{\partial \tilde{\mu}_k}{\partial P} \right)_{T,\underline{w}} \nabla P + \sum_{j=1}^{nc-1} \left(\frac{\partial \tilde{\mu}_k}{\partial w_j} \right)_{T,P,w_{i \neq j}} \nabla w_j \\ & - \frac{1}{w_{nc}} \left[-\sum_{j=1}^{nc-1} w_j \left(\frac{\partial \tilde{\mu}_j}{\partial P} \nabla P + \sum_{l=1}^{nc-1} \frac{\partial \tilde{\mu}_j}{\partial w_l} \nabla w_l \right) + \frac{\nabla P}{\rho} \right] \end{aligned} \tag{A.86}$$

Let us rearrange Eq. (A.86) to obtain

$$\begin{aligned} \nabla(\tilde{\mu}_i - \tilde{\mu}_{nc}) = {} & \sum_{j=1}^{nc-1} \left[\left(\frac{\partial \tilde{\mu}_k}{\partial w_j} \right)_{T,P,w_{i \neq j}} \nabla w_j + \frac{w_j}{w_{nc}} \sum_{l=1}^{nc-1} \left(\frac{\partial \tilde{\mu}_j}{\partial w_l} \right)_{T,P,w_{i \neq l}} \nabla w_l \right] \\ & + \left[\left(\frac{\partial \tilde{\mu}_k}{\partial P} \right)_{T,\underline{w}} + \sum_{j=1}^{nc-1} \frac{w_j}{w_{nc}} \left(\frac{\partial \tilde{\mu}_j}{\partial P} \right)_{T,\underline{w}} - \frac{1}{w_{nc} \rho} \right] \nabla P \end{aligned} \tag{A.87}$$

Rewriting the first term of the right-hand side of Eq. (A.87), we can write the compositional gradient only in terms of ∇w_l:

$$\sum_{j=1}^{nc-1} \left(\frac{\partial \tilde{\mu}_k}{\partial w_j} \right)_{T,P,w_{i \neq j}} \nabla w_j = \sum_{l=1}^{nc-1} \left(\frac{\partial \tilde{\mu}_k}{\partial w_l} \right)_{T,P,w_{i \neq l}} \nabla w_l = \sum_{j=1}^{nc-1} \sum_{l=1}^{nc-1} \left(\frac{\partial \tilde{\mu}_j}{\partial w_l} \right)_{T,P,w_{i \neq l}} \delta_{jk} \nabla w_l \tag{A.88}$$

reaching to

$$\nabla(\tilde{\mu}_i - \tilde{\mu}_{nc}) = \sum_{j=1}^{nc-1}\left[\left(\frac{w_j}{w_{nc}} + \delta_{jk}\right)\sum_{l=1}^{nc-1}\left(\frac{\partial\tilde{\mu}_j}{\partial w_l}\nabla w_l\right)\right] + \left[\left(\frac{\partial\tilde{\mu}_k}{\partial P}\right)_{T,\underline{w}} + \sum_{j=1}^{nc-1}\frac{w_j}{w_{nc}}\left(\frac{\partial\tilde{\mu}_j}{\partial P}\right)_{T,\underline{w}} - \frac{1}{w_{nc}\rho}\right]\nabla P \tag{A.89}$$

Comparing Eqs. (A.80) and (A.89), the diffusive flux might be given by

$$\underline{j}_i = -\frac{L_{i0}}{T^2}\nabla T - \frac{1}{T}\sum_{k=1}^{nc-1}L_{ik}\sum_{j=1}^{nc-1}\left[\left(\frac{w_j}{w_{nc}} + \delta_{jk}\right)\sum_{l=1}^{nc-1}\left(\frac{\partial\tilde{\mu}_j}{\partial w_l}\nabla w_l\right)\right] - \frac{1}{T}\sum_{k=1}^{nc-1}L_{ik}\left[\left(\frac{\partial\tilde{\mu}_k}{\partial P}\right)_{T,\underline{w}} + \sum_{j=1}^{nc-1}\frac{w_j}{w_{nc}}\left(\frac{\partial\tilde{\mu}_j}{\partial P}\right)_{T,\underline{w}} - \frac{1}{w_{nc}\rho}\right]\nabla P \tag{A.90}$$

Now, let us redefine some auxiliary variables useful to write the diffusive flux in a more compact form:

$$k_{Ti} \equiv \frac{M w_i w_{nc} L_{i0}}{RTL_{ii}}, \quad \text{for } i = 1, 2, \ldots, nc-1 \tag{A.91}$$

$$D_{ic} \equiv \frac{RL_{ii}}{\rho w_i w_{nc} a_{ic} M}, \quad \text{for } i = 1, 2, \ldots, nc-1 \tag{A.92}$$

$$a_{ic} \equiv \frac{M_i M_{nc}}{M^2}, \quad \text{for } i = 1, 2, \ldots, nc-1 \tag{A.93}$$

Let us do some algebra to introduce these quantities in Eq. (A.90). Combining Eqs. (A.91)–(A.93), one observes that the term of the thermal gradient can be written as

$$-\frac{L_{i0}}{T^2}\nabla T = -\rho D_{ic} a_{ic} k_{Ti}\frac{\nabla T}{T} \tag{A.94}$$

Now, let us review the concentration-gradient term. Rewriting $\left(\frac{w_j}{w_{nc}} + \delta_{jk}\right)$ as $\frac{1}{w_{nc}}(w_j + \delta_{jk}w_{nc})$, the term $\frac{1}{w_{nc}}$ will get off the summations in j, k, and l. Manipulating Eq. (A.92), one can write:

$$-\frac{1}{Tw_{nc}} = -\frac{\rho M w_i a_{ic} D_{ic}}{RT L_{ii}} \tag{A.95}$$

We still can use the following relation:

$$\left(\frac{\partial\tilde{\mu}_j}{\partial w_l}\right)_{T,P,w_{i\neq l}} = \left(\frac{\partial\mu_j/M_j}{\partial w_l}\right)_{T,P,w_{i\neq l}} = \frac{RT}{M_j}\left(\frac{\partial \ln f_j}{\partial w_l}\right)_{T,P,w_{i\neq l}} \tag{A.96}$$

Now, let us treat the pressure-gradient term:

$$
\begin{aligned}
-\frac{1}{T}\sum_{k=1}^{nc-1} L_{ik}\left[\left(\frac{\partial\tilde{\mu}_k}{\partial P}\right)_{T,\underline{w}} + \sum_{j=1}^{nc-1}\frac{w_j}{w_{nc}}\left(\frac{\partial\tilde{\mu}_j}{\partial P}\right)_{T,\underline{w}} - \frac{1}{w_{nc}\rho}\right]\nabla P = \\
-\frac{1}{w_{nc}T}\sum_{k=1}^{nc-1} L_{ik}\left[w_{nc}\tilde{v}_k + \sum_{j=1}^{nc-1} w_j\tilde{v}_j - \frac{1}{\rho}\right]\nabla P
\end{aligned}
\tag{A.97}
$$

where we use the following relations:

$$
\left(\frac{\partial\tilde{\mu}_k}{\partial P}\right)_{T,\underline{w}} = \tilde{v}_k \quad \text{and} \quad \left(\frac{\partial\tilde{\mu}_j}{\partial P}\right)_{T,\underline{w}} = \tilde{v}_j \tag{A.98}
$$

Replacing Eqs. (A.94) into (A.98) in Eq. (A.90), we finally get

$$
\underline{j}_i = -\rho D_{ic} a_{ic}\left[\begin{aligned}&\frac{k_{Ti}}{T}\nabla T + \frac{w_i M}{L_{ii}}\sum_{k=1}^{nc-1} L_{ik}\sum_{j=1}^{nc-1}\sum_{l=1}^{nc-1}\left(\frac{w_j + w_{nc}\delta_{jk}}{M_j}\right)\left(\frac{\partial \ln f_j}{\partial w_l}\right)\nabla w_l \\ &+ \frac{M w_i}{RTL_{ii}}\sum_{k=1}^{nc-1} L_{ik}\left(w_{nc}\tilde{v}_k + \sum_{j=1}^{nc-1} w_j\tilde{v}_j - \frac{1}{\rho}\right)\nabla P\end{aligned}\right] \tag{A.99}
$$

Let us define some auxiliary matrices in order to simplify Eq. (A.99):

$$
\underline{\underline{D}} \equiv [D_{ij}] = [a_{ic} D_{ic}\delta_{ij}] \tag{A.100}
$$

$$
\underline{K}_T \equiv [k_{Ti}] = \left[\frac{k_{Ti}}{T}\right] \tag{A.101}
$$

$$
\underline{\underline{M}} \equiv [M_{ij}] = \left[\frac{M w_i}{L_{ii}}\delta_{ij}\right] \tag{A.102}
$$

$$
\underline{\underline{W}}_{kj} \equiv [W_{kj}] = \left[\frac{w_j + w_{nc}\delta_{jk}}{M_j}\right] \tag{A.103}
$$

$$
\underline{\underline{F}} \equiv [F_{jl}] = \left[\left(\frac{\partial \ln f_j}{\partial w_l}\right)_{T,P,w_{m\neq l}}\right] \tag{A.104}
$$

$$
\underline{V} \equiv [V_k] = \frac{1}{RT}\left(w_{nc}\tilde{v}_k + \sum_{j=1}^{nc-1} w_j\tilde{v}_j - \frac{1}{\rho}\right) \tag{A.105}
$$

$$
\underline{\underline{L}} \equiv [L_{ij}] = \begin{bmatrix} L_{11} & L_{12} & \cdots & L_{1,nc-1} \\ L_{21} & L_{22} & \cdots & L_{2,nc-1} \\ \vdots & & & \\ L_{nc-1,1} & L_{nc-1,2} & \cdots & L_{nc-1,nc-1} \end{bmatrix} \tag{A.106}
$$

$$\nabla T = \begin{bmatrix} \dfrac{\partial T}{\partial x} \\ \dfrac{\partial T}{\partial y} \\ \dfrac{\partial T}{\partial z} \end{bmatrix}; \quad \nabla P = \begin{bmatrix} \dfrac{\partial P}{\partial x} \\ \dfrac{\partial P}{\partial y} \\ \dfrac{\partial P}{\partial z} \end{bmatrix}; \quad \nabla w_i = \begin{bmatrix} \dfrac{\partial w_i}{\partial x} \\ \dfrac{\partial w_i}{\partial y} \\ \dfrac{\partial w_i}{\partial z} \end{bmatrix} \tag{A.107}$$

$$\underline{\underline{\nabla w^t}} = \begin{bmatrix} \dfrac{\partial w_1}{\partial x} & \dfrac{\partial w_2}{\partial x} & \cdots & \dfrac{\partial w_{nc-1}}{\partial x} \\ \dfrac{\partial w_1}{\partial y} & \dfrac{\partial w_2}{\partial y} & \cdots & \dfrac{\partial w_{nc-1}}{\partial y} \\ \dfrac{\partial w_1}{\partial z} & \dfrac{\partial w_2}{\partial z} & \cdots & \dfrac{\partial w_{nc-1}}{\partial z} \end{bmatrix}; \quad \underline{\underline{j}} = \begin{bmatrix} j_{1,x} & j_{2,x} & \cdots & j_{nc-1,x} \\ j_{1,y} & j_{2,y} & \cdots & j_{nc-1,y} \\ j_{1,z} & j_{2,z} & \cdots & j_{nc-1,z} \end{bmatrix} \tag{A.108}$$

So we get

$$\underline{\underline{j}}^t = -\rho\left[\underline{\underline{D}}\,\underline{\underline{M}}\,\underline{\underline{L}}\,\underline{\underline{W}}\,\underline{\underline{F}}\left(\nabla w^t\right)^t + \underline{\underline{D}}\,\underline{K_T}(\nabla T)^t + \underline{\underline{D}}\,\underline{\underline{M}}\,\underline{\underline{L}}\,\underline{V}(\nabla P)^t\right] \tag{A.109}$$

or

$$\underline{\underline{j}}^t = -\rho\left[\underline{\underline{D}}^m\left(\nabla w^t\right)^t + \underline{\underline{D}}^{T,m}(\nabla T)^t + \underline{\underline{D}}^{P,m}(\nabla P)^t\right] \tag{A.110}$$

where $\underline{\underline{D}}^m = \underline{\underline{D}}\,\underline{\underline{M}}\,\underline{\underline{L}}\,\underline{\underline{W}}\,\underline{\underline{F}}$, $\underline{\underline{D}}^{T,m} = \underline{\underline{D}}\underline{K}_T$, and $\underline{\underline{D}}^{P,m} = \underline{\underline{D}}\,\underline{\underline{M}}\,\underline{\underline{L}}\,\underline{V}$.

Comparing Eq. (A.110) to Eq. (A.99), we have

$$D_{il}^m = a_{ic}D_{ic}\frac{w_i M}{L_{ii}}\sum_{k=1}^{nc-1} L_{ik}\sum_{j=1}^{nc-1}\sum_{l=1}^{nc-1}\left(\frac{w_j + w_{nc}\delta_{jk}}{M_j}\right)\left(\frac{\partial \ln f_j}{\partial w_l}\right)_{T,P,w_{m\neq l}} \tag{A.111}$$

$$D_i^{T,m} = a_{ic}D_{ic}\frac{k_{Ti}}{T} \tag{A.112}$$

$$D_i^{P,m} = a_{ic}D_{ic}\frac{Mw_i}{RTL_{ii}}\sum_{k=1}^{nc-1} L_{ik}\left(w_{nc}\bar{v}_k + \sum_{j=1}^{nc-1} w_j\bar{v}_j - \frac{1}{\rho}\right) \tag{A.113}$$

$$\underline{j}_i = -\rho\left[\sum_{j=1}^{nc-1} D_{ij}^m \nabla w_j + D_i^{T,m}\nabla T + D_i^{P,m}\nabla P\right] \tag{A.114}$$

A.5 DERIVING THE CHANGES OF AXIS FOR THE DIFFUSIVE FLUXES

With the expression for the mass-based diffusive flux, that is considering the barometric velocity axis $\underline{v} = \sum_{i=1}^{nc} w_i \underline{v}_i$, it is possible to obtain the flux in molar basis, which is relative to the molar-averaged velocity axis $\underline{v} = \sum_{i=1}^{nc} x_i \underline{v}_i$. Let us see

$$\underline{v}_i = \frac{\underline{j}_i}{\rho_i} + \sum_{k=1}^{nc} w_k \underline{v}_k = \frac{\underline{J}_i}{c_i} + \sum_{k=1}^{nc} x_k \underline{v}_k \tag{A.115}$$

$$\underline{J}_i = \frac{\underline{j}_i}{M_i} - \frac{\rho_i}{M_i} \sum_{k=1}^{nc} \underline{v}_k (x_k - w_k) \tag{A.116}$$

Applying the sum over index i in Eq. (A.116), we have

$$\sum_{i=1}^{nc} \underline{J}_i = \sum_{i=1}^{nc} \frac{\underline{j}_i}{M_i} - \sum_{i=1}^{nc} \frac{\rho_i}{M_i} \sum_{k=1}^{nc} \underline{v}_k (x_k - w_k) = 0 \tag{A.117}$$

Knowing also that $\sum_{i=1}^{nc} \frac{\rho_i}{M_i} = \sum_{i=1}^{nc} c_i = c$, we have

$$\sum_{k=1}^{nc} \underline{v}_k (x_k - w_k) = \frac{\sum_{i=1}^{nc} \frac{\underline{j}_i}{M_i}}{c} \tag{A.118}$$

Replacing Eq. (A.118) into Eq. (A.116), we have

$$\underline{J}_i = \frac{\underline{j}_i}{M_i} - x_i \sum_{k=1}^{nc} \frac{\underline{j}_k}{M_k} \tag{A.119}$$

Separating the last (nc^{th}) term of the sum on the right-hand side of Eq. (A.119), we have

$$\underline{J}_i = \frac{\underline{j}_i}{M_i} - x_i \sum_{k=1}^{nc-1} \frac{\underline{j}_k}{M_k} - x_i \frac{\underline{j}_{nc}}{M_{nc}} \tag{A.120}$$

Using the relation $\underline{j}_{nc} = -\sum_{k=1}^{nc-1} \underline{j}_k$ in Eq. (A.120), one can write

$$\underline{J}_i = \frac{\underline{j}_i}{M_i} - x_i \sum_{k=1}^{nc-1} \frac{\underline{j}_k}{M_k} + x_i \sum_{k=1}^{nc-1} \frac{\underline{j}_k}{M_{nc}} \tag{A.121}$$

Multiplying the last term on the right-hand side of Eq. (A.121) by $\left(\frac{M_k}{M_k}\right)$:

$$\underline{J}_i = \frac{\underline{j}_i}{M_i} - x_i \sum_{k=1}^{nc-1} \frac{\underline{j}_k}{M_k} + x_i \sum_{k=1}^{nc-1} \frac{\underline{j}_k}{M_k} \frac{M_k}{M_{nc}} \tag{A.122}$$

and writing the mass fraction in terms of the mole fraction:

$$w_k = \frac{M_k x_k}{M} \quad and \quad w_{nc} = \frac{M_{nc} x_{nc}}{M} \tag{A.123}$$

We can replace Eq. (A.123) on the last term of Eq. (A.122) to obtain

$$\underline{J}_i = \frac{\underline{j}_i}{M_i} - x_i \sum_{k=1}^{nc-1} \frac{\underline{j}_k}{M_k} + x_i \sum_{k=1}^{nc-1} \frac{\underline{j}_k}{M_k} \frac{w_k}{w_{nc}} \frac{x_{nc}}{x_k} \tag{A.124}$$

Grouping all the terms on the right-hand side of Eq. (A.124) in the same summation, using Kronecker delta (δ_{ik}) function for the first one, we finally get

$$\underline{J}_i = \sum_{k=1}^{nc-1} \frac{\underline{j}_k}{M_k} \left[\delta_{ik} - \frac{x_i}{x_k} \left(x_k - \frac{w_k}{w_{nc}} x_{nc} \right) \right] \tag{A.125}$$

Obtaining the mass-based flux from molar-based flux can be done by matrix-inversion techniques or applying the same thought of the derivation above, though isolating $\underline{j}_i$ in Eq. (A.115):

$$\underline{j}_i = \underline{J}_i \frac{\rho_i}{c_i} + \rho_i \sum_{k=1}^{nc} \underline{v}_k (x_k - w_k) \tag{A.126}$$

or

$$\underline{j}_i = \underline{J}_i M_i + \rho_i \sum_{k=1}^{nc} \underline{v}_k (x_k - w_k) \tag{A.127}$$

Applying the sum over index i in Eq. (A.127), we have

$$\sum_{i=1}^{nc} \underline{j}_i = \sum_{i=1}^{nc} \underline{J}_i M_i + \sum_{i=1}^{nc} \rho_i \sum_{k=1}^{nc} \underline{v}_k (x_k - w_k) = 0 \tag{A.128}$$

where

$$\sum_{k=1}^{nc} \underline{v}_k (x_k - w_k) = - \frac{\sum_{i=1}^{nc} \underline{J}_i M_i}{\rho} \tag{A.129}$$

Replacing Eq. (A.129) in Eq. (A.127), we have

$$\underline{j}_i = \underline{J}_i M_i - w_i \sum_{k-1}^{nc} \underline{J}_k M_k \tag{A.130}$$

Separating the last term of the summation:

$$\underline{j}_i = \underline{J}_i M_i - w_i \sum_{k=1}^{nc-1} \underline{J}_k M_k - w_i \underline{J}_{nc} M_{nc} \tag{A.131}$$

Applying $\underline{J}_{nc} = -\sum_{k=1}^{nc-1} \underline{J}_k$, we can write

$$-w_i \underline{J}_{nc} M_{nc} = w_i \sum_{k=1}^{nc-1} \underline{J}_k M_{nc} \tag{A.132}$$

Now, multiplying Eq. (A.132) by $\frac{M_k}{M_k}$ and recognizing that $w_k = \frac{x_k M_k}{M}$ and $w_{nc} = \frac{x_{nc} M_{nc}}{M}$, we have

$$w_i \sum_{k=1}^{nc-1} \underline{J}_k M_k \frac{M_{nc}}{M_k} = w_i \sum_{k=1}^{nc-1} \underline{J}_k M_k \frac{w_{nc}}{w_k} \frac{x_k}{x_{nc}} \tag{A.133}$$

Replacing Eq. (A.133) in Eq. (A.131), we get

$$\underline{j}_i = \underline{J}_i M_i - w_i \sum_{k=1}^{nc-1} \underline{J}_k M_k + w_i \sum_{k=1}^{nc-1} \underline{J}_k M_k \frac{w_{nc}}{w_k} \frac{x_k}{x_{nc}} \tag{A.134}$$

Rearranging Eq. (A.134), using Kronecker delta (δ_{ik}) function in the first term of the right-hand side, we finally get

$$\underline{j}_i = \sum_{k=1}^{nc-1} \underline{J}_k M_k \left[\delta_{ik} + \frac{w_i}{w_k} \left(-w_k + \frac{w_{nc}}{x_{nc}} x_k \right) \right] \tag{A.135}$$

As observed in Eqs. (A.125) and (A.135), the computation of diffusive fluxes in terms of mass- or molar-averaged velocity *is not merely a unit transformation involving exclusively the molar mass*. The sum of the diffusive fluxes only vanishes if we are in the same reference axis. So, note that $\underline{J}_i \neq \underline{J}_i^* = \frac{\underline{j}_i}{M_i}$ and $\underline{j}_i \neq \underline{j}_i^* = \underline{J}_i M_i$. Note also that, obviously, $\sum_{i=1}^{nc} \underline{J}_i^* \neq 0$ and $\sum_{i=1}^{nc} \underline{j}_i^* \neq 0$.

Now, let us obtain the molecular-, thermal-, and pressure-diffusion coefficients from one basis to the other, using Eq. (A.125) or (A.135). Defining the following auxiliary matrices:

$$\underline{\underline{K}} \to K_{ik} \equiv \delta_{ik} + \frac{w_i}{w_k} \left(-w_k + \frac{w_{nc}}{x_{nc}} x_k \right); \quad \text{for } i, k = 1, \ldots, nc-1 \tag{A.136a}$$

$$\underline{\underline{M}}^M = \begin{bmatrix} M_1 & 0 & \ldots & 0 \\ 0 & M_2 & \ldots & 0 \\ 0 & 0 & \ldots & M_{nc-1} \end{bmatrix} \tag{A.136b}$$

$$\underline{\underline{J}} = \begin{bmatrix} J_{1,x} & J_{2,x} & \ldots & J_{nc-1,x} \\ J_{1,y} & J_{2,y} & \ldots & J_{nc-1,y} \\ J_{1,z} & J_{2,z} & \ldots & J_{nc-1,z} \end{bmatrix} \tag{A.136c}$$

$$\underline{\underline{j}} = \begin{bmatrix} j_{1,x} & j_{2,x} & \ldots & j_{nc-1,x} \\ j_{1,y} & j_{2,y} & \ldots & j_{nc-1,y} \\ j_{1,z} & j_{2,z} & \ldots & j_{nc-1,z} \end{bmatrix} \tag{A.136d}$$

we can write

$$\underline{j}^t = \underline{\underline{K}}\,\underline{\underline{M}}^M \underline{\underline{J}}^t \tag{A.137}$$

We know that $\underline{j}_i$ and $\underline{J}_i$ are proportional to $\nabla \underline{w}$ and $\nabla \underline{x}$, respectively. That is why the basis change of the molecular-diffusion coefficient is a little bit more complex than the other coefficients, because ∇T and ∇P are common to both basis, which makes the transformation of D^T and D^P considerably simpler.

Conventioning operator ∇ to be a column vector, we must transpose $\underline{w}$ and $\underline{x}$ to generate the matrices of the concentration gradients. Therefore, we have

$$\nabla \underline{w}^t = \begin{bmatrix} \dfrac{\partial}{\partial x} \\ \dfrac{\partial}{\partial y} \\ \dfrac{\partial}{\partial z} \end{bmatrix} [w_1\ w_2 \ldots w_{nc-1}] = \begin{bmatrix} \dfrac{\partial w_1}{\partial x} & \dfrac{\partial w_2}{\partial x} & \cdots & \dfrac{\partial w_{nc-1}}{\partial x} \\ \dfrac{\partial w_1}{\partial y} & \dfrac{\partial w_2}{\partial y} & \cdots & \dfrac{\partial w_{nc-1}}{\partial y} \\ \dfrac{\partial w_1}{\partial z} & \dfrac{\partial w_2}{\partial z} & \cdots & \dfrac{\partial w_{nc-1}}{\partial z} \end{bmatrix} \tag{A.138a}$$

$$\nabla \underline{x}^t = \begin{bmatrix} \dfrac{\partial}{\partial x} \\ \dfrac{\partial}{\partial y} \\ \dfrac{\partial}{\partial z} \end{bmatrix} [x_1\ x_2 \ldots x_{nc-1}] = \begin{bmatrix} \dfrac{\partial x_1}{\partial x} & \dfrac{\partial x_2}{\partial x} & \cdots & \dfrac{\partial x_{nc-1}}{\partial x} \\ \dfrac{\partial x_1}{\partial y} & \dfrac{\partial x_2}{\partial y} & \cdots & \dfrac{\partial x_{nc-1}}{\partial y} \\ \dfrac{\partial x_1}{\partial z} & \dfrac{\partial x_2}{\partial z} & \cdots & \dfrac{\partial x_{nc-1}}{\partial z} \end{bmatrix} \tag{A.138b}$$

Writing also the matrices of the diffusion coefficients:

$$\underline{\underline{D}}^m = \begin{bmatrix} D_{11}^m & D_{12}^m & \cdots & D_{1,nc-1}^m \\ D_{21}^m & D_{22}^m & \cdots & D_{2,nc-1}^m \\ & & \cdots & \\ D_{nc-1,1}^m & D_{nc-1,2}^m & \cdots & D_{nc-1,nc-1}^m \end{bmatrix} \tag{A.139a}$$

$$\underline{\underline{D}}^M = \begin{bmatrix} D_{11}^M & D_{12}^M & \cdots & D_{1,nc-1}^M \\ D_{21}^M & D_{22}^M & \cdots & D_{2,nc-1}^M \\ & & \cdots & \\ D_{nc-1,1}^M & D_{nc-1,2}^M & \cdots & D_{nc-1,nc-1}^M \end{bmatrix} \tag{A.139b}$$

We can write only the molecular terms of each reference axis:

$$\underline{\underline{J}}^t{}_{\text{molecular}} = -\rho \underline{\underline{D}}^m (\nabla \underline{w}^t)^t; \quad \underline{J}^t{}_{\text{molecular}} = -c \underline{\underline{D}}^M (\nabla \underline{x}^t)^t \tag{A.140}$$

So, we can combine Eqs. (A.137) and (A.140):

$$-\rho \underline{\underline{D}}^m (\nabla \underline{w}^t)^t = -\underline{\underline{K}}\,\underline{\underline{M}}^M c \underline{\underline{D}}^M (\nabla \underline{x}^t)^t \tag{A.141}$$

Writing also the mass fractions in terms of mole fractions, we have

$$\begin{bmatrix} w_1 \\ w_2 \\ \cdots \\ w_{nc-1} \end{bmatrix} = \frac{1}{M} \begin{bmatrix} x_1 M_1 \\ x_2 M_2 \\ \cdots \\ x_{nc-1} M_{nc-1} \end{bmatrix} \quad \text{(A.142)}$$

where $1/M = c/\rho$ and $M = \sum_{i=1}^{nc} x_i M_i$ is the overall molar mass (which is a scalar, not to be confused with matrix $\underline{\underline{M}}^M$).

Using $\underline{\underline{M}}^M$ matrix, we can rewrite Eq. (A.142):

$$\begin{bmatrix} w_1 \\ w_2 \\ \cdots \\ w_{nc-1} \end{bmatrix} = \frac{c}{\rho} \begin{bmatrix} M_1 & 0 & \cdots & 0 \\ 0 & M_2 & \cdots & 0 \\ 0 & 0 & \cdots & M_{nc-1} \end{bmatrix} \begin{bmatrix} x_1 \\ x_2 \\ x_{nc-1} \end{bmatrix} \quad \text{(A.143)}$$

We can also rewrite Eq. (A.143) using diagonal matrices $\underline{\underline{W}}$ and $\underline{\underline{X}}$:

$$\underbrace{\begin{bmatrix} w_1 & 0 & \cdots & 0 \\ 0 & w_2 & \cdots & 0 \\ 0 & 0 & \cdots & w_{nc-1} \end{bmatrix}}_{\underline{\underline{w}}} = \frac{c}{p} \underbrace{\begin{bmatrix} M_1 & 0 & \cdots & 0 \\ 0 & M_2 & \cdots & 0 \\ 0 & 0 & \cdots & M_{nc-1} \end{bmatrix}}_{\underline{\underline{M}}} \underbrace{\begin{bmatrix} x_1 & 0 & \cdots & 0 \\ 0 & x_2 & \cdots & 0 \\ 0 & 0 & \cdots & x_{nc-1} \end{bmatrix}}_{\underline{\underline{x}}} \quad \text{(A.144)}$$

Now, let us isolate $\underline{\underline{D}}^m$ on the left-hand side of Eq. (A.141). We must post-multiply both sides by $\nabla \underline{w}^t$ in order to obtain a square matrix, suitable to be canceled out by its inverse:

$$\underline{\underline{D}}^m = \left(\frac{c}{\rho}\right) \underline{\underline{K}}\,\underline{\underline{M}}\,\underline{\underline{D}}^M (\nabla \underline{x}^t)^t (\nabla \underline{w}^t) \left[(\nabla \underline{w}^t)^t (\nabla \underline{w}^t)\right]^{-1} \quad \text{(A.145)}$$

From Eq. (A.144), we can write

$$\frac{c}{\rho}\underline{\underline{M}} = \underline{\underline{W}}\,\underline{\underline{X}}^{-1} \quad \text{(A.146)}$$

And, from Eq. (A.143), one has

$$\nabla \underline{w}^t = \nabla \left(\frac{c}{\rho}\underline{\underline{M}}\,\underline{x}\right)^t = \nabla \left(\underline{x}^t \frac{c}{\rho}\underline{\underline{M}}\right) = (\nabla \underline{x}^t)\frac{c}{\rho}\underline{\underline{M}} \quad \text{(A.147)}$$

Replacing Eqs. (A.146) and (A.147) into Eq. (A.145), using matrices properties $(AB)^{-1} = B^{-1}A^{-1}$ and $(AB)^t = B^t A^t$, we have

$$\underline{\underline{D}}^m = \underline{\underline{K}}\,\underline{\underline{W}}\,\underline{\underline{X}}^{-1} \underline{\underline{D}}^M (\nabla \underline{x}^t)^t \left[\frac{c}{\rho}\underline{\underline{M}} (\nabla \underline{x}^t)^t\right]^{-1} \quad \text{(A.148)}$$

$$\underline{\underline{D}}^m = \underline{\underline{K}\,\underline{W}\,\underline{X}}^{-1} \underline{\underline{D}}^M \left(\frac{c}{\rho}\underline{\underline{M}}\right)^{-1} \quad \text{(A.149)}$$

$$\underline{\underline{D}}^m = \underline{\underline{KWX}}^{-1}\underline{\underline{D}}^M(\underline{\underline{WX}}^{-1})^{-1} \tag{A.150}$$

$$\underline{\underline{D}}^m = \underline{\underline{KWX}}^{-1}\underline{\underline{D}}^M\underline{\underline{XW}}^{-1} \tag{A.151}$$

Starting again from Eq. (A.140), but considering only thermal diffusion, we have

$$\underline{j}^t_{\text{thermal}} = -\rho\underline{D}^{T,m}(\nabla T)^t; \quad \underline{J}^t_{\text{thermal}} = -c\underline{D}^{T,M}(\nabla T)^t \tag{A.152}$$

$$-\rho\underline{D}^{T,m}(\nabla T)^t = -\underline{\underline{KM}}^M c\underline{D}^{T,M}(\nabla T)^t \tag{A.153}$$

$$\underline{D}^{T,m} = \frac{c}{\rho}\underline{\underline{KM}}^M\underline{D}^{T,M} \tag{A.154}$$

And, analogously, for the pressure term

$$\underline{D}^{P,m} = \frac{c}{\rho}\underline{\underline{KM}}^M\underline{D}^{P,M} \tag{A.155}$$

A.6 OBTAINING ONSAGER PHENOMENOLOGICAL COEFFICIENTS. THE CONCEPT OF "HEAT OF TRANSPORT"

The fundamental step to obtain the diffusion coefficients as functions of T, P, and $\underline{x}$ is the calculation of the Onsager phenomenological coefficients, L_{i0} and L_{ik}.

Consider again Eq. (A.80), which provides the diffusive flux as a consequence of cross phenomena. Performing a matrix inversion, for the isothermal case, to isolate the chemical potential gradients, we get

$$-\frac{1}{T}\nabla_T(\tilde{\mu}_k - \tilde{\mu}_{nc}) = \sum_{i=1}^{nc-1}[\underline{\underline{L}}]^{-1}_{ki}\underline{j}_i \tag{A.156}$$

Replacing Eq. (A.156) in Eq. (A.79), we have

$$\underline{q} = \sum_{i=1}^{nc-1} L_{0i}\sum_{l=1}^{nc-1}[\underline{\underline{L}}]^{-1}_{il}\underline{j}_l \equiv \sum_{l=1}^{nc-1}(Q^*_l - Q^*_{nc})\underline{j}_l \tag{A.157}$$

where $(Q^{*,m}_l - Q^{*,m}_{nc}) \equiv \sum_{i=1}^{nc-1} L_{0i}[\underline{\underline{L}}]^{-1}_{il}$ and $Q^{*,m}_l$ is the *heat of transport of component* l, that is the heat which is transported by isothermal diffusion per unit mass of component l per unit cross-section area (see Appendix B). So, we have

$$\underline{q} = \sum_{l=1}^{nc-1}\hat{Q}^{*,m}_l\underline{j}_l, \quad \text{for } \nabla T = 0 \tag{A.158}$$

where

$$\hat{\underline{Q}}^{*,m} = \underline{\underline{L}}^{-1}\underline{L}_0 = \begin{bmatrix} (Q_1^{*,m} - Q_{nc}^{*,m}) \\ \vdots \\ (Q_{nc-1}^{*,m} - Q_{nc}^{*,m}) \end{bmatrix} \quad \text{(A.159)}$$

From Eq. (A.159), one can realize that each position of the vector $\hat{\underline{Q}}^*$ is called *Net Heat of Transport* of the component $l = 1, 2, \ldots, nc - 1$, since it is always compared with the property for the reference component *nc* (Fitts, 1962). The constitutive equation for the *heat of transport* in binary mixtures proposed by Shukla and Firoozabadi (1998), based on the kinetic viscosity theories presented by Glasstone et al. (1941) and Dougherty and Drickamer (1955), was generalized by Firoozabadi et al. (2000) for multicomponent mixtures, considering as reference the molar-averaged velocity axis:

$$Q_i^{*,M} = -\frac{\overline{U}_i}{\tau_i} + \left[\sum_{j=1}^{nc} x_j \frac{\overline{U}_j}{\tau_j}\right] \frac{\overline{V}_i}{\sum_{j=1}^{nc} x_j \overline{V}_j} \quad \text{(A.160)}$$

where $Q_i^{*,M}$ is the heat of transport of component i related to the molar basis axis (hence the superscript *M*), $\overline{U}_i$ and $\overline{V}_i$ are, respectively, the partial molar internal energy and the partial molar volume of component i (which can be calculated with an equation of state, e.g.), and $\tau_i = \frac{\Delta U_i^{\text{vap}}}{\Delta U_i^{\text{vis}}}$, where ΔU_i^{vap} and ΔU_i^{vis} are, respectively, the vaporization internal energy and the so-called *energy of viscous flux* of pure component i (Glasstone et al., 1941). Generally, τ_i is considered an adjustable parameter of the model. For hydrocarbons, $\tau_i = 4.0$ is a good estimate recommended by Shukla and Firoozabadi (1998). The complete deduction of Eq. (A.160) is presented on Appendix B. We should emphasize that, as the heat of transport is written in molar basis ($Q_i^{*,M}$) in Eq. (A.160), it is necessary to convert it to the mass basis one ($Q_i^{*,m}$) before applying it on Eq. (A.159). This conversion is straightforward, since one should apply Eq. (A.135), with $\underline{j}_i = Q_i^{*,m}$ and $\underline{J}_k = Q_k^{*,M}$.

The question one can ask now is, how to obtain the terms of the diffusive flux from the *heat of transport* expression of Shukla and Firoozabadi (1998)?

The answer is simple, but nontrivial: these authors introduced an important approximation for the calculation of the Onsager coefficients, equating the expression of the isothermal heat flux, depending only on heats of transport (Eq. (A.157) or (A.158)), to the general expression given by Eq. (A.79) to get

$$\sum_{i=1}^{nc-1} (Q_i^{*,m} - Q_{nc}^{*,m})\underline{j}_i = L_{00}\nabla(1/T) + \sum_{k=1}^{nc-1} L_{0k}\left[-\frac{1}{T}\nabla_T(\tilde{\mu}_k - \tilde{\mu}_{nc})\right] \quad \text{(A.161)}$$

Replacing Eq. (A.82) on the left-hand side of Eq. (A.161), one gets

$$\sum_{i=1}^{nc-1}\left(Q_i^{*,m}-Q_{nc}^{*,m}\right)\left\{L_{i0}\nabla(1/T)+\sum_{k=1}^{nc-1}L_{ik}\left[-\frac{1}{T}\nabla_T\left(\tilde{\mu}_k-\tilde{\mu}_{nc}\right)\right]\right\}$$
$$=L_{00}\nabla(1/T)+\sum_{k=1}^{nc-1}L_{0k}\left[-\frac{1}{T}\nabla_T\left(\tilde{\mu}_k-\tilde{\mu}_{nc}\right)\right] \tag{A.162}$$

Comparing the equivalent terms, one can conclude that

$$L_{00}=\sum_{i=1}^{nc-1}\left(Q_i^{*,m}-Q_{nc}^{*,m}\right)L_{i0} \tag{A.163}$$

$$L_{0k}=\sum_{i=1}^{nc-1}(Q_i^{*,m}-Q_{nc}^{*,m})L_{ik} \tag{A.164}$$

Eq. (A.164) being identical to the already deduced Eq. (A.159). In matrix notation:

$$\underline{L_0}=\underline{\underline{L}}\cdot\underline{\hat{Q}}^{*,m} \tag{A.165}$$

According to Eq. (A.91) and the definition of vector $\underline{K_T}$, we can write

$$\underline{K_T}=\frac{w_{nc}}{RT^2}\underline{\underline{M}}\cdot\underline{\underline{L}}\cdot\underline{\hat{Q}}^{*,m} \tag{A.166}$$

So $\underline{D}^T$, in Eq. (A.110), reads

$$\underline{D}^{T,m}=\frac{w_{nc}}{RT^2}\underline{\underline{DM}}\cdot\underline{\underline{L}}\cdot\underline{\hat{Q}}^{*,m} \tag{A.167}$$

and the diffusive flux of Eq. (A.109) is then

$$\underline{\underline{j}}^t=-\rho\left[\underline{\underline{D}}\,\underline{\underline{M}}\,\underline{\underline{L}}\,\underline{\underline{W}}\,\underline{\underline{F}}(\nabla\underline{w}^t)^t+\frac{w_{nc}}{RT^2}\underline{\underline{D}}\,\underline{\underline{M}}\,\underline{\underline{L}}\,\underline{\hat{Q}}^{*,m}(\nabla T)^t+\underline{\underline{D}}\,\underline{\underline{M}}\,\underline{\underline{L}}\,\underline{V}(\nabla P)^t\right] \tag{A.168}$$

Finally, the molecular-diffusion coefficient $\underline{\underline{D}}^m$, defined in Eq. (A.111), is directly calculated, for example, by the empirical correlation of Leahy-Dios and Firoozabadi (2007), presented in detail in Appendix C, so that matrix $\underline{\underline{L}}$ might be obtained by

$$\underline{\underline{L}}=(\underline{\underline{D}}\,\underline{\underline{M}})^{-1}\underline{\underline{D}}^m(\underline{\underline{W}}\,\underline{\underline{F}})^{-1} \tag{A.169}$$

and then replaced in Eq. (A.165) to give $\underline{L_0}$, or, more straightforwardly, in Eq. (A.167) to give $\underline{D}^{T,m}$.

In summary, given the temperature (T), pressure (P), and composition ($\underline{w}$), the diffusive flux is calculated through the following steps:

1. Obtain the vector of Net Heats of Transport $\underline{\hat{Q}}^{*,m}$, using Eqs. (A.135), (A.159), and (A.160).
2. Obtain the molecular-diffusion coefficient $\underline{\underline{D}}^{m}$ by the correlation of Leahy-Dios and Firoozabadi (2007)—see procedure in Appendix C.
3. Obtain matrix $\underline{\underline{L}}$ by Eq. (A.169).
4. With $\underline{\hat{Q}}^{*,m}$, obtain $\underline{D}^{T,m}$ by Eq. (A.167).
5. Obtain $\underline{D}^{P,m}$ by Eq. (A.113).

REFERENCES

De Groot, S.R., Mazur, P., 1962. Nonequilibrium Thermodynamics. North-Holland Publishing Co., Amsterdam.

Dougherty, E.L., Drickamer, H.G., 1955. A theory of thermal diffusion in liquids. J. Chem. Phys. 23 (2), 295.

Firoozabadi, A., Ghorayeb, K., Shukla, K., 2000. Theoretical model of thermal diffusion factors in multicomponent mixtures. AIChE J. 46 (5), 892–900.

Fitts, D.D., 1962. Nonequilibrium Thermodynamics. A Phenomenological Theory of Irreversible Processes in Fluid Systems. McGraw-Hill, New York.

Ghorayeb, K., Firoozabadi, A., 2000. Modeling multicomponent diffusion and convection in porous media. SPE J. 5 (2), 158–171.

Glasstone, S., Laidler, K.J., Eyring, H., 1941. The Theory of Rate Processes. McGraw-Hill Book Co, New York.

Haase, R., 1969. Thermodynamics of Irreversible Processes. Addison-Wesley, London.

Leahy-Dios, A., Firoozabadi, A., 2007. Unified model for nonideal multicomponent molecular diffusion coefficients. AIChE J. 53 (11), 2932–2939.

Onsager, L., 1931a. Reciprocal relations in irreversible processes I. Phys. Rev. E 37, 405–426.

Onsager, L., 1931b. Reciprocal relations in irreversible processes II. Phys. Rev. E 38, 2265–2279.

Shukla, K., Firoozabadi, A., 1998. A new model of thermal diffusion coefficients in binary hydrocarbon mixtures. Ind. Eng. Chem. Res. 37, 3331–3342.

Appendix B

Heat of Transport

In this Appendix, we will derive the semiempirical heat-of-transport expression proposed by Shukla and Firoozabadi (1998) for binary systems, and generalized by Firoozabadi et al. (2000) for multicomponent mixtures. The concepts involved in these equations date back to the 1940s and still lack for a more consistent theoretical basis. According to Dougherty and Drickamer (1955), the *Heat of Transport* Q_i *is the amount of energy carried through a plane normal to the diffusive flux of i per mole of this component in the absence of a thermal gradient.*

The molecular motion within a liquid mixture was modeled initially by Glasstone et al. (1941). These authors stated that a liquid may be considered as a system which spaces might be vacant or occupied by molecules from time to time. The flow itself would be the act of creating empty spaces by displacement of molecules which leave a certain region of the fluid, demanding energy, and filling them up by the displacement of other molecules that arrive to the region, releasing energy into the medium. Density fluctuations in a certain point are related to the Brownian motion of the fluid. The *Net Heat of Transport of component i* is the amount of energy required by a certain region of the fluid in order that a determined number of molecules of type i leave this region by diffusion, minus the energy released into the region when other molecules of type i occupy the spaces left empty by the leaving ones.

Fig. B.1 shows a sketch of some molecules of different types leaving a certain region of the fluid, creating empty spaces in this region, and reaching to other regions to occupy empty spaces previously created.

Consider then a region arbitrarily chosen in the fluid. Suppose that dn_i moles diffuse out of this region crossing the limits of its surface. According to the definition of Q_i given above, this diffusive flux will be followed up by an energy flux through the boundary given by $Q_i dn_i$. To keep the temperature constant, an amount of heat dq will be absorbed and, still during this process, a certain amount of work dw will be exerted quasi-statically by the

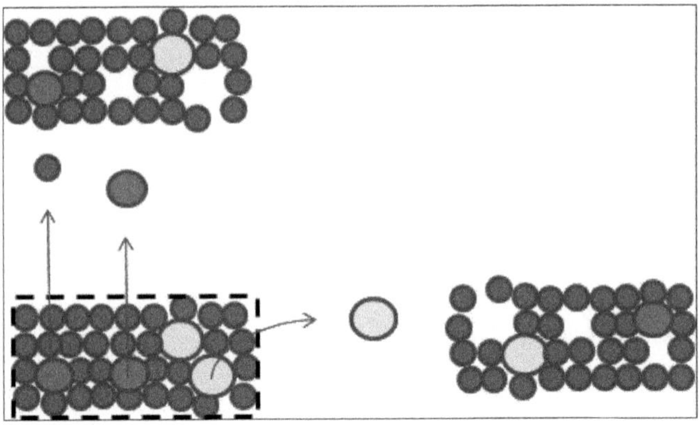

FIGURE B.1 Molecules of different types leaving a certain region of the fluid (generating vacancies and demanding energy) to occupy empty spaces in other regions (filling up vacancies and releasing energy into the medium).

fluid that remains in the region. So, the variation of internal energy of the region is given by

$$dU = -Q_i dn_i + dq - dw \tag{B.1}$$

Since temperature and pressure are constant, the variation of internal energy can be written as $-\overline{U}_i dn_i$ and the variation in volume, as $-\overline{V}_i dn_i$. Therefore, the work performed on the boundary of the region is $-P\overline{V}_i dn_i$, so that Eq. (B.1) reads:

$$-\overline{U}_i dn_i = -Q_i dn_i + dq + P\overline{V}_i dn_i, \quad \text{for } i = 1, 2, \ldots, nc \tag{B.2}$$

So, the heat absorbed by the region will be

$$dq = (Q_i - \overline{U}_i - P\overline{V}_i)dn_i = (Q_i - \overline{H}_i)dn_i = Q_i^* dn_i, \quad \text{for } i = 1, 2, \ldots, nc \tag{B.3}$$

Then, $Q_i^* = (Q_i - \overline{H}_i) = \dfrac{dq}{dn_i}$ is the net amount of energy that must be absorbed by the region per mole of component i diffusing outwards through its limits to keep its temperature and pressure constant.

In a binary mixture, the liquid heat of transport of each one of the two components might be expressed in terms of two main quantities: (1) the energy to detach a molecule from its neighbors, creating the vacancy. Let us call it W_H, according to nomenclature of Dougherty and Drickamer (1955); and (2) the energy released in the region when a molecule fills up a vacancy, let us call it W_L. Assuming a totally random molecular distribution, Dougherty and Drickamer (1955) state that the total energy W_L needed to fill up vacancies already left behind by molecules of types 1 and 2 will be given by:

$$W_L = x_1 W_{H1} + x_2 W_{H2} \tag{B.4}$$

For molecules of different sizes and shapes, the volume and contours of the vacancies occupied by the two types of molecules will also be different. It is reasonable to suppose that, for mixtures of big and small molecules, roughly both types of molecules will move towards the empty space left by the bigger molecule, which does not occur for vacancies left by the smaller type. Assuming that ψ_1 molecules migrate toward the vacancy left by a molecule of type 1 and similarly for $\psi_2\psi_2$, we have the following expressions for the net heats of transport:

$$Q_1^* = W_{H1} - \psi_1 W_L \tag{B.5}$$

$$Q_2^* = W_{H2} - \psi_2 W_L \tag{B.6}$$

Applying Gibbs–Duhem relation to Q_i^*, we get

$$x_1\psi_1 + x_2\psi_2 = 1 \tag{B.7}$$

Eq. (B.7) states that, on average, the number of molecules that leave the region is equal to the number of molecules that get into it, which keeps its volume constant, still obeying the limits:

$$x_1 \to 1, \quad \psi_1 \to 1 \text{ and } x_2 \to 1, \quad \psi_2 \to 1$$

Let us detail a little bit more the concept of parameter ψ_1. It represents the number of molecules (they being of type 1 or 2) that fill up the vacancies left behind by molecules of type 1 when those left the region:

$$\psi_1 = \frac{N_{1,s1} + N_{2,s1}}{N_1} \tag{B.8}$$

where $N_{1,s1}$ and $N_{2,s1}$ are, respectively, the number of molecules of type 1 and 2 getting into the region to occupy the sites (vacancies) of type 1 created by the N_1 molecules of type 1 that left it. Analogously,

$$\psi_2 = \frac{N_{1,s2} + N_{2,s2}}{N_2} \tag{B.9}$$

Note that $\psi_1 + \psi_2 \neq 1$.

The total number of molecules that get into the region is

$$N = N_{1,s1} + N_{2,s1} + N_{1,s2} + N_{2,s2} = N_1\psi_1 + N_2\psi_2 \tag{B.10}$$

Consequently, due to the random distribution, the number of molecules of type 1 getting into the region is given by

$$N_{1,s1} + N_{1,s2} = x_1 N = x_1(N_1\psi_1 + N_2\psi_2) = x_1^2 N\psi_1 + x_1 x_2 N\psi_2 \tag{B.11}$$

Analogously, for type 2, we have

$$N_{2,s1} + N_{2,s2} = x_2 N = x_2(N_1\psi_1 + N_2\psi_2) = x_2 x_1 N\psi_1 + x_2^2 N\psi_2 \tag{B.12}$$

The total volume of molecules getting into the region (V_t^{in}) might be calculated from Eqs. (B.11) and (B.12), using partial molar volumes:

$$V_t^{in} = (x_1^2 N\psi_1 + x_1 x_2 N\psi_2)\overline{V}_1 + (x_2 x_1 N\psi_1 + x_2^2 N\psi_2)\overline{V}_2 \tag{B.13}$$

The total volume of molecules that leave the region is given by

$$V_t^{out} = x_1 N\overline{V}_1 + x_2 N\overline{V}_2 \tag{B.14}$$

Now consider only the molecules of type 1. The volume of vacancies left by them ($x_1 N\overline{V}_1$) will be occupied by the total number of molecules that move towards this site: $x_1^2 N\psi_1\overline{V}_1 + x_2 x_1 N\psi_1\overline{V}_2$. Equating these two quantities, we have

$$\psi_1 = \frac{\overline{V}_1}{x_1\overline{V}_1 + x_2\overline{V}_2} \tag{B.15}$$

Analogously, for molecules of type 2:

$$\psi_2 = \frac{\overline{V}_2}{x_1\overline{V}_1 + x_2\overline{V}_2} \tag{B.16}$$

Generalizing for multicomponent mixtures, we have

$$\psi_i = \frac{\overline{V}_i}{\sum_{j=1}^{nc} x_j\overline{V}_j}, \quad \text{for } i = 1, 2, \ldots, nc \tag{B.17}$$

Now, let us apply the concept of ψ_i, returning to Eqs. (B.5) and (B.6) to build up the final expression for the heat of transport: according to Shukla and Firoozabadi (1998), it is reasonable to suppose that the energy W_{H1} required to detach a type-1 molecule from its neighbors to generate a vacancy is proportional to its partial molar internal energy, by its own definition: $\overline{U}_1 = \left(\frac{\partial U}{\partial n_1}\right)_{T,P,n_{j\neq 1}}$. Let us call this proportionality constant $\frac{1}{\tau_1}$:

$$W_{H1} = -\frac{1}{\tau_1}\overline{U}_1 \tag{B.18}$$

So, for a binary mixture, we can replace Eqs. (B.4), (B15), and (B.18) in Eq. (B.5) to get

$$Q_1^* = -\frac{1}{\tau_1}\overline{U}_1 + \frac{\overline{V}_1}{x_1\overline{V}_1 + x_2\overline{V}_2}\left(x_1\frac{1}{\tau_1}\overline{U}_1 + x_2\frac{1}{\tau_2}\overline{U}_2\right) \tag{B.19}$$

Generalization of Eq. (B.19) is straightforward for multicomponent mixtures:

$$Q_i^{*,M} = -\frac{1}{\tau_i}\overline{U}_i + \frac{\overline{V}_i}{\sum_{j=1}^{nc} x_j\overline{V}_j}\sum_{k=1}^{nc} x_k\frac{1}{\tau_k}\overline{U}_k, \quad \text{for } i = 1, 2, \ldots, nc \tag{B.20}$$

where the index M was now added to emphasize the reference axis of molar average bulk velocity, keeping the coherence with the same net heats of transport used both in Chapter 4 and Appendix A.

Except for the adjustable parameter τ_i, all other variables in Eq. (B.20) might be obtained through a well-fitted equation of state based on a PVT analysis.

REFERENCES

Dougherty, E.L., Drickamer, H.G., 1955. A theory of thermal diffusion in liquids. J. Chem. Phys. 23 (2), 295.

Firoozabadi, A., Ghorayeb, K., Shukla, K., 2000. Theoretical model of thermal diffusion factors in multicomponent mixtures. AIChE J. 46 (5), 892–900.

Glasstone, S., Laidler, K.J., Eyring, H., 1941. The Theory of Rate Processes. McGraw-Hill Book Co, New York.

Shukla, K., Firoozabadi, A., 1998. A new model of thermal diffusion coefficients in binary hydrocarbon mixtures. Ind. Eng. Chem. Res. 37, 3331–3342.

Appendix C

Calculation of Molecular Diffusion Coefficient

This Appendix presents the calculation procedure for the molecular diffusion coefficient $\underline{\underline{D}}^M$ according to the correlation of Leahy-Dios and Firoozabadi (2007). We recommend the reader to go over the text of Taylor and Krishna (1993) to get familiar with the so-called Stefan–Maxwell approach for the diffusive flux computation, which, although being absolutely equivalent to the Fickian approach we presented in Chapter 4 is based in the separation of the EoS-dependent terms of nonideality. Here, we present only the equations necessary to the computational implementation of the correlation. On the approach of Stefan–Maxwell, the diffusive flux relative to the reference axis of molar average bulk velocity (already defined in Chapter 4) is given by

$$\underline{\underline{J}}^t = -c(\underline{\underline{B}}^M)^{-1}\underline{\underline{\Gamma}}\nabla\underline{x}^t \tag{C.1}$$

where

$$B_{ii}^M = \frac{x_i}{\mathcal{D}_{i,nc}} + \sum_{\substack{k=1 \\ i \neq k}}^{nc} \frac{x_k}{\mathcal{D}_{ik}}, \quad \text{for } i = 1, \ldots, nc-1 \tag{C.2}$$

$$B_{ij}^M = -x_i\left(\frac{1}{\mathcal{D}_{ij}} - \frac{1}{\mathcal{D}_{i,nc}}\right), \quad \text{for } i,j = 1, \ldots, nc-1 \ (i \neq j) \tag{C.3}$$

$$\Gamma_{ij} = x_i\left(\frac{\partial \ln f_i}{\partial x_j}\right)_{T,P,x_{k\neq j}}, \quad \text{for } i,j = 1, \ldots, nc-1 \tag{C.4}$$

By direct comparison with Eq. (4.17a) (Chapter 4), one can write

$$\underline{\underline{D}}^M = (\underline{\underline{B}}^M)^{-1}\underline{\underline{\Gamma}} \tag{C.5}$$

At infinite dilution, that is, for a binary mixture where component i is infinitely diluted in component j ($x_i \rightarrow 0$ and $x_j \rightarrow 1$), it can be shown that $\mathcal{D}_{ij}$, D_{ij}^{M}, and D_{ij}^{m} tend to the same value D_{ij}^{∞}, independently of the reference axis.

The correlation of Leahy-Dios and Firoozabadi (2007) actually provides the diffusion coefficients for the pairs ($i,j = 1, 2 \ldots, nc$) of the mixture at infinite dilution, D_{ij}^{∞}:

$$\frac{cD_{ij}^{\infty}}{(cD)^{0}} = A_0 \left(\frac{Tr_j Pr_i}{Tr_i Pr_j}\right)^{A_1} \left(\frac{\mu}{\mu^0}\right)^{[A_2(\omega_1,\omega_2)+A_3(P_r,T_r)]} \tag{C.6}$$

where Tr and Pr are reduced temperature and pressure, and

$$A_0 = e^{a_1} \tag{C.7a}$$

$$A_1 = 10a_2 \tag{C.7b}$$

$$A_2 = a_3(1 + 10\omega_j - \omega_i + 10\omega_j\omega_i) \tag{C.7c}$$

$$A_3 = a_4(Pr_j^{3a_5} - 6Pr_i^{3a_5} + 6Tr_j^{10a_6}) + a_7 Tr_i^{-a_6} + a_2\left(\frac{Tr_j Pr_i}{Tr_i Pr_j}\right) \tag{C.7d}$$

with:

$a_1 = -0.0472$
$a_2 = 0.0103$
$a_3 = -0.0147$
$a_4 = -0.0053$
$a_5 = -0.3370$
$a_6 = -0.1852$
$a_7 = -0.1914$

Let $(cD)^{0}$ be the so-called *diffusion coefficient-diluted gas density product* given by Fuller's relation (Poling et al., 2001):

$$(cD)^0 = (0.0101)T^{0.75} \frac{\left(\frac{1}{M_j} + \frac{1}{M_i}\right)^{0.5}}{R\left[(\Sigma v_j)^{1/3} + (\Sigma v_i)^{1/3}\right]^2} \tag{C.8}$$

where (Σv_j) are the so-called *volume increments for diffusion* of the component j, reported by Poling et al. (2001) and reproduced here in Table C.1.

In relation to the viscosities $\left(\frac{\mu}{\mu^0}\right)$, the correlation of Stiel and Thodos (1961) might be used for the numerator at the conditions of temperature (T), pressure (P), and composition ($\underline{x}$) of the mixture:

$$\left[(\mu - \mu^0)\xi + 10^{-4}\right]^{0.25} = a_0 + a_1\rho_r + a_2\rho_r^2 + a_3\rho_r^3 + a_4\rho_r^4 \tag{C.9}$$

where $\rho_r = \dfrac{\rho_{\text{mix}}}{\rho_{c,\text{mix}}}$ is the reduced density of the pair i,j.

TABLE C.1 Volume Increments for Diffusion (Σv_j) According to Poling et al. (2001)

Functional Groups		Simple Molecules	
C	15.90	He	2.67
H	2.31	H_2	6.12
O	6.11	O_2	16.30
N	4.54	N_2	18.50
F	14.70	Ar	19.70
CL	21.00	CO	18.00
Br	21.90	CO_2	26.90
I	29.80	NH_3	20.70
S	22.90	SO_2	41.80
Aromatic ring	−18.30	H_2O	13.10
Heterocyclic ring	−18.30	SF_6	71.30

with

$$\xi = \frac{Tc_{\text{mix}}^{1/6}}{M_{\text{mix}}^{1/2} Pc_{\text{mix}}^{2/3}} \tag{C.10}$$

where the index *mix* refers to the pair *i,j*.

The low-pressure viscosity (μ^0) is given by

$$\mu_i^0 \xi_i = 34 \times 10^{-8} Tr_i^{0.94}, \quad \text{if } Tr_i < 1.5, \quad \text{for } i = 1, \ldots, nc \tag{C.11}$$

$$\mu_i^0 \xi_i = 17.78 \times 10^{-8} (4.58\, Tr_i - 1.67)^{5/8}, \ \text{if } Tr_i > 1.5, \quad \text{for } i = 1, \ldots, nc \tag{C.12}$$

where

$$\xi_i = \frac{Tc_i^{1/6}}{M_i^{1/2} Pc_i^{2/3}}, \quad \text{for } i = 1, \ldots, nc \tag{C.13}$$

$$\mu^0 = \frac{\mu_j^0 M_j^{1/2} + \mu_i^0 M_i^{1/2}}{M_j^{1/2} + M_i^{1/2}} \tag{C.14}$$

where Tc_i is in Kelvin (K) and Pc_i is in atm.

Once calculated the infinite dilution coefficient for every pair *i,j* of the mixture, Leahy-Dios and Firoozabadi (2007) have used the so-called

Vignes Relation (Vignes, 1966) to obtain the Stefan–Maxwell diffusion coefficient at the actual composition:

$$\mathcal{D}_{ij} = (D_{ij}^{\infty})^{x_j}(D_{ji}^{\infty})^{x_i} \prod_{\substack{k=1 \\ k \neq i,j}}^{nc} (D_{ik}^{\infty} D_{jk}^{\infty})^{x_k/2}, \quad \text{for } i,j = 1,\ldots,nc \ (i \neq j) \quad \text{(C.15)}$$

Therefore, the matrix $\underline{\underline{B}}^M$ might be calculated by Eqs. (C.2) and (C.3) and, finally, $\underline{\underline{D}}^M$ is calculated by Eq. (C.5), using the fitted EoS to obtain matrix $\underline{\underline{\Gamma}}$.

REFERENCES

Leahy-Dios, A., Firoozabadi, A., 2007. Unified model for nonideal multicomponent molecular diffusion coefficients. AIChE J. 53 (11), 2932–2939.

Poling, B.E., Prausnitz, J.M., O'Connell, J.P., 2001. The Properties of Gases and Liquids, 5th Edition McGraw Hill, New York.

Stiel, L.I., Thodos, G., 1961. The viscosity of nonpolar gases at normal pressures. AICHE J. 7, 611–615.

Taylor, R., Krishna, R., 1993. Multicomponent Mass Transfer. John Wiley & Sons, Inc., New York.

Vignes, A., 1966. Diffusion in binary mixtures. Ind. Eng. Chem. Fund 5, 189–199.

Appendix D

Another Approach for Heat of Transport

Another approach to the calculation of the diffusive-flux terms in steady state, derived from the heat of transport, follows the formulation initially proposed by Haase (1969) and effectively implemented by Pedersen and Lindeloff (2003). Let us get back to Eq. (A.29) (Appendix A), which defines the total heat flux $\underline{j}_E$ as a function of the conductive term and of the enthalpy carried by diffusion:

$$\underline{j}_E = \underline{q} + \sum_{i=1}^{nc} \underline{j}_i \tilde{h}_i \tag{A.29}$$

For an isothermal system, one can rewrite Eq. (A.29) relative to the molar bulk average velocity axis, with the conductive term already in terms of the *Net Heat of Transport* $\hat{Q}_i^{*,M}$, defined in Appendix A:

$$\underline{J}_E = \sum_{i=1}^{nc-1} \hat{Q}_i^{*,M} \underline{J}_i + \sum_{i=1}^{nc} \underline{J}_i \overline{H}_i \tag{D.1}$$

Separating the last term of the second sum, one has

$$\underline{J}_E = \sum_{i=1}^{nc-1} \hat{Q}_i^{*,M} \underline{J}_i + \sum_{i=1}^{nc-1} \underline{J}_i \overline{H}_i + \underline{J}_{nc} \overline{H}_{nc} \tag{D.2}$$

Knowing that $\sum_{i=1}^{nc} \underline{J}_i = 0$ and, therefore, $\underline{J}_{nc} = -\sum_{i-1}^{nc-1} \underline{J}_i$, one can write

$$\underline{J}_E = \sum_{i=1}^{nc-1} (\hat{Q}_i^{*,M} + \overline{H}_i - \overline{H}_{nc}) \underline{J}_i \tag{D.3}$$

According to Pedersen and Lindeloff (2003), in an isothermal system, the total heat flux $\underline{J}_E$ is zero because the conduction stream $\underline{q}$ is counterbalanced by the thermal energy flux carried by diffusion, i.e.:

$$\underline{J}_E = 0, \quad \text{for} \quad \nabla T = 0 \Rightarrow \hat{Q}_i^{*,M} = Q_i^{*,M} - Q_{nc}^{*,M} = \overline{H}_{nc} - \overline{H}_i, \quad \text{for} \quad i = 1, \ldots, nc \tag{D.4}$$

Here is the great difference between the approach of Firoozabadi et al. (2000) and Pedersen and Lindeloff (2003): the latter don't need additional models like the one in Eq. (A.160) for the calculation of the heat of transport. As long as one sets pre-fitted values of the individual enthalpies in arbitrary reference states (e.g., for ideal gas at 298 K), the absolute enthalpies of Eq. (D.4) can be straightforwardly obtained through the integral $\int_{T_{ref}}^{T} C_p^{ig}(T)dT$ and the residual enthalpy expression provided by the tuned EoS. It is important to mention that both approaches are similar on the matter of using an expression derived for a particular case (that is the heat of transport in an isothermal case) in a general situation where temperature is no longer constant. As the heat of transport is an intrinsic property of the fluid, the expression is generally valid.

As discussed in Appendix A, Firoozabadi and collaborators calculate the Onsager coefficients after obtaining $Q_i^{*,M}$ and only then build up the expression of the diffusive flux. Let us see further what Pedersen and Lindeloff (2003) do: rewriting Eq. (A.80) at steady state when $\underline{j}_i$ is zero, one has

$$L_{i0}\nabla(1/T) = -\sum_{k=1}^{nc-1} L_{ik}\left[-\frac{1}{T}\nabla_T\left(\frac{\mu_k}{M_k} - \frac{\mu_{nc}}{M_{nc}}\right)\right] \tag{D.5}$$

Replacing Eq. (A.165)

$$\underline{L}_0 = \underline{\underline{L}} \cdot \underline{\hat{Q}}^{*,m} \tag{A.165}$$

in the left-hand side of Eq. (D.5) and assuming molar basis for the heats of transports and chemical potentials, one has

$$\sum_{k=1}^{nc-1} L_{ik}^M (Q_k^{*,M} - Q_{nc}^{*,M})\nabla(1/T) = -\sum_{k=1}^{nc-1} L_{ik}^M\left[-\frac{1}{T}\nabla_T(\mu_k - \mu_{nc})\right] \tag{D.6}$$

So, by comparison between the summations in Eq. (D.6) (supposing valid the relation for each component, since this is true for a binary system):

$$(Q_k^{*,M} - Q_{nc}^{*,M})\nabla(1/T) = \frac{1}{T}\nabla_T(\mu_k - \mu_{nc}), \quad \text{for} \quad k = 1, \ldots, nc-1 \tag{D.7}$$

Substituting Eq. (D.4) into Eq. (D.7), changing the index k by i, and knowing that $\nabla(1/T) = -\frac{1}{T^2}\nabla T$, we obtain

$$-(\overline{H}_{nc} - \overline{H}_i)\frac{\nabla T}{T} = -\nabla_T(\mu_i - \mu_{nc}), \quad \text{for } i = 1, \ldots, nc \tag{D.8}$$

Applying Eq. (D.8) to a 1D column, one has

$$-(\overline{H}_{nc} - \overline{H}_i)\frac{dT/dz}{T} = -\left(\frac{d}{dz}\mu_i - \frac{d}{dz}\mu_{nc}\right)_T, \quad \text{for } i = 1, \ldots, nc \tag{D.9}$$

From this point, Pedersen and Lindeloff (2003) manipulate Eq. (D.9) to obtain a general expression that relates the chemical potential of component i to the composition profile along the entire non-isothermal column, eliminating the need of calculating the Onsager coefficients. Separating the derivatives of chemical potential at constant temperature,

$$-(\overline{H}_{nc} - \overline{H}_i)\frac{dT/dz}{T} = -\left[\left(\frac{d\mu_i}{dz}\right)_T - \left(\frac{d\mu_{nc}}{dz}\right)_T\right], \quad \text{for } i = 1, \ldots, nc \tag{D.10}$$

The authors redefine the driving force (F_i) for the entropy generation in an isothermal system, already presented in mass basis in Eqs. (A.64) to (A.68) as $\nabla_T\tilde{\mu}_i - \underline{X}_i$ (see Appendix A). Rewriting it in molar basis, one has

$$M_iF_i \equiv M_ig + \left(\frac{d\mu_i}{dz}\right)_T, \quad \text{for } i = 1, \ldots, nc \tag{D.11}$$

Replacing Eq. (D.11) in Eq. (D.10), one has

$$M_iF_i - M_{nc}F_{nc} = -(\overline{H}_i - \overline{H}_{nc})\frac{dT/dz}{T}, \quad \text{for } i = 1, \ldots, nc \tag{D.12}$$

Using Gibbs–Duhem relation:

$$-S\frac{dT}{dz} + V\frac{dP}{dz} - \sum_{i=1}^{nc} x_i\frac{d\mu_i}{dz} = 0 \tag{D.13}$$

which, at constant temperature, reduces to

$$V\frac{dP}{dz} = \sum_{i=1}^{nc} x_i\left(\frac{d\mu_i}{dz}\right)_T \tag{D.14}$$

And using the hydrostatic relation $\frac{dP}{dz} = -\rho g$, substitution of Eq. (D.11) into Eq. (D.14) gives

$$-V\rho g = \sum_{i=1}^{nc} x_iM_i(-g + F_i) \tag{D.15}$$

Knowing that ρV is the overall molar mass of the mixture ($M = \sum_{i=1}^{nc} x_i M_i$), one can write

$$V\rho g = g\sum_{i=1}^{nc} x_i M_i \tag{D.16}$$

Replacing Eq. (D.16) in Eq. (D.15), we reach to

$$\sum_{i=1}^{nc} x_i M_i F_i = 0 \tag{D.17}$$

Therefore, if one multiplies Eq. (D.12) by x_i and sum over all $i = 1,\ldots,$ nc components, one has

$$\underbrace{\sum_{i=1}^{nc} x_i M_i F_i}_{\text{zero}} - \sum_{i=1}^{nc} x_i M_{nc} F_{nc} = -\frac{dT/dz}{T}\sum_{i=1}^{nc} x_i(\overline{H}_i - \overline{H}_{nc}) \tag{D.18}$$

or

$$M_{nc} F_{nc} = \frac{dT/dz}{T}\sum_{i=1}^{nc} x_i(\overline{H}_i - \overline{H}_{nc}) \tag{D.19}$$

Using Euler's theorem in terms of the partial molar enthalpy:

$$H = \sum_{i=1}^{nc} x_i \overline{H}_i \tag{D.20}$$

where H is the molar enthalpy of the mixture, Eq. (D.19) reduces to

$$M_{nc} F_{nc} = \frac{dT/dz}{T}(H - \overline{H}_{nc}) \tag{D.21}$$

Replacing Eq. (D.21) in Eq. (D.12), one has

$$M_i F_i = \frac{dT/dz}{T}(H - \overline{H}_i), \quad \text{for } i = 1,\ldots,nc \tag{D.22}$$

And, replacing Eq. (D.22) into the definition of F_i itself (i.e., Eq. (D.11)), one has

$$M_i F_i \equiv M_i g + \left(\frac{d\mu_i}{dz}\right)_T = \frac{dT/dz}{T}(H - \overline{H}_i), \quad \text{for } i = 1,\ldots,nc \tag{D.23}$$

Rearranging Eq. (D.23) to isolate the chemical potential on the left-hand side, one has

$$d\mu_i = -M_i g dz + (H - \overline{H}_i)\frac{dT}{T}, \quad \text{for } i = 1, \ldots, nc \tag{D.24}$$

Replacing chemical potential by fugacity and integrating Eq. (D.24) from a reference depth z_{ref} to any other depth z, we reach to the final equation of Pedersen and Lindeloff (2003), which has the advantage of being absolutely analogue to Eq. (3.27) (see Chapter 3), but now with a term that accounts for the existence of a thermal gradient:

$$\hat{f}_i^z = \hat{f}_i^{z_{\text{ref}}} \exp\left[\frac{M_i g(z - z_{\text{ref}})}{RT}\right] \exp\left[\frac{1}{RT}\int_{T_{z_{\text{ref}}}}^{T_z} (H - \overline{H}_i)\frac{dT}{T}\right], \quad \text{for } i = 1, \ldots, nc \tag{D.25}$$

Pedersen and Lindeloff (2003) suggested also to calculate the enthalpies of Eq. (D.25) in an average condition of temperature and pressure between z_{ref} and z, allowing them to be taken off the integral and making the resolution of the system easier.

$$\hat{f}_i^z = \hat{f}_i^{z_{\text{ref}}} \exp\left[\frac{M_i g(z - z_{\text{ref}})}{RT}\right] \exp\left[\frac{(H - \overline{H}_i)}{RT}\ln\left(\frac{T_z}{T_{z_{\text{ref}}}}\right)\right], \quad \text{for } i = 1, \ldots, nc \tag{D.26}$$

The closer z_{ref} and z are, the better this approximation is, which is up to the user. It is important to emphasize two inconvenients of this method: the first one is a consequence of the fact that the calculation of the heat of transport has been done in some particular isothermal case: the derivative of the chemical potential with height used in Eq. (D.23) is, by hypothesis, taken at constant temperature. However, when integrated to obtain Eq. (D.26), it encompasses a changing-temperature term. In this case, the second term of the right-hand side of Eq. (D.25) could have been considered constant, at least between z_{ref} and z. This is a slightly stronger approximation than the one which generated Eq. (D.26), as there will not be any integration to be done on T.

Assuming that $\dfrac{dT}{dz}$ is constant, we can integrate Eq. (D.24) in the following way:

$$d\mu_i = -M_i g dz + (H - \overline{H}_i)\frac{dT/dz}{T}dz, \quad \text{for } i = 1, \ldots, nc \tag{D.27}$$

$$\hat{f}_i^z = \hat{f}_i^{z_{\text{ref}}} \exp\left[\frac{M_i g(z - z_{\text{ref}})}{RT}\right] \exp\left[\frac{(H - \overline{H}_i)}{RT^2}\frac{dT}{dz}(z - z_{\text{ref}})\right], \quad \text{for } i = 1, \ldots, nc \tag{D.28}$$

Eq. (D.28) is more coherent than Eq. (D.25) because it keeps the temperature of the chemical potential derivative constant in all terms, even the enthalpies. In our opinion, Eq. (D.28) should be the final equation presented by Pedersen and Lindeloff (2003), instead of Eq. (C14) of Appendix C of their paper.

The second inconvenient is the set of adjustable parameters of this model, made up completely by the enthalpies in the reference state of ideal gas. By the definition of these terms themselves as intrinsic properties of each component, it would not be coherent that they have to change from case to case to match experimental data of compositional grading. It is interesting to note that this is not a problem on flowing processes, because the energy balance calculates enthalpy variations from one point to another in the system, always relative to the same reference state, which cancels out inevitably.

REFERENCES

Firoozabadi, A., Ghorayeb, K., Shukla, K., 2000. Theoretical model of thermal diffusion factors in multicomponent mixtures. AIChE J. 46 (5), 892–900.

Haase, R., 1969. Thermodynamics of Irreversible Processes. Addison-Wesley, London.

Pedersen, K.S., Lindeloff, N., *Simulations of compositional gradients in hydrocarbon reservoirs under the influence of a temperature gradient*, SPE 84364, SPE Annual Technical Conference and Exhibition, Denver, Colorado, October/2003.

Appendix E

Parameter Estimation

In this Appendix, one presents briefly the procedure of parameter estimation employed in this book. As one remarked in Chapters 4–6, the sensitivity of the thermal diffusion model to the parameter τ_i (see Eq. (4.30)) leads to the necessity of a robust parameter estimation procedure to improve the model predictions on matching the experimental data. The nonlinearity of the model derived in Chapter 4 (and detailed in Appendix A) turn difficult to use conventional strategies of parameter estimation by derivative-based methods, which frequently lead to singularities in the Jacobian matrix and local extrema (both minimum or maximum). Same thought holds for the EoS fitting based on Pressure-Volume-Temperature (PVT) data presented along this book. One observes limitations on these latter regressions in commercial thermodynamic simulators.

Therefore, we used a hybrid optimization algorithm, built up by coupling a heuristic method in the initial iterations, followed by a deterministic procedure. In this way, one avoids both singularities—since heuristic methods do not compute derivatives—and local minima problems, since the initial scanning by the heuristic approach turns the problem independent of the parameters' initial guesses. Next, we present the general guidelines of this methodology. For a more detailed comprehension, we refer the reader to Bard (1974), Schwaab and Pinto (2007), and Schwaab et al. (2008).

E.1 THE PARAMETER ESTIMATION PROBLEM

Parameter estimation techniques are basic tools that establish the links of a given physical problem and help its interpretation. On trying to describe the physical reality, one often relies on models developed from experimental observations and hypothesis assumed by the investigator. Models will be always approximations to physical reality, since the available experimental data are always subject to uncertainties, some variables being not always easily measured or observed (Schwaab and Pinto, 2007).

Formally, the estimation procedure of some model's parameters consists on changing the parameters values till the model predictions get as close as they can to the experimental data, **respecting the experimental uncertainties**. The highlighting on the latter phrase is not casual, since neglecting the experimental errors upcoming from measurements may lead to overestimated models and statistically mistaken interpretations.

The parameter estimation procedure can be divided into three steps: the synthesis of the objective function, the minimization of the objective function, and the statistical analysis of the results (Severo Jr., 2011). In the petroleum industry, however, experimental data, even when available, are rare, expensive, and frequently subject to imprecisions. Analysis replications that would allow one to infer such uncertainties are seldom reported. So, in the procedures we carried out in this book, due to the absence of experimental information enough to establish such uncertainties, rigorous experimental-error-based statistical analysis about the quality of the adjustments are not reported.

E.2 OBJECTIVE FUNCTION SYNTHESIS

A statistically rigorous definition of the objective function is a fundamental step of a suitable parameter estimation procedure. The experimental variables, as we discussed, are always subject to errors and might be considered as aleatory variables, following a determined probability density function $\wp(z^e; z^*, V_e)$, which describes the probability of obtaining the experimental values (z^e), given the unknown actual ones z^* and a preview of the errors shown in the covariance matrix V_e (Severo Jr., 2011).

The maximum likelihood method consists in maximizing this probability, honoring the restrictions imposed by the model. Mathematically, one has

$$f(z, \theta) = 0 \tag{E.1}$$

where z is the vector of both dependent and independent variables, f represents the vector with the model equations, whereas θ is the vector of the model's parameters.

Assuming, by hypothesis, that the model is perfect and that the experiments are well performed, it seems reasonable to assume that the experimental data obtained are in fact the most probable ones, statistically (Bard, 1974). Therefore, the perfect-model hypothesis allows us to admit that the values calculated by the model (z^m) are equal to the actual and unknown ones (z^*).

Normally, one admits that the experimental data follow a normal Gaussian probability distribution. Then, the method of maximum likelihood consists in maximizing the following function (Severo Jr., 2011):

$$L(z^m, \theta) = \frac{(2\pi)^{-N/2}}{\sqrt{\det(V_e)}} \exp\left[-\frac{1}{2}(z^e - z^m)^T V_e^{-1}(z^e - z^m)\right] \tag{E.2}$$

where N is the total number of measurements.

Maximizing Eq. (E.2) means minimizing the following function:

$$S(z^m, \theta) = (z^e - z^m)^T V_e^{-1} (z^e - z^m) \tag{E.3}$$

If the experiments are independent, the off-diagonal elements of the covariance matrix V_e are null, resulting in

$$S(z^m, \theta) = \sum_{i=1}^{NE} (z^e - z^m)^T V_e^{-1} (z^e - z^m) \tag{E.4}$$

where NE is the total number of experiments.

Frequently, one divides the vector of variables z in both dependent (y) and independent (x). In general, the latter are not correlated, which allows us to separate their respective covariance matrices, resulting in

$$S(x^m, \theta) = \sum_{i=1}^{NE} \left[y_i^e - y_i^m(x_i^e, \theta)\right]^T V_{y_i}^{-1} \left[y_i^e - y_i^m(x_i^e, \theta)\right] + \sum_{i=1}^{NE} (x_i^e - x_i^m)^T V_{x_i}^{-1} (x_i^e - x_i^m) \tag{E.5}$$

where V_{x_i} and V_{y_i} are the covariance matrices for experiment i, associated, respectively to dependent and independent variables.

Normally, the experimental errors in the independent variables are neglected ($x^m = x^e$), since these are precisely controlled in laboratory experiments. So, the second sum on the right-hand side of Eq. (E.5) can be neglected. Plus, if the dependent-variable measurements are performed so that there is no correlation among them (i.e., the deviations of the dependent variables are not correlated), Eq. (E.5) might be rewritten as

$$S(\theta) = \sum_{i=1}^{NE} \sum_{j=1}^{NY} \frac{\left[y_i^e - y_i^m(x_i^e, \theta)\right]^2}{\sigma_{ij}^2} \tag{E.6}$$

where NY is the number of dependent variables, and σ_{ij}^2 is the variance associated to the dependent-variable j in experiment i. Eq. (E.6) is known as the weighted least squares function. If the errors in the dependent variables are constant (i.e., independent of the variable j), one reaches the well-known ordinary least squares function:

$$S(\theta) = \sum_{i=1}^{NE} \left[y_i^e - y_i^m(x_i^e, \theta)\right]^2 \tag{E.7}$$

The ordinary least squares objective function is much used in optimization and parameter estimation procedures. However, one has to be careful to the hypothesis assumed when deriving Eq. (E.7) starting from a maximum likelihood function, so one can avoid the indiscriminate use of the least squares method, which may lead to misgiven statistical interpretations and unsuccessful estimations. Same thought holds for the cases in which the variables and experimental deviations are strongly correlated, as it occurs very often in the evaluation of catalytic processes (Cerqueira et al., 1999; Rawet et al., 2001).

E.3 MINIMIZING THE OBJECTIVE FUNCTION

With the experimental data, after selecting the model and defining the objective function, the next step is the latter's minimization, manipulating the parameters' values. Such a procedure is nothing but an optimization problem, which, as one knows, might not be trivial, especially when one deals with nonlinear models. Schwaab and Pinto (2007) point down a series of common issues in optimization problems: excess of experimental data, nonlinear models, objective functions with many local minima, strongly correlated parameters, the number of the latter being very big sometimes. Plus, the analytical solution in optimization problems is only possible in very particular cases, when the objective function does not consider deviations in the independent variables, and the model is linear regarding the parameters.

So, in the majority of the engineering problems (frequently nonlinear in the parameters), one must recall the numerical methods. For the purposes of this book, where the experimental data are expensive and sparse, turning difficult the information obtaining about the measurements' variance, the use of the ordinary (or weighted) least squares objective function is justified. So, in every example exposed here, one does the minimization of Eq. (E.6) or (E.7), from a hybrid optimization methodology, which consists initially in a global-optimum seek through a heuristic method, followed by a refinement using a classic derivative-based method.

E.3.1 Particle Swarm Optimization (PSO) Algorithm

The PSO, or *Particle Swarm Optimization* algorithm, originally developed by Kennedy and Eberhart (1995) to simulate the gregarious behavior of animals (fishes, birds, bees, etc.), is a heuristic optimization algorithm that consists in the minimization of an objective function from an information interchange among the elements (particles) of the group that move in the seek space with changing velocity. It is a proved-efficiency method, also robust and simply implemented (Severo Jr., 2011).

This method does not require an initial estimate for the parameters values, only the determination of a finite domain for the parameter space. The motion of each particle at each iteration is induced by the sum of three contributions: (1) the inertia of the particle relative to its moving in the given "parameter seek" space; (2) the attraction of the particle to the lowest value of the objective function found by the particle itself; (3) the attraction of the particle to the lowest value of the objective function found by the whole set of particles (or part of it). Shi and Eberhart (1998) introduced a slight modification to the initial algorithm, which consisted in an inertia factor that weighs the term relative to the previous velocity of the particles, resulting in the following recursive procedure:

$$v_{i,d}^{k+1} = w \cdot v_{i,d}^{k} + c_1 \cdot r_1 \cdot \left(p_{i,d}^{k} - x_{i,d}^{k}\right) + c_2 \cdot r_2 \cdot \left(p_{g,d}^{k} - x_{i,d}^{k}\right) \tag{E.8}$$

where v is the particle velocity; x is its position in the seek space; p_i is the position correspondent to the lowest value of the objective function for particle i; and p_g is the position correspondent to the lowest value of the objective function among all the particles. Indexes i, d, and k indicate the particle, the seek direction, and the interaction counter, respectively; c_1 and c_2 are two arbitrary positive constants, known as *cognitive* and *social* parameter, respectively; w is the inertia weight; r_1 and r_2 are two aleatory numbers with uniform distribution between 0 and 1, distinct for each direction, interaction and particle. The position of each particle is updated at each iteration in the following way:

$$x_{i,d}^{k+1} = x_{i,d}^{k} + v_{i,d}^{k+1} \tag{E.9}$$

An interesting feature of the particle swarm algorithm is its initially global seeking character, which goes local along the iterations, allowing one to reach the global minimum after a number of iterations big enough. In addition, the algorithm allows one to fit both local and global characters of the seek, manipulating the weights c_1 and c_2, as well as the inertia parameter w. Finally, conversely to the derivative-based algorithms, the heuristic methods are insensitive to the initial guesses of the parameters due to their global seek character.

A problem common to the heuristic methods is associated to the relatively big number of objective function evaluations necessary to reach the global minimum. It can turn this methods significantly lower compared to the derivative-based ones. However, this problem can be attenuated when one uses a hybrid strategy: the heuristic method is used only in the initial iterations to perform a global seek, the derivative-based method being applied to refine the optimal solution close to the convergence in the final iterations. In this book, one uses a Gauss–Newton type method, described below.

E.3.2 Gauss–Newton Method

The method of Gauss–Newton consists in approximating the objective function by a Taylor series truncated on the quadratic term, as it would be the natural way for the case of a linear model in the parameters, with normal fluctuations in the dependent variables and absence of errors in the independent ones:

$$S(\theta) \approx S(\theta_0) + (\theta - \theta_0)\nabla S\big|_{\theta_0} + \frac{1}{2}(\theta - \theta_0)^T H\big|_{\theta_0}(\theta - \theta_0) \tag{E.10}$$

where $\theta = (\theta_1, \ldots, \theta_N)$ and $\theta_0 = (\theta_{0,1}, \ldots, \theta_{0,N})$ are the perturbed- and original-parameter vectors, respectively; the gradient vector (∇S) and the Hessian matrix (H) are given by

$$\nabla S = \begin{bmatrix} \dfrac{\partial S}{\partial \theta_1} \\ \vdots \\ \dfrac{\partial S}{\partial \theta_N} \end{bmatrix} \tag{E.11}$$

$$H = \begin{bmatrix} \dfrac{\partial^2 S}{\partial {\theta_1}^2} & \cdots & \dfrac{\partial^2 S}{\partial \theta_N \theta_1} \\ \vdots & \ddots & \vdots \\ \dfrac{\partial^2 S}{\partial \theta_1 \partial \theta_N} & \cdots & \dfrac{\partial^2 S}{\partial {\theta_N}^2} \end{bmatrix} \tag{E.12}$$

Therefore, both gradient vector and Hessian matrix can be evaluated after successive derivatives of the objective function with respect to the model parameters. As a necessary condition for the optimization, the gradient of the objective function at the optimum point must be null, i.e., $\nabla S = 0$. Therefore, neglecting the second-order term in Eq. (E.10) and differentiating it with respect to the parameters, one has

$$\nabla S\big|_{\theta_0} + H\big|_{\theta_0}(\theta - \theta_0) = 0 \tag{E.13}$$

Proceeding this way, the parameter vector can be updated at each iteration. The solution is exact when the function is quadratic. Validating this approach, the procedure can be executed iteratively, so the parameters obtained are used as the starting point for the next series expansion and numerically calculated derivatives, obtaining a new vector of parameters, till one reaches convergence. That means: within a determined tolerance, either the new vector of parameters generated is equal to the previous one or the objective function calculated with the new-parameter set is equal to the one obtained with the previous set. So, one defines the recursive procedure for the update of the parameters at each iteration:

$$\theta^{k+1} = \theta^k - \lambda^k \left(H\big|_{\theta_k}\right)^{-1} \cdot \nabla S\big|_{\theta_k} \tag{E.14}$$

where subscript k is the iteration counter, θ is the vector of parameters, and λ is the step-size given on the direction defined by the product $\left(H\big|_{\theta_k}\right)^{-1} \cdot \nabla S\big|_{\theta_k}$.

Moreover, the algorithm uses a step-size control to avoid losing track of the seek direction towards the optimal point. It also monitors the derivative signal in order to avoid the attraction to a maximum point, which equally satisfies the null-derivative criterion.

An approximation for the calculation of the Hessian matrix was proposed by Gauss, being valid for problems of parameter estimation where the independent variables are not subject to errors and also in the absence of correlation between the experiments. So, the Hessian matrix can be approximated by

$$H \approx 2\sum_{i=1}^{NE}\left(\frac{\partial y_i^m}{\partial \theta}\right)^T V_{y_i}^{-1}\left(\frac{\partial y_i^m}{\partial \theta}\right) \tag{E.15}$$

So, when the approximation of Eq. (E.15) is used, the Newton method described above is known as the method of Gauss–Newton. Its main advantage is the possibility of calculating the Hessian matrix only with first-order derivatives with respect to the parameters, which are already necessarily calculated when lining up the gradient vector of the objective function.

E.4 MOTIVATING EXAMPLE

To exemplify the just described parameter estimation procedure, let us consider the PR-EoS fitting from the PVT analysis of the fluid used in the example presented in Section 2.2 (Chapter 2).

Although the particle swarm algorithm is independent of the initial guesses for the parameters values, it is necessary to define a search interval for each parameter. In this sense, their upper and lower limits are always around the reference values given by the correlations presented in Section 2.5 (Chapter 2). Table E.1 shows the type of experimental data and

TABLE E.1 Experimental Data and Their Respective Weights in the Optimization of the PR-EoS Parameters

Variable	Weighs
Saturation pressure (P_{sat})	10
Bo[a]	0.1
Rs[b]	0.1
Oil density[c]	0.1
V_{rel}[d]	0.1
Liq_{Vtotal}[e]	0.1

[a]*Oil volume at both pressure and temperature of the assay per residual oil volume at standard conditions.*
[b]*Dissolved gas volume (measured at standard conditions) at both pressure and temperature of the assay per residual oil volume at standard conditions.*
[c]*Density of the live oil at both pressure and temperature of the assay.*
[d]*Relative volume: total volume at both pressure and temperature of the assay per volume at saturation pressure.*
[e]*Liquid volume percent related to the total volume of the cell at both pressure and temperature of the assay.*

their respective weights in the objective function. These weights are used to define the relevance of the variables in the estimation procedure. If all weights are the same, the deviations of all the variables in relation to the experimental data contribute equally to the objective function. Variables with higher weights contribute more to objective function, whereas the variables with lower weight contribute less. Therefore, the algorithm prioritizes adjusting variables with greater weight. In the example presented in Table E.1, adjusting the saturation pressure has priority over the other variables. The weights come in the objective function as multipliers, corresponding to the inverse of the variances in the method of weighted least squares.

The experimental data of this example come from a differential liberation (DL) and five constant-composition expansions (CCE's), the latter carried out during the swelling test with 0%, 5.03%, 10.02%, 20.01%, and 35.02% of injected gas. Among the DL properties, one took Bo, Rs, and oil density. As to the CCE's, the relative volume (V_{rel}) and the liquid volume percent below P_{sat} $\left(\mathrm{Liq}_{V\mathrm{total}}\right)$ were also used in the regression. Also, one has to add the saturation pressures measured in each one of these assays. The total number of experimental data makes up 273.

Table E.2 shows the information about the main parameters of the estimation procedure employed here.

Parameters c_1, c_2, w_0, and w_f are used in the equation proposed by Shi and Eberhart (1998)—Eq. (E.8)—described in Section E-3.1. The particle swarm optimization algorithm allows updating the value of the inertia parameters at each iteration. This variation is linear from the first (w_0) to the last iteration $\left(w_f\right)$.

Table 2.3 shows the approximate compositions of both reservoir fluid and injected gas. The optimized parameters were the pure-component properties (T_c, P_c, ω, M, and volume shift) and the binary interaction coefficients $\left(k_{ij}\right)$

TABLE E.2 Parameters of the Particle Swarm Optimization Method Applied in Example of Section 2.2 (Chapter 2)

Description	Value
Number of minimizations (inside each one repeats all the procedure)	1
Maximum number of iterations	50
Number of particles	10
Constants of the PSO algorithm (c_1; c_2)	1.5; 1.5
Inertia weighs of the PSO algorithm (w_o; w_f)	0.75; 0.2
Number of parameters	40

for CO_2 and N_2-CH_4 (the only nonzero ones) with all other pseudocomponents, making up 40 parameters to be estimated.

Figs. 2.13–2.23 present the graphs with both experimental and calculated PVT properties of this fluid. The adjustment might be considered reasonable for both DL and CCE's with low contents (up to 10%) of injected gas. Discrepancies observed in the CCE's for injected gas contents higher than 10% (Figs. 2.20 and 2.21) are likely due to the limitations of the thermodynamic model.

REFERENCES

Bard, Y., 1974. Non-linear Parameter Estimation. Academic Press Inc, San Diego.

Cerqueira, H.S., Rawet, R., Pinto, J.C., 1999. The influence of experimental errors during laboratory evaluation of FCC catalysts. Appl. Catal. A, v 181, 209–220.

Kennedy, J. and Eberhart, R.C., 1995. "Particle swarm optimization". In: Proc. IEEE International Conference on Neural Networks, Perth, Australia, pp. 1942–1948.

Rawet, R., Cerqueira, H.S., Pinto, J.C., 2001. The influence of covariances during laboratory evaluation of FCC catalyst. Appl. Catal. A, v 207, 199–209.

Schwaab, M., Pinto, J.C., 2007. Análise de Dados Experimentais, I: Fundamentos de Estatística e Estimação de Parâmetros. E-papers, Rio de Janeiro.

Schwaab, M., Biscaia Jr., E.C., Monteiro, J.L., Pinto, J.C., 2008. Nonlinear parameter estimation through particle swarm optimization. Chem. Eng. Sci. 63, 1542–1552.

Severo J.B. Jr., 2011. Avaliação de Técnicas de Planejamento de Experimentos no Reconhecimento do Equilíbrio de Adsorção em Sistemas Cromatográficos, Tese de D.Sc., PEQ/COPPE, Universidade Federal do Rio de Janeiro, Rio de Janeiro, RJ, Brasil.

Shi, Y. and Eberhart, R., 1998. "A modified particle swarm optimizer". In: Proc. Conference on Evolutionary Computation, Anchorage, Alaska, pp. 69–73.

FURTHER READING

Noronha, F.B., Pinto, J.C., Monteiro, J.L., Lobão and M.W., Santos, T.J., (1993), ESTIMA—Um Pacote Computacional para Estimação de Parâmetros e Projeto de Experimentos. Guia de Usuários, PEQ/COPPE/UFRJ, Rio de Janeiro.

Appendix F

Fundamentals of Statistical Thermodynamics and Perturbation Theory

The objective of this Appendix is to present a brief description of the fundamentals of Statistical Thermodynamics and Perturbation Theory, building up the framework for the development of the molecular association theories used in Chapter 7. One also derives the configurational part of the cubic equations of state on a molecular basis. Going through a full explanation of these tools is beyond the scope of this book. Here, we intend to provide the basic concepts needed to understand the origin of the thermodynamic models in a straight way. For a deeper discussion on Statistical Mechanics concepts, see, for example, McQuarrie (2000). A wider view of applied Statistical Thermodynamics is shown by Sandler (2011). The equation-of-state derivations based on statistical mechanics are presented in an explaining way by two papers that date back to the 1990s: One refers to pure substances (Sandler, 1990a); the other considers mixtures (Sandler, 1990b). One uses much of these papers' contents hereby.

F.1 GENERALIZED VAN DER WAALS PARTITION FUNCTION: FROM STATISTICAL MECHANICS TO CLASSIC THERMODYNAMICS

The molecular theory from which it is possible to derive thermodynamic models is statistical mechanics. For the case in which temperature T, volume V and the number of elements N are the independent variables, the canonic partition function is given by

$$Q(T,V,N) \equiv \sum_{i}^{\text{states}} e^{-\beta E_i(V,N)} \tag{F.1}$$

where index i refers to the quantum state of the N elements within a volume V; $\beta = (kT)^{-1}$, k is the Boltzmann constant; and E_i is the energy of the quantum state i.

The relation with Classic Thermodynamics is essential for the generation of macroscopic thermodynamic models, like the equations of state, from the microscopic-level information coming out the canonical partition function. For example, one can write the Helmholtz energy as

$$A(T,V,N) = -kT \ln Q(T,V,N) \tag{F.2}$$

Therefore, the equation of state may be written as

$$P = -\left(\frac{\partial A}{\partial V}\right)_{T,N} = kT\left(\frac{\partial \ln Q}{\partial V}\right)_{T,N} \tag{F.3}$$

All other thermodynamic properties can be obtained according to their definitions. So, one has, for example:

$$S = -\left(\frac{\partial A}{\partial T}\right)_{V,N} = k\ln Q + kT\left(\frac{\partial \ln Q}{\partial T}\right)_{V,N} \tag{F.4}$$

$$\mu = \left(\frac{\partial A}{\partial N}\right)_{T,V} = -kT\left(\frac{\partial \ln Q}{\partial N}\right)_{T,V} \tag{F.5}$$

$$U = kT^2\left(\frac{\partial \ln Q}{\partial T}\right)_{V,N} \tag{F.6}$$

Defining the exact partition function for complex molecules is a hard task, except for special cases, like the ideal gas. For simple molecules (apart from long-chain hydrocarbons or polymers), the total energy in a group of molecules might be splitted into its internal (electronic, translation, rotation, and vibration) and external (intermolecular interactions) contributions, each one of them being independent of the others (Sandler, 1990a). The so-called internal degrees of freedom may be considered separately for each individual molecule. For a pure fluid made up by N identical molecules, the canonical partition function is given by

$$Q(T,V,N) = \frac{q^N(T)}{N!}\Lambda^{-3N} \cdot Z(T,V,N) \tag{F.7}$$

where $q^N(T)$ is the partition function of the molecule representing the degrees of freedom of rotation, vibration, and electronic; the wavelength of De Broglie Λ is related to the translation movement and is defined as

$$\Lambda \equiv \left(\frac{h^2}{2m\pi kT}\right)^{1/2} \tag{F.8}$$

where m is the mass of the particle, and h is the Planck constant

The last term in Eq. (F.7) is remarkably defined as the *configuration integral* $Z(T,V,N)$, which is related to the intermolecular interactions (McQuarrie, 2000):

$$Z(T,V,N) \equiv \int \ldots \int e^{-\frac{\varphi(\underline{r}_1,\underline{r}_2,\ldots,\underline{r}_N)}{kT}} d\underline{r}_1 d\underline{r}_2 \ldots d\underline{r}_N \tag{F.9}$$

where $\varphi(\underline{r}_1,\underline{r}_2,\ldots,\underline{r}_N)$ is the interaction potential energy for a molecule between the position vectors $\underline{r}_1$ and $\underline{r}_1 + d\underline{r}_1$; a second molecule between $\underline{r}_2$ and $\underline{r}_2 + d\underline{r}_2$, and so on. The multiple integrals refer to all possible position vectors in volume V.

An interesting particular case is the ideal-gas equation of state, which comes when one sets the intermolecular interaction potential to zero, that is,

$$Z(T,V,N) \equiv \int \ldots \int d\underline{r}_1 d\underline{r}_2 \ldots d\underline{r}_N = V^N, \tag{F.10}$$

for the integral of the position vector $d\underline{r}_i$ over the space gives the volume V itself. Then the canonical partition function reduces to

$$Q(T,V,N) = \frac{q^N(T)}{N!}\Lambda^{-3N}V^N \tag{F.11}$$

Applying Eq. (F.3) to Eq. (F.11) gives

$$P = kT\left(\frac{\partial \ln Q}{\partial V}\right)_{T,N} = kT\left(\frac{\partial \ln V^N}{\partial V}\right) = kT\frac{N}{V} \rightarrow PV = NkT \tag{F.12}$$

For any other equation of state, it is necessary to assume some kind of interaction energy that represents the potential $\varphi(\underline{r}_1,\underline{r}_2,\ldots,\underline{r}_N)$. According to Sandler (1990a), it is the derivative of $Z(T,V,N)$ with temperature that provides the average interaction energy among the molecules. He refers to this global average energy as the *Configuration Energy* E^{conf}. In fact, recognizing the configurational part of U and applying Eq. (F.6) to Eq. (F.7), one has

$$E^{conf} = U^{conf} = kT^2\left(\frac{\partial \ln Z}{\partial T}\right)_{V,N} \tag{F.13}$$

Now, we will define the premises necessary to derive the van der Waals equation of state, the first semiempirical cubic one, which formed the basis for the proposal of the Peng–Robinson EoS, among others. One assumes that the interaction energy for the entire ensemble at any particular configuration can be calculated as the sum of the interaction energies of all possible pairs of molecules, following the pairwise additivity hypothesis:

$$\varphi(\underline{r}_1,\underline{r}_2,\ldots,r_N) = \sum_{\substack{i \\ i>j}}\sum_{j}\varphi(r_{ij}) \tag{F.14}$$

where r_{ij} is the distance between the centers of mass of molecules i and j.

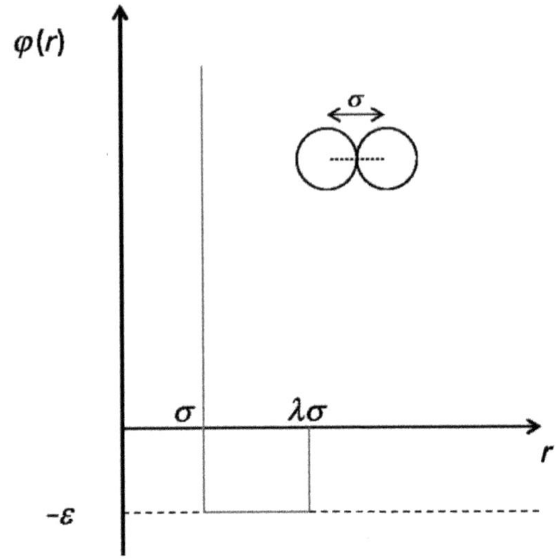

FIGURE F.1 Square-well potential as a function of the distance between the centers of mass of two molecules. As they are impenetrable, the potential diverges for distances lower than the diameter.

Let us assume that each pair of molecules interacts under a square-well type potential, as it is represented in Fig. F.1, although other potentials can be used. Dropping the indexes i and j, this potential has three noticeable values, depending on distance r:

$$\varphi(r) = \begin{cases} \infty, r < \sigma \\ -\varepsilon, \sigma \leq r \leq \lambda\sigma \\ 0, \lambda\sigma < r \end{cases} \tag{F.15}$$

where σ is the molecule diameter, since it is considered spherical.

This simple model gathers the essential characteristics of a real interaction: it has a repulsive region in $r < \sigma$ (molecule being impenetrable); an attractive region in $\sigma \leq r \leq \lambda\sigma$; and vanishes at long distances. Once the potential is adopted, the average global energy, or configuration energy, might be obtained by a simple analysis: let $N_C(\rho, T)$ be the so-called Coordination Number, that is, the average number of molecules between σ and $\lambda\sigma$, given a central molecule when the fluid has a number density ρ at the temperature T. The interaction energy of the central molecule with all others is $-N_C(\rho, T)\varepsilon$. As there are N possible choices for the central molecule, the total configuration energy is given by

$$E^{\text{conf}} = -\frac{NN_C(\rho, T)\varepsilon}{2} \tag{F.16}$$

where factor 2 arises from the fact that each interaction is counted twice, for each member of the interacting pair is considered as being the central molecule.

Given the expression for E^{conf}, one can integrate Eq. (F.13) from $T \to \infty$ to the actual temperature T of the system to obtain the numerical result of the configuration integral:

$$\ln Z(T, V, N) = \ln Z(T \to \infty, V, N) + \int_{\infty}^{T} \frac{E^{\text{conf}}}{kT^2} dT \tag{F.17}$$

As to the first term of the right-hand side of Eq. (F.17), one knows that, in the limit of infinite temperature, only repulsion forces are important and therefore the potential φ also diverges ($\varphi \to \infty$). Under these conditions, the result of the configuration integral is finite, being a fraction of the total volume to the Nth power:

$$Z(T \to \infty, V, N) = (\alpha V)^N = V_f^N \tag{F.18}$$

This fraction, known as *free volume* (V_f), is, by definition, the space available in the system when one discounts, from the total volume, that one which is not only occupied by the resident molecules but also voided for interaction. The latter is known as *compact volume* or *co-volume of van der Waals*. Fig. F.2 shows that the compact volume is not simply the sum of the molecules' volumes, since there is also the interstitial space that depends on the kind of fluid. Therefore, defining the extensive covolume as being Nb, one has

$$V_f = V - Nb \tag{F.19}$$

The second term on the right-hand side of Eq. (F.17) may be integrated using Eq. (F.16), assuming (like in the case of van der Waals equation) that the Coordination Number N_C is not a function of temperature, being only proportional to density ($N_C = \alpha\rho$). So, one has

$$\int_{\infty}^{T} \frac{E^{\text{conf}}}{kT^2} dT = -\frac{N\alpha\rho\varepsilon}{2k}\left[-\frac{1}{T}\right]_{\infty}^{T} = \frac{N\alpha\rho\varepsilon}{2kT} \tag{F.20}$$

Replace Eq. (F.20) in Eq. (F.17), to get

$$\ln Z(T, V, N) = N\ln(V - Nb) + \frac{N^2\alpha\varepsilon}{2kTV} \tag{F.21}$$

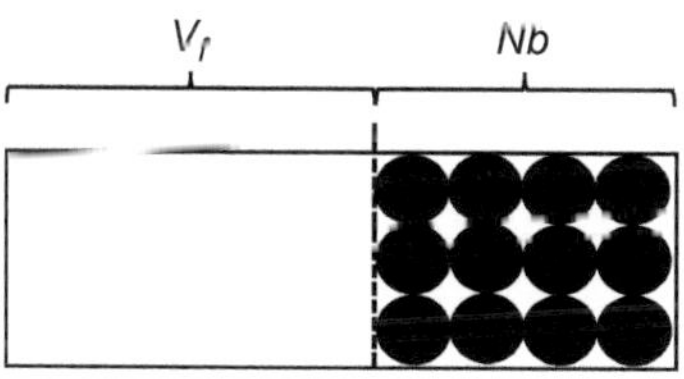

FIGURE F.2 Free volume, defined as a function of the compact volume of the molecules present in the system. Simplified model by van der Waals.

Finally, the equation of state might be obtained using Eq. (F.3):

$$P = kT\left(\frac{\partial \ln Q}{\partial V}\right)_{T,N} = kT\left(\frac{\partial \ln Z}{\partial V}\right)_{T,N} = \frac{NkT}{V - Nb} - \frac{\alpha\varepsilon N^2}{2V^2} = \frac{RT}{\overline{V} - b} - \frac{a}{\overline{V}^2} \tag{F.22}$$

where $a = \alpha\varepsilon/2$ is the attractive parameter of van der Waals EoS.

Sandler (1990a) presents some other choices for both free volume and coordination number, which give rise to other equations of state. Table F.1 shows some of these expressions, which correspond to the cubic models most commonly used in the petroleum industry.

F.2 DERIVATION VIA PERTURBATION THEORY

We proceed now to the derivation of the same van der Waals equation, using the Perturbation Theory approach, which is also useful to obtain the association term discussed in Chapter 7. Starting from an arbitrary reference fluid, for example, rigid spheres with exclusively repulsive potential, one can obtain the configurational Helmholtz energy by adding an attractive term, which is responsible for long-range interactions. The derivative of this Helmholtz energy with respect to volume provides pressure as a function of $T, V,$ and N, characterizing the equation of state. Once we are given the configurational energy, we can use the Gibbs–Helmholtz equation to obtain A^{conf}:

$$\left[\frac{\partial(A/T)}{\partial T}\right]_{V,N} = -\frac{U}{T^2} \tag{F.23}$$

Therefore, we have

$$\frac{A^{\text{conf}}}{NkT} - \lim_{T\to\infty}\frac{A^{\text{conf}}}{NkT} = \int_0^{1/T}\frac{E^{\text{conf}}}{Nk}d(1/T) \tag{F.24}$$

The right-hand side of Eq. (F.24) is the so-called Perturbation Helmholtz energy A^{pert}, or the work needed to bring the molecules from infinite temperature up to the configuration at the actual temperature, where the attractive potentials become relevant. The average configurational energy E^{conf}, although is already defined, might be rewritten as the integral, in spherical coordinates, of the attractive potential around a central molecule. The occurrence of such a central molecule disturbs locally the probability of other molecules to be around it, in relation to the aleatory distribution. We define this probability of finding molecules at a distance r from the central-molecule center of mass as the *radial distribution function*, $g(r, T, \rho)$.

$$E^{\text{conf}} = 4\pi\frac{N}{2}\int_0^{\infty}\rho g(r, T, \rho)\varphi^{\text{pert}}(r)r^2dr \tag{F.25}$$

TABLE F.1 Approximations for Both Free Volume (V_f) and Coordination Number (N_C) for the Cubic Equations of State Most Commonly Used in the Petroleum Industry (Sandler, 1990a)

Equation of State		V_f	N_C
van der Waals	$P = \frac{RT}{V-b} - \frac{a}{V^2}$	$V - Nb$	$\alpha\rho$
Redlich–Kwong	$P = \frac{RT}{V-b} - \frac{a/\sqrt{T}}{V(V+b)}$	$V - Nb$	$\frac{\alpha}{\sqrt{T}}\ln(1 + b\rho)$
Soave–Redlich–Kwong	$P = \frac{RT}{V-b} - \frac{a(T)}{V(V+b)}$	$V - Nb$	$\alpha(T)\ln(1 + b\rho)$
Peng–Robinson	$P = \frac{RT}{V-b} - \frac{a(T)}{V(V+b)+b(V-b)}$	$V - Nb$	$\alpha(T)\ln\left[\frac{1+(1+\sqrt{2})b\rho}{1+(1-\sqrt{2})b\rho}\right]$

where the factor 4π refers to the full integration over both angles in spherical coordinates; $N/2$ is the correction on counting all molecules at the central position; ρ is the probability of occurrence of any molecule at the central position (considering aleatory distribution); $g(r, T, \rho)$ is the probability of other molecules occurrence around the central one in relation to the aleatory distribution; and $\varphi^{\text{pert}}(r)$ is the intermolecular interaction potential, not necessarily of square-well type.

Substituting Eq. (F.25) into Eq. (F.24) and rearranging, we have

$$\frac{A^{\text{pert}}}{NkT} = \frac{2\pi\rho}{k}\int_{\sigma}^{\infty}\left[\int_{0}^{1/T} g(r, T, \rho)d(1/T)\right]\varphi^{\text{pert}} r^2 dr \qquad \text{(F.26)}$$

The strongest approximation of van der Waals equation, independently of the interaction-potential choice, is to assume that the radial distribution function is equal to 1, that is, the probability of finding molecules around the central-position one is compatible with the aleatory distribution:

$$g(r, T, \rho) = \frac{p(r)}{\rho} = 1 \qquad \text{(F.27)}$$

So,

$$\frac{A^{\text{pert}}}{NkT} = \frac{2\pi\rho}{kT}\int_{\sigma}^{\infty}\varphi^{\text{pert}}(r) r^2 dr \qquad \text{(F.28)}$$

Defining

$$a \equiv 2\pi\int_{\sigma}^{\infty}\varphi^{\text{pert}}(r) r^2 dr \qquad \text{(F.29)}$$

we have

$$A^{\text{pert}} = a\frac{N^2}{V} \tag{F.30}$$

Therefore, the attractive term of the equation of state is

$$P^{\text{pert}} = -\left(\frac{\partial A^{\text{pert}}}{\partial V}\right)_{T,N} = a\frac{N^2}{V^2} \tag{F.31}$$

The approaches of the canonical partition function (configuration integral) and the Perturbation Theory are equivalent. Assuming the square-well potential for $\varphi^{\text{pert}}(r)$, though relaxing the hypothesis of the aleatory distribution for $g(r,T,\rho)$, we can define the coordination number as a function of the integral of Eq. (F.28):

$$\frac{A^{\text{pert}}}{NkT} = -\frac{2\pi\rho\varepsilon}{k}\int_{\sigma}^{\lambda\sigma}\left[\int_{0}^{1/T} g(r,T,\rho)d\left(1/T\right)\right]r^2dr \equiv \frac{\varepsilon N_C(T,\rho)}{2k} \tag{F.32}$$

In other words, depending on the functional relation of $g(r,T,\rho)$, the expression for the coordination number obtained from the integration of Eq. (F.32) will lead to the various equations of state used in the literature.

REFERENCES

McQuarrie, D.A., 2000. Statistical Mechanics. University Science Books.

Sandler, S.I., 1990a. From molecular theory to thermodynamic models. Part I. Pure fluids. Chem. Eng. Edu 24 (1), 12–19.

Sandler, S.I., 1990b. From molecular theory to thermodynamic models. Part 2: Mixtures. Chem. Eng. Edu. 24 (2), 80–87.

Sandler, S.I., 2011. An Introduction to Applied Statistical Thermodynamics. John Wiley & Sons, Inc, Hoboken, NJ.

Index

Note: Page numbers followed by "*f*" and "*t*" refer to figures and tables, respectively.

compositional grading enhanced by, 150–159

reservoir not much influenced by, 159–161

CPI Antony Rowe
Chippenham, UK
2017-06-07 21:13